Constructive Engineering of Large Reverse Osmosis Desalination Plants

Constructive Engineering of Large Reverse Osmosis Desalination Plants

By Pedro Maria Gonzalez Olabarria

ISBN: 978-0-8206-0200-4
e-Book ISBN: 978-0-8206-0208-0

First Edition:

Chemical Publishing Company:
www.chemical-publishing.com
Printed in the United States of America

TABLE OF CONTENTS

1

REVERSE OSMOSIS BASIC CONCEPTS.

1.1 Basic concepts.

- **Permeate:** Low-saline and/or purified product.
- **Brine:** High saline or concentrated brine, called concentrate or reject.
- **Raw and Feed water:** Raw and Feed water.
- **Seawater:** Water with a salinity higher than 30,000 ppm.
- **Brackish water**: Water with a salinity less than 15,000 ppm.
- **Recovery:** It is the relation between production and raw water flow.
- **Osmotic pressure:** Difference of pressure between two liquids of different salinity separated with a semipermeable membrane.
- **Salinity:** Total dissolved solids in water.
- **SS (Suspended solids):** Suspended solids in water.
- **Fouling:** Fouling of membranes which can be organic or inorganic.
- **TDS (Total dissolved solids):** Total dissolved solids in water.
- **TOC (Total organic carbon):** Total organic carbon.
- **DOC (Dissolved organic carbon):** Dissolved organic carbon.
- **SWRO (Sea water reverse osmosis):** Sea water reverse osmosis.

1.2 Units.

- **Production:** m^3/day or MGD (mega gallons per day).
- **Recovery:** %.
- **Salinity:** mg/L, gr/m^3 or ppm (parts per million).
- **TDS**: mg/L or ppm.
- **SS**: mg/L or ppm.
- **Flow:** m^3/h, m^3/day, gpm or gpd (gallons per minute or gallons per day)
- **Pressure:** bar, mcw, psig or ft.
- **Power:** kW or hp (horse power).
- **Conductivity:** mS/cm or µS/cm (mili siemens/cm or micro siemens/cm).
- **Turbidity:** NTU (nephelometric turbidity unit).
- **TOC**: mg/L.
- **DOC**: mg/L.

2

FEED WATER TYPE AND ANALYSIS.

The major water types being treated by SWRO can be roughly characterized from the total dissolved solids (TDS) content and the organic (TOC).

- High salinity: 15,000-45,000 ppm, Sea water.
- High salinity-brackish water: 5000-15,000 ppm.
- Medium salinity-brackish water: 500-5000 ppm .
- Medium-salinity tertiary effluent with high TOC and biological oxygen demand (BOD) levels and TDS up to 5000 mg/L.
- Low salinity tap waters with TDS up to 500 mg/L.
- Very-low-salinity, high-purity waters (HPW) coming from the first RO systems (double-pass RO system) or the polishing stage in ultrapure water (UPW) systems with TDS up to 50 mg/L.

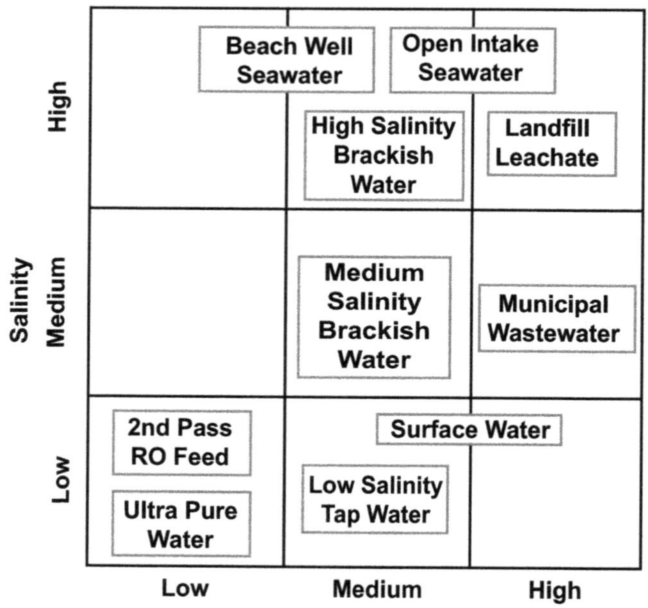

Figure 2.1. Type of water in function of salinity.

Depending on the type of water source:

1. Open intake water.
2. Well water.
3. Microfiltration and Ultrafiltration water.
4. Water from reverse osmosis.

Elements of feed water characteristics for reverse osmosis

Cations	Anions	Metals	Other elements
Calcium	Carbonates	Iron	Silica
Magnesium	Bicarbonate	Aluminum	Boron
Potassium	Sulfates	Manganese	Turbidity
Sodium	Chlorines		TSS
Estroncium	Nitrates		TOC
Barium	Fluoride		PH
Amonium			Temperature
			Conductivity

Seawater with TDS of 35,000 mg/L is considered standard seawater, constituting, the largest amount of water worldwide. The composition is nearly same all over the world. The actual TDS content may, however, vary within wide limits.

In the next table, water salinity in different areas are shown:

	Mediterranean	Red sea	Atlantic
Na^+	11,177	12,300	11,520
Ca^{++}	471	570	450
K^+	478	470	460
Mg^{++}	1355	1700	1320
Cl^-	20,530	34,000	20,600
SO_4^{--}	2488	3400	2910
CO_3H^-	208	185	134
pH	7.9	8.2	8.1
TDS	36,720	47,000	37,830

As a consequence of the high salinity of seawater involving a high osmotic pressure, the recovery of the system is limited to typically 40 to 50%.

2.1 Characteristic parameters.

It is necessary to analyze independently the parameters and characteristics of water which affect the reverse osmosis process.

1. Salinity.

TDS, in water treatment, is the inorganic residue left after the filtration of colloidal and suspended solids and then the evaporation of a known volume of water. Feed or permeate TDS, in RO design projections, can also be estimated by applying a conversion factor to the conductivity of the solution. TDS can also be determined in the field by use of a TDS meter. TDS meter measures the conductivity of the water and then apply a conversion factor that reports TDS to a known reference solution (e.g. ppm sodium chloride or ppm potassium chloride). The user is cautioned that TDS levels for water with a mixture of ions and determined from conductivity measurements may not agree with TDS calculated as a sum of the ions. As a rough rule of thumb, relationship between conductivity and salinity ranges between 1.38 for a salinity of 44,000 ppm and 1.43 for a salinity of 36,000 ppm.

It affects directly to operation pressure of reverse osmosis, increasing when increasing salinity of feedwater. It also affects to the quality of permeate, getting worse with the increase of feed water salinity.

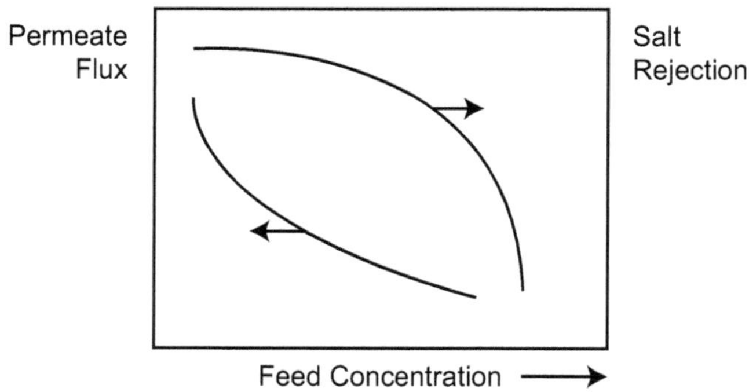

Figure 2.2. Effect of feed concentration on permeate flow.

2. Scaling.

Scaling of RO membranes may occur when sparingly soluble salts are concentrated within the element beyond their solubility limit. For example, if a reverse osmosis plant is operated at 50% recovery, the concentration in the concentrate stream will be almost double the concentration in the feed stream. As the recovery of a plant is increased, so it is the risk of scaling. The most common sparingly soluble salts encountered are:

- Calcium carbonate.
- Calcium sulfate.
- Strontium Sulfate.
- Barium Sulfate.
- Silica.
- Calcium Fluoride.

Solubility products of soluble inorganic components are encountered in table below:

Substance	Formula	Temp. °C	Solubility product	Negative log K_{sp}
Aluminum hydroxide	$Al(OH)_3$	25	3×10^{-34}	33.5
Aluminum phosphate	$AlPO_4$	25	9.84×10^{-21}	20
Barium carbonate	$BaCO_3$	25	2.58×10^{-9}	8.6
Barium Sulfate	$BaSO_4$	25	1.1×10^{-10}	10
Calcium carbonate	$CaCO_3$	25	Calcite: 3.36×10^{-9}	8.5
			Aragonite: 6×10^{-9}	8.2
Calcium fluoride	CaF_2	25	3.45×10^{-11}	10.5
Calcium phosphate	$Ca_3(PO_4)_2$	25	2.07×10^{-33}	32.7
Calcium Sulfate	$CaSO_4$	25	4.93×10^{-5}	4.3
Iron(II) hydroxide	$Fe(OH)_2$	25	4.87×10^{-17}	16.3
Iron(II) sulfide	FeS	25	8×10^{-19}	18.1
Iron(III) hydroxide	$Fe(OH)_3$	25	2.79×10^{-39}	38.6
Iron(III) phosphate dehydrate	$FePO_4{\cdot}2H_2O$	25	9.91×10^{-16}	15
Lead carbonate	$PbCO_3$	25	7.4×10^{-14}	13.1
Lead fluoride	PbF_2	25	3.3×10^{-8}	7.5
Lead sulfate	$PbSO_4$	25	2.53×10^{-8}	7.6
Magnesium ammonium phosphate	$MgNH_4PO_4$	25	2.5×10^{-13}	12.6
Magnesium carbonate	$MgCO_3$	12	2.6×10^{-6}	4.58
		25	6.82×10^{-6}	5.17
Magnesium fluoride	MgF_2	18	7.1×10^{-9}	8.15
		25	5.16×10^{-11}	10.3
Magnesium hydroxide	$Mg(OH)_2$	18	1.2×10^{-11}	10.9
		25	5.61×10^{-12}	11.25
Magnesium phosphate	$Mg_3(PO_4)_2$	25	1.04×10^{-24}	24
Manganese hydroxide	$Mn(OH)_2$	18	4.0×10^{-14}	13.4
		25	2×10^{-13}	12.7
Strontium carbonate	$SrCO_3$	25	5.6×10^{-10}	9.25
Strontium Sulfate	$SrSO_4$	17.4	3.8×10^{-7}	6.42
Zinc carbonate	$ZnCO_3$	25	1.46×10^{-10}	9.84

Reverse osmosis design software inform in detail about the saturation levels of different salts. $CaCO_3$, $BaSO_3$ and $CaSO_4$ scaling criteria are very similar. It is located in the last membranes of the vessels, and a reduction of flux and an increase differential head loss will be detected. A photograph of a membrane with calcium carbonate scaling is shown below.

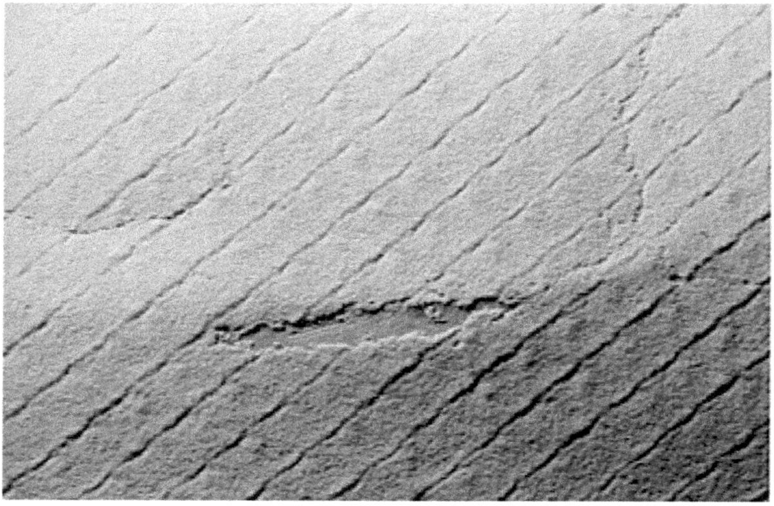

Figure 2.3. Calcium carbonate scaling.

3. Langelier and Stiff & Davis index.

$CaCO_3$ tends to dissolve in the concentrate stream rather than precipitate. This tendency can be expressed by the Langelier Saturation Index (LSI) for brackish waters and the Stiff and Davis Stability Index (S&DSI) for seawaters.

The Langelier Saturation Index is calculated according to the relation LSI = pH - pHs (pHs = pH of saturation) where pH is the actual pH of the water and pHs is pH that corresponds to saturation concentration of ions forming calcium carbonate. Water solution has potential for $CaCO_3$ scaling at LSI higher than 0 and it is assumed that saturation prediction using LSI is reliable up to salinity of about 5000 ppm TDS.

For high salinity and seawater applications the LSI was modified by Stiff and Davis Index. It is determined by the expression pH – pCa - pALK - K, where pCa and pALK are negative logarithms of the molar concentration of Ca and bicarbonate and K is a constant function of temperature and ionic strength.

Example 1

Seawater system is design to operate at a recovery rate of 50%.

Feed water feed has TDS = 40,000 ppm ionic strength 0.69 and the following concentration of the relevant ions:

Ca = 450 ppm; HCO_3 = 150; pH = 8.1; Temp = 21°C

After acidification and pH adjustment to 6.5 HCO_3 = 115 ppm; CO_2 = 42 ppm

Concentration factor = 2

Approximate concentrations in the concentrate.

TDS = 80,000 ppm, ionic strength 1.39; Ca = 900 ppm; HCO_3 *= 230;* CO_2 *= 42; pH = 7.2*

Calculation of concentrate pH

$$\text{Concentrate } pH = 6.37 + \log((900/42) \times (44/61)) = 6.37 + 0.6 = 6.97$$

Calculation of pHs and LSI

$$LSI = pH - pH_s$$
$$pH_s = (9.3 + A + B) - (C + D)$$
$$A = (\log_{10}(TDS) - 1)/10$$
$$B = -13.12 \times \log_{10}(°C + 273) + 34.55$$
$$C = \log_{10}(Ca \text{ as } CO_3) - 0.4$$
$$D = \log_{10}(Alcalinity \text{ as } CaCO_3)$$

pH_s *(feed) = 9.3 + 0.36 + 2.15 - 2.65 - 1.97 = 7.19*

LSI (feed) = 6.5 - 7.19 = -0.69

pH_s *(concentrate) = 9.3 + 0.39 + 2.15 - 2.98 - 2.28 = 6.58*

LSI (concentrate) = 6.97 - 6.58 = 0.39

An LSI calculation is shown in the figure below.

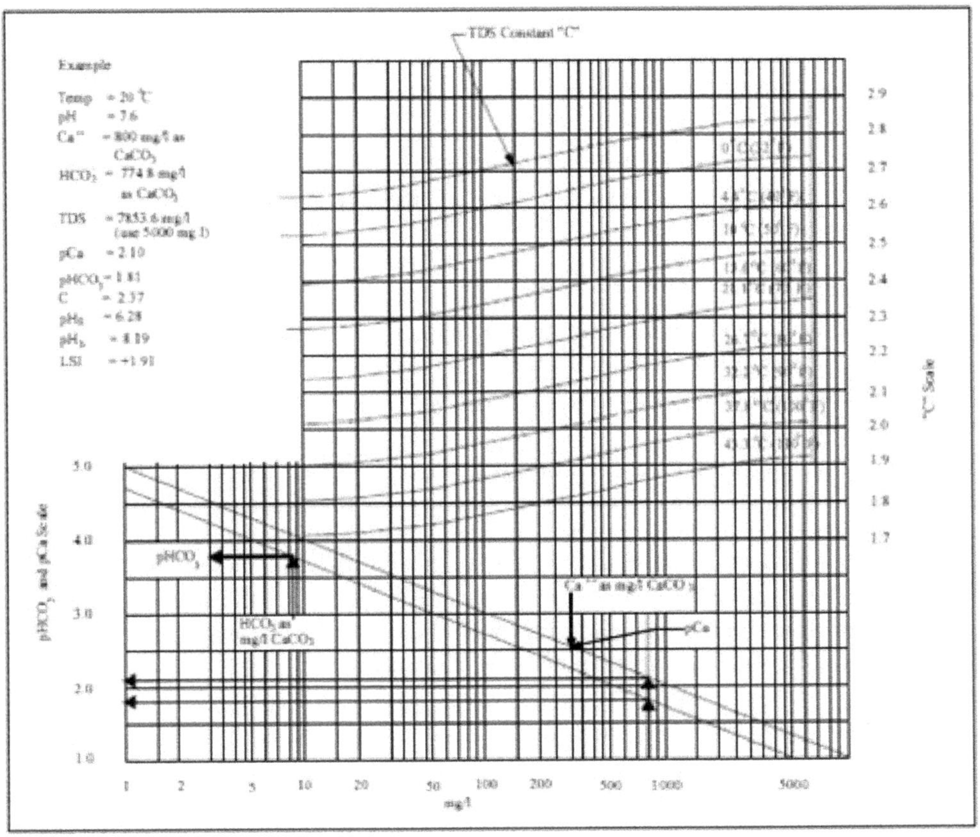

Figure 2.4. LSI table.

4. pH.

The pH of the feed water measures the acidity of basicity. A pH of 7.0 is considered neutral. A pH between 0.0 and 7.0 is acidic. A pH between 7.0 and 14.0 is basic. To the water chemist, pH is important in defining the alkalinity equilibrium levels of carbon dioxide, bicarbonate, carbonate and hydroxide ions, so it affects the solubility of carbonates and regulates scaling. Usually seawater pH ranges between 7.5 to 8. Lowering the feed pH with acid results in a lower LSI value which reduces the scaling potential for calcium carbonate.

The concentrate pH is typically higher than the feed due to the higher concentration of carbonates/bicarbonates ions relative to the concentration of carbon dioxide.

Feed and concentrate pH can also affect the rejection of ions. For example, fluoride, boron and silica rejection are lower when the pH becomes more acidic.

Lowering the pH, the bactericidal activity of disinfectants improve, so it tends to work in pH between 6.5-7. Sometimes, operators form a short (1/2 to 1 hour) abrupt drop of pH.

5. Suspended solids and turbidity.

Turbidity is an expression of the optical property of water that causes light to be scattered and absorbed rather than transmitted in straight lines through the simple. Turbidity is caused by suspended and colloidal particulate matter such as clay, silt, finely divided organic and inorganic matter, plankton and other microscopic organisms.

In theory, there is relationship between suspended solids content and turbidity. However, there are some solids which produce more turbidity than others.

Suspended solids must be removed before the reverse osmosis in order to prevent clogging. The turbidity of feed water to reverse osmosis should be less than 1 NTU as one of the minimum requirements of feed water. Usually, filtration processes guarantee suspended solids removal except colloidal particles.

6. Colloidal and Particulate Fouling.

Colloidal fouling of RO elements can seriously impair performance by lowering productivity and sometimes salt rejection. It affects the first stage and a sign of colloidal fouling is a production reduction, an increased differential pressure across the system and an increase of permeate salinity.

The source of silt or colloids in reverse osmosis feed water is varied and often includes bacteria, clay, colloidal silica and iron corrosion products. Pretreatment chemicals used in a clarifier such as aluminum sulfate, ferric chloride or cationic polyelectrolyte are materials that can be used to combine these fine particle size colloids resulting in an agglomeration or large particles that then can be removed more easily by media or cartridge filters. Such agglomeration, consequently, can reduce the performance criteria of media filtration or the pore size of cartridge filtration where these colloids are present in the feed water.

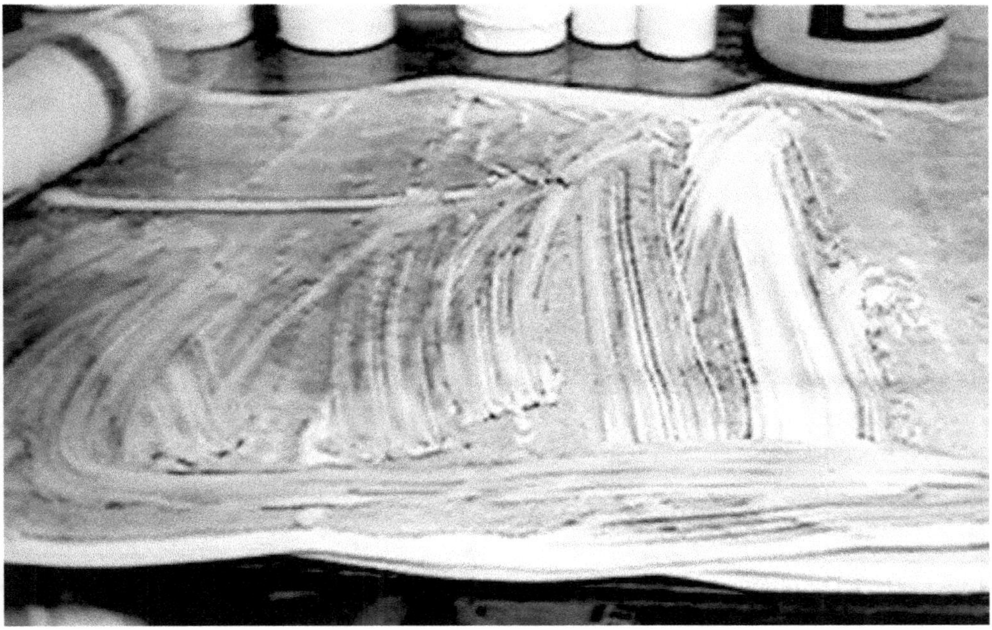

Figure 2.5. Colloidal fouling. Clay.

7. Aluminum fouling.

Aluminum, based on its low solubility, is typically not found in any significant concentrations in well or surface waters. Aluminum, when present in an RO feed water, is typically colloidal in nature (not ionic) and is the result of alum carry over by an on-site or municipal clarifier or lime softener. Alum (aluminum sulfate) is a popular coagulant that is effective in the absorption and precipitation of naturally occurring, negatively charged colloidal material Aluminum, when introduced into water, disassociates into trivalent aluminum and sulfate. Fouling by aluminum based colloid carry-over can occur, with alert levels for the RO designer ranging from 0.1 to 1 ppm aluminum in the feed water. Aluminum at high pH can exist as a negatively charged anionic compound. Typically the range of least solubility for aluminum compounds is in the pH range of 5-5.5. So for the pH values the RO systems operate (between 6.5 and 7.5), the risk aluminum fouling is minimum .

8. Silica Fouling.

Dissolved silica (SiO_2) is naturally present in most feed waters in the range of 1-100 mg/L. The prevailing forms of silica are meta silica acids as $(H_2SiO_3)_n$ with low *n* numbers. Since silicic acid is a weak acid, it is mostly in the undissociated form at or below a neutral pH. Supersaturated silicic acid can further polymerize to form insoluble colloidal silica or silica gel, which can cause membrane scaling. The maximum allowable SiO_2 concentration in the concentrate stream is based on the solubility of SiO_2.

The scaling potential for the concentrate stream will be quite different from that of the feed solution because of the increase in the concentration of SiO_2 and the change in pH. It can be calculated from the feed water analysis and the RO operating parameters.

As the pH exceeds neutral, silicic acid dissociates into the silicate anion. This can react with calcium, magnesium, iron, manganese or aluminum to form insoluble silicates.

The aluminum is the most powerful precipitant of silicic acid, and the occurrence of silica scaling is mostly correlated with the occurrence of aluminum or iron. It has been reported that, when Al^{3+} and Fe^{3+} coexist in the pretreated feed water, silica is precipitated even below its saturation. Both Al^{3+} and Fe^{3+}, therefore, must be less than 0.05 mg/L in the feed water, even if the silica level is below saturation. Since Al^{3+} and Fe^{3+} salts are used for coagulation in municipal and other industrial water processing, frequent and accurate measurements of these ions are needed even though the feed water itself does not contain high levels of aluminum and iron ions. Fouling with metal silicates may occur from a chemical reaction and precipitation process (scaling), and also from colloidal fouling with submicron particles entering the membrane system. Feed water acidification and preventive acid cleanings are possible measures in cases of a metal silica scaling potential. If colloidal silica and silicates are present in the feed water, a flocculation/filtration process and/or a fine grade prefilter (1 µm or less) should be chosen.

9. Manganese and Fe fouling.

Iron is a water contaminant that takes two major forms. The water soluble form is known as the ferrous state and has a +2 valence state. In non aerated well waters ferrous iron behaves much like calcium or magnesium hardness in that it can be removed by softeners or its precipitation in the back of the RO can be controlled by dispersants. The insoluble form is known as the ferric state and has a +3 valence. Typically reverse osmosis system manufacturers will recommend that the combined form, iron levels up to 0.5 ppm in the feed, can be tolerated if pH is less than 7. The introduction of air into water with soluble ferrous iron will result in the oxidation to insoluble ferric iron. Insoluble ferric iron oxides on ferric hydroxides, being colloidal in nature, will foul the front end of reverse osmosis.

Sources of insoluble iron are aerated well waters, surface sources, and iron scale from unlined pipe and tanks.

Iron as foulant will quickly increase RO feed pressure requirements and increase permeate TDS. In some cases, the presence of iron can create a bio-fouling problem by being the energy source of iron reducing bacteria. Iron reducing bacteria can cause the formation of slimy biofilm than can plug the RO.

We must take into account that filters sand media has approximately 1% of Fe_2O_3, that sometimes can precipitate in the filter when it gets oxidized. For that reason, it is recommended to work for a certain period of time at a pH of 6-6.5 to dissolve Fe.

The usual treatment to remove iron with a content lower than 5 mg/L from water is to oxidize the minerals with air as rapidly as possible at a proper pH and then to remove the oxidized material through filtration. If Fe content is higher than 5 mg/L, it will require a settling process before filtration.

Iron removal processes are no useful for manganese removal. Precipitation as hydroxides or the oxidation with oxygen would only be possible at pH 9-9.5. However, a quick oxidation with chlorine dioxide or potassium permanganate is possible.

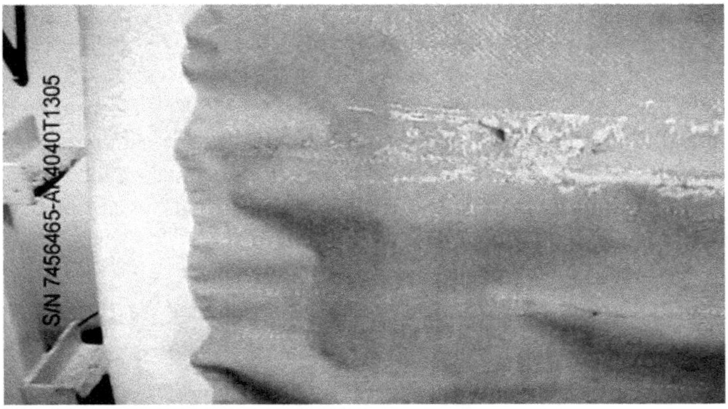

Figure 2.6. Fe fouling.

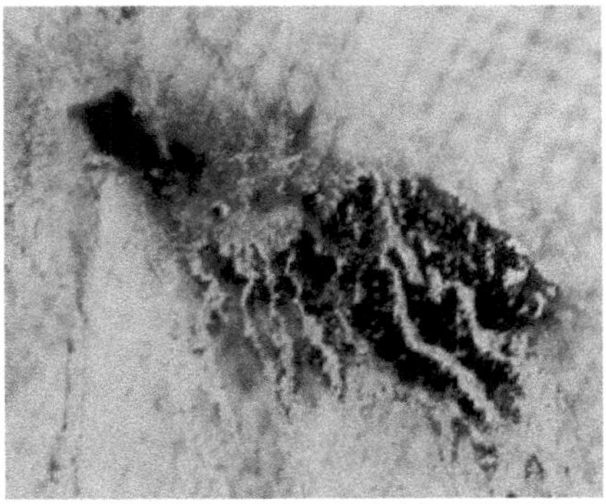

Figure 2.7. Deposit of MnO_2.

10. Organic Fouling (biofouling).

All raw waters contain microorganisms such as bacteria, algae, fungi, viruses and higher organisms. The microorganisms entering a RO/NF system find a large membrane surface where dissolved nutrients from the water are enriched due to concentration polarization, thus creating an ideal environment for the formation of a biofilm. Biological fouling of the membrane may affect seriously the performance of the RO system.

The biofouling is located in the first stages, and the symptoms are an increase in the differential pressure from feed to concentrate, finally leading to telescoping and mechanical damages of membranes, and a decline in membrane flux.

"Biofouling" prevention can be obtained with a proper arrangement of system and operation:

- Direct hydraulic connections between water intake and reverse osmosis. Do not install intermediate tanks or elements where water could be aerated. If intermediate open basins or tanks are used, provisions should be made to allow for proper sanitization at that open source and the part of the system downstream from it. If intermediate sealed tanks are used, their air breathing or ventilation systems should be equipped with bacteria-retaining devices (example HEPA filters).
- Blind, long pieces of piping should be avoided by design.

- Avoid direct exposition of water to sun.
- Avoid adding chemical products which may be metabolized by microorganisms.
- Control organic elements which can boost biological activity.
- It should be made possible to physically isolate the RO/NF section from the pretreatment by using a flange. This allows one to use chlorine for sanitizing the pretreatment section while the membranes are protected from chlorine contact. A drain valve should be installed at the lowest point close to the flange, to allow complete drainage of the chlorine solution.
- Do not chlorinate continuously.

Figure 2.8. Biological fouling.

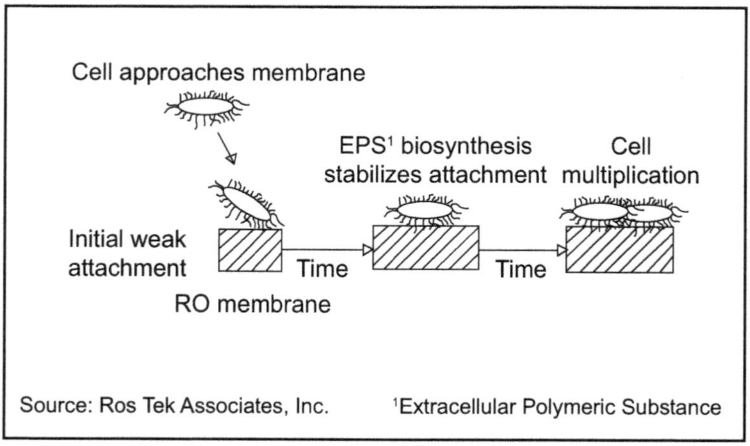

Figure 2.9. Biological fouling.

11. Organic fouling - TOC.

Organics occurring in natural waters are usually humic substances in concentrations between 0.5 mg/L in deep well waters to 6 mg/L TOC in surface open intakes. Pretreatment should be considered when TOC is higher than 3 mg/L. TOC (total organic carbon) is the value measured to determine organic content. DOC (dissolved organic carbon) or AOC are other analysis which could be used to evaluate the organic matter.

Typically, as indicated, well water will have a lower content of organic matter than open intakes. The temperature and sun light will boost and increase TOC so direct contact of sea water with sun light should be avoided.

Adsorption of organic substances on the membrane surfaces causes flux loss, which is irreversible in serious cases. The adsorption process is favored with high molecular mass compounds when these compounds are hydrophobic or positively charged. Organics present as a emulsion may form an organic film on the membrane surface which will increase the head loss and reduce production.

The microorganisms are difficult to eliminate completely with the membrane cleaning, so the organics must, therefore, be removed in pretreatment.

Organic fouling occurs in all the stages. A flux and permeate salinity reduction will be detected, without a significant increase in differential pressure.

Figure 2.10. Oil fouling. Courtesy of DOW water.

12. SDI , MFI and MEB.

Several methods or indices have been proposed to predict a colloidal fouling potential of feed waters, including turbidity, Silt Density Index (SDI) and Modified Fouling Index (MFI). The SDI is the most commonly used fouling index. The guideline is to maintain SDI less than 5. To minimize the fouling, however, SDI_{15} of less than three (3) is recommended. A number of pretreatment technologies have proven effective in SDI reduction, including media filtration (such as sand/anthracite), ultrafiltration and cross-flow microfiltration.

Usually well water will have an SDI of less than 1 and an open intake between 3 and 4.

The SDI is calculated from the rate of plugging of a 0.45 μm membrane filter, when water is passed through at a constant applied gauge pressure.

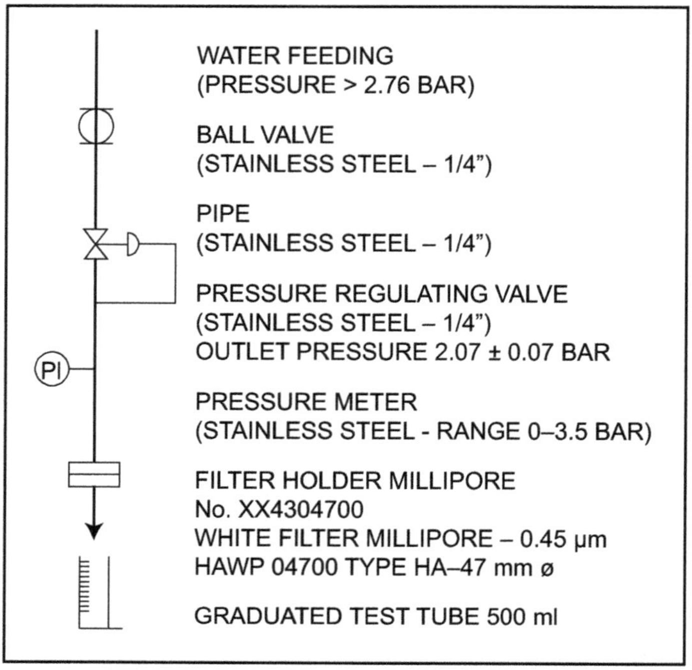

Figure 2.11. SDI measurement

The SDI is calculated with the following expression:

$$SDI = \frac{\%P_{30}}{T} = \frac{100\left(1 - \frac{T_0}{T_{15}}\right)}{T}$$

Assembling sketch of the elements necessary to measure the SDI, in which:

P_{30} = Clogging with a feeding pressure of 2.07 bar (psig). To exactly measure the SDI, % P_{30}, it should not surpass 75%. If the % P_{30} is superior to that quantity, it would be necessary to carry out the test again, obtaining the T_{15} in a shorter time (T).

T = Total duration of the test in minutes, usually 15, being possible to have a shorter duration if 75% of the clogging is produced in less than 15 minutes.

T_0 = Initial time, in seconds, necessary to collect a sample of 500 ml.

T_{15} = Time, in seconds, necessary to collect a sample of 500 ml, when the testing time T has passed.

The Modified Fouling Index (MFI) is proportional to the concentration of suspended matter and is a more accurate index than the SDI for predicting the tendency of a water to foul RO/NF membranes. The method is the same as for the SDI except that the volume is recorded every 30 seconds over a 15 minute filtration period. The MFI is obtained graphically as the slope of the straight part of the curve when t/V is plotted against V (t is the time in seconds to collect a volume of V in liters).

A MFI value lower than one (1) corresponds to a SDI value of about less than three (3) and can be considered as sufficiently low to control colloidal and particulate fouling.

13. Feed Water Temperature.

Feed water temperature affects the performance of membranes so it is required to know the range of temperatures during the year:

- If the temperature increases and all other parameters are kept constant, the permeate flux and the salt passage will increase. An increase of feed water temperature enables production of a given permeate flow at a reduced feed pressure. The RO systems are designed to operate at constant permeate capacity. Therefore, increase of feed water temperature will result in increased permeate salinity. The increase is about 3% per degree C.
- Solubility of some salts increases with temperature. For that reason, at high temperature, scaling potentially decreases.
- Temperature increase will result in more microbiological activity and fouling associated.

3

RAW WATER REQUIREMENTS.

The first design data is the desalted water production (net production). Designers must consider the internal services requirements (reactives dilution, services water, flushing and chemical washing), which could be approximately 0.5% to 1% of the total production depending on the size of plant.

The second value is the recovery which usually ranges from 42% to 48%. The typical value is 45%.

If global recovery is 45%, seven (7) elements should be installed per vessel not to exceed the recommended values by manufacturers for membrane recovery. So, a single stage with seven (7) membranes installed in each vessel will be the right arrangement.

If water characteristics are very good (deep well water), a global recovery of up to 50% could be obtained with only one stage.

If a higher recovery is required, it will be necessary to install two (2) stages.

The recovery rate or conversion rate is the ratio of membrane permeate flow rate to the membrane feed, so raw water requirements will be obtained from desalted water production and RO recovery.

Filters backwashing needs and any additional pretreatment losses must be added to feed water flow, Water intake mode, i.e. indirect or direct. The first mode requires few treatments to achieve the required turbidity and SDI for membranes feed water, whereas the second mode may require more intensive treatments (classical like coagulation, flocculation, clarification or flotation, filtration or membrane treatments like micro or ultrafiltration). All those pretreatment create water losses for sludge discharge or wash water. Water losses may vary from 2% for a classical filtration, to 10% for ultrafiltration.

If backwashing is carried out with brine, this concept will be 0 m^3. However, if backwashing is carried out with filtered water, 200 to 300 m^3/h should be added depending on the n° of filters.

In all the cases, maturation flow will be added. This flow goes from 400 to 500 m^3/h (raw water flow / nº of filters -1)

Raw water volumes will be consequently calculated with the following relation

$$Q_{sw} = \frac{Q_{wtp} \times 100}{(Y \times l \times d)}$$

With:

Q_{sw} : daily seawater capacity (m^3/day).

Q_{wtp} : daily fresh water production (m^3/day).

Y : recovery ratio (%) = (freshwater flowrate x 100) : feed water flowrate.

L : pretreatment water losses compared to seawater flowrate (%).

d : other losses in water pipes and network (%).

4

SEA WATER INTAKE.

Seawater intake types range from screened wells onshore, to large surface water intakes along the shore and to offshore intake structures. Each type varies in design, power consumption and environmental considerations. Intake types are generally divided into two main groups, which are direct intakes and indirect intakes as shown in Figure 4.1.

Selecting the most feasible type of intake structure will mainly depend on the following:

- The required feed water volume and quality.
- Site-specific coastal and physical processes.
- The seabed bathymetry.
- Geophysical conditions.
- Environmental considerations.
- Cost of construction; and
- Future maintenance costs.

A direct seawater intake refers to an intake structures which extracts seawater directly from the ocean. This type of structure can either be constructed below the surface (sub-surface) at an offshore location, or as an open channel (surface intake) protected from waves by groynes or in a natural bay as illustrated in Figure 4.7.

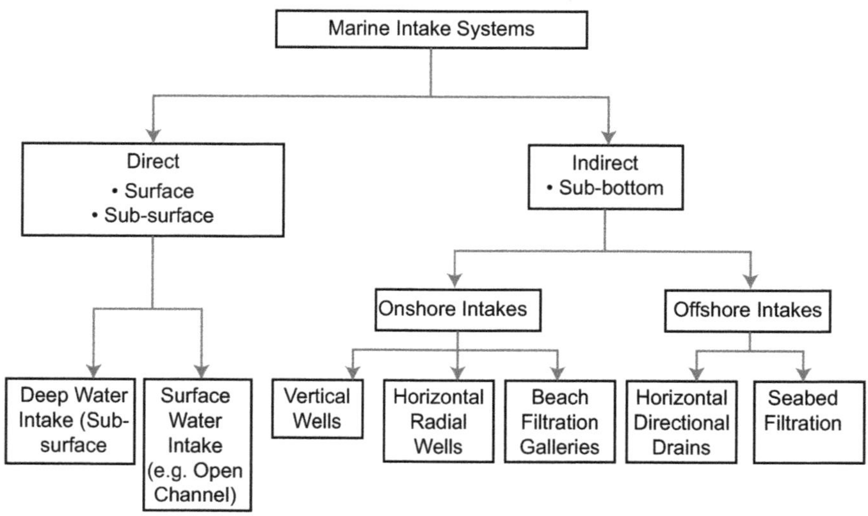

Figure 4.1. Types of intakes

Indirect intakes, also referred to as sub-bottom intake structures, extract seawater from a point below the seafloor or beach. Various sub-bottom intake types are available, depending on the local geotechnical conditions and specific project specifications. The indirect intake types include horizontal and vertical wells, beach filtration galleries, horizontal drains and seabed filtrations systems.

We will focus on the four major alternatives of intakes for seawater desalination plants, that are:

1. Open intakes.
2. Well water intakes.
3. Horizontal drains.
4. Natural pit.

Vertical well intakes are used in brackish water desalination plants

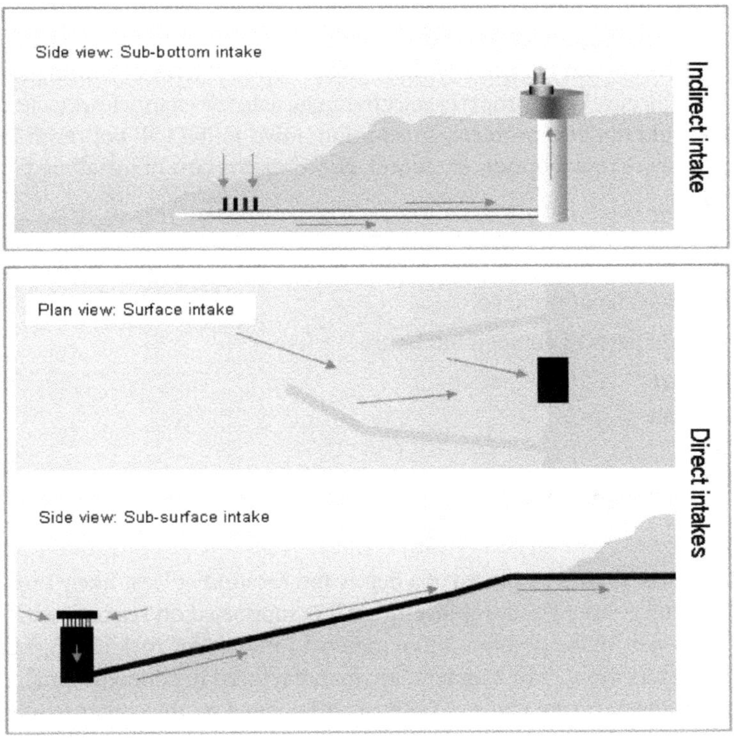

Figure 4.2. View of different types of intakes.

4.1 Open intake.

Open intakes usually have a singular collection structure that is located in a convenient position, at a sufficient sea depth, where the water enters into the underwater pipelines, which duly transport the same to land, whereby a pump station pumps the water to its destination.

Knowledge of the geotechnical characteristics of the underlying seabed where the tower and pipeline is to be positioned is of upmost importance.

Most desalination plants withdraw water from one to six meters below the water surface. However, at these depths, the presence of sand, fish, seaweed, algae, jellyfish, and microorganisms might be a concern. Better water quality can be obtained at depths of greater than 35 meters below the surface where debris loads can be 20 times lower than at the surface. However, water depths of 35 meters are usually not available within 500 meters from the shoreline. Minimum water depth should be between 15-20 meters and distance to coast should be 400 meters to guarantee good water quality.

Desalination plants are often designed to take shallow water to avoid the high cost of pipeline or other transmission methods.

For sandy and silt bottoms, wave action could result in sand and silt being placed in suspension causing higher turbidity and worse water quality. A detailed investigation of sediment gradation, settling rates, littoral transport patterns, currents, and wave activity is required for the design of intake in silty or sandy bottoms.

For rocky bottoms, the problems of littoral transport, erosion, and suspension of particles due to wave activity are not significant considerations in designing channel type intakes and water quality is good. The bottom is generally more stable and problems of shifting sand and sediments should not occur. With properly selected materials for channel protection, erosion of channel walls should not be a problem. In addition, wave action will not result in the stirring up of sediments where rock bottoms are found. However, the cost of installing piping in rocky bottoms is higher.

It must be taken into consideration that pumping facilities must be designed to work when the intakes operates at maximum capacity and the source water elevation is at its minimum, so piping head loss must be minimized.

The most important parts of an open intake are:

1. **Intake structure**
2. **Pipe connection**
3. **Shore well**
4. **Reactives piping**

Intake structure

The feed water intake should be placed at a depth and location of least likely impact to reduce the potential for entrainment and impingement. It is supported on sea bed, with an elevation from bed to the water intake point of 3 to 6 meters (3 meters for rock and 5-6 for sand). Big sea intakes (diameters more than 2 meters) are manufactured in concrete. Small and medium intakes (less than 2 meters) are made of GRP (It will depend on the construction company).

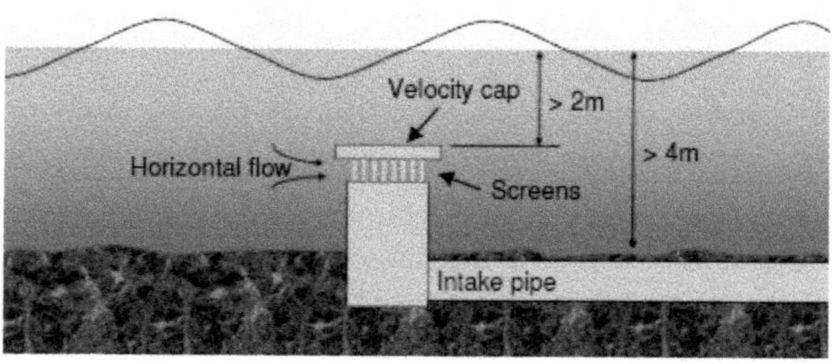

Figure 4.3. Details of intake structure.

The velocity of water entering an intake structure exerts a direct physical force against which fish and other organisms must act to avoid impingement or entrainment. The appropriate velocity thresholds should be based on the maximum current velocities of which the local fish species can tolerate without getting forced into the intake pipes. EPA recommends a threshold of 0.15 m/sec.

Two velocities are of importance in the design of intake structures: the approach velocity and the through-screen velocity. The approach velocity is the velocity measured just in front of the screen face. This velocity has the most influence on an aquatic organism and its ability to escape from being impinged or entrained by the seawater intake structure.

As commented, diameter of towers must guarantee a velocity lower than 0.15 m/sec (example: for a flow of 25000 $m^{3/}$day, the diameter would be 3.2 meters).

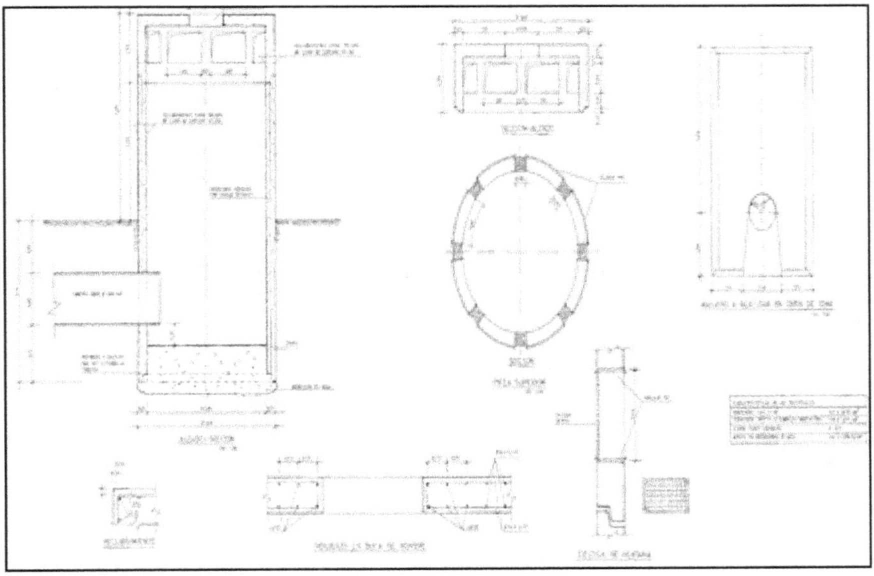

Figure 4.4. Constructive drawing. Open intake.

Glass reinforced plastic (GRP) screens with a pass which will range from 15-20 mm and a water pass velocity of 0.25 m/sec will be installed. If biological growth is possible, it will be necessary to clean the area at least once a year.

It should be possible to have access to check piping through opening and manhole (An opening diameter of 800 mm/1000 mm at the top).

Passive metallic Screens is an alternative. They comprise slotted screens, which are constructed of trapezoidal-shaped "wedgewire" and aligned on a horizontal axis with openings from 0.5 mm to 10 mm. The cylindrical shape dissipates the through slot-velocity allowing organisms to escape the flow field. The diameter of these passive screens ranges from 0.4 m to 2.1 m and the length ranges from 0.5 m to 8 m.

These screens can reduce impingement and entrainment (up to 80%) of marine organisms effectively. The screens are normally cleaned with an air backwash system when debris accumulates and adequate counter-current flow is needed to transport organisms away from the screen. The materials used will be 2205 Duplex and Z-Alloy with a high biological resistance. Water pass velocity is approximately 0.135 m/s. Due to low inlet velocity of seawater, impingement and entrainment of marine organisms is reduced by 80%. These screens are cleaned by automated air backwash system.

There are four main components of the air back wash which are listed below

(Johnson Screens, 2010):

1. A compressor is used to recharge the air in the receiver tank.
2. A receiver tank which then stores the compressed air supplied by the compressor.
3. A distributor system, required for multi-screen application with air-actuated valves and piping for the air to flow from the tank to the desired passive screen.
4. A control panel for either manual or automated service to carry out the cleaning process.

This short burst of compressed air is released after every few hours to repel debris and zebra mussels from the screens. Also, the screens slots are designed in a V-shaped wire specifically to prevent clogging as the wire allows only two-point particle contact.

Figure 4.5. Johnson screens. Photograph.

Johnson Screens is the main manufacturer of these type of screens.

Another interesting screening option for desalination intakes is proposed by the company Taprogge. It combines the functions of the classic multi-stage pre-screening systems in a single stage. It is a system for the mechanical treatment of water at the place of extraction by making use of hydrophysical principles. The material of filter elements is modified polyamide with the housing in stainless steel.

Figure 4.6. Taprogge screens. Photograph. Courtesy of Taprogge.

Pipe connections and Marine pipelines.

The following section summarizes the general considerations which have to be taken into account when designing sea pipeline.

Usually one or two pipes are installed depending on material and size and must be designed for low velocities between 1 to 1.5 m/sec.

Marine growth inside a seawater intake pipeline could have a major impact on the hydraulics and subsequently the operation of a desalination plant. Therefore pipe materials which are more resilient to marine growth such as HDPE, GRP or concrete should be considered. Material selection is purely economic.

Usually for diameters up to 1200 mm, HDPE is used. NP6 or NP8 is enough to handle the pipe pressure in the sea. Less pressure can lead to a bigger insecurity. It is a material with a high chemical and corrosion resistance, easy to fabricate, flexible and has many jointing options. The welding process is carried out off shore. Next phase of the process is to install, concrete weights. These are fixed to the pipeline at a certain centre distance. The distance can vary along the pipeline, depending on the calculated forces to act at a special depth. The weights can be installed on shore or off shore. When all pipe sections are fitted together, the pipeline is ready for the sinking process. The pipeline is equipped with blind flanges in each end. At the outmost end, the blind flange is also equipped with pipes and valves for air evacuation and air filling.

The total pipeline is positioned in the correct route by boats, barges and small boats. The inmost end is connected to the flange in the Outfall Shaft. There must be a pipe through the wall in the shaft, so that seawater can enter the shaft during the sinking.

The sinking starts by opening the air valve in the outmost end carefully and controlling the inside pressure by a manometer, charge the pipe with compressed air if required. Beforehand there will be a calculated curve showing the necessary air pressure as a function of the sinking depth. By regulating the inside pressure according to this curve, we will get a controlled sinking with a nearly constant speed. The sinking velocity may be approximately 0.3 m/sec.

The critical factor is the radius of curvature at the sea surface. If this radius is less than approximately 50 meters in this case, the pipeline runs the risk for buckling (safety factor of 2). It is necessary to carry out the sinking operation as a continuous process.

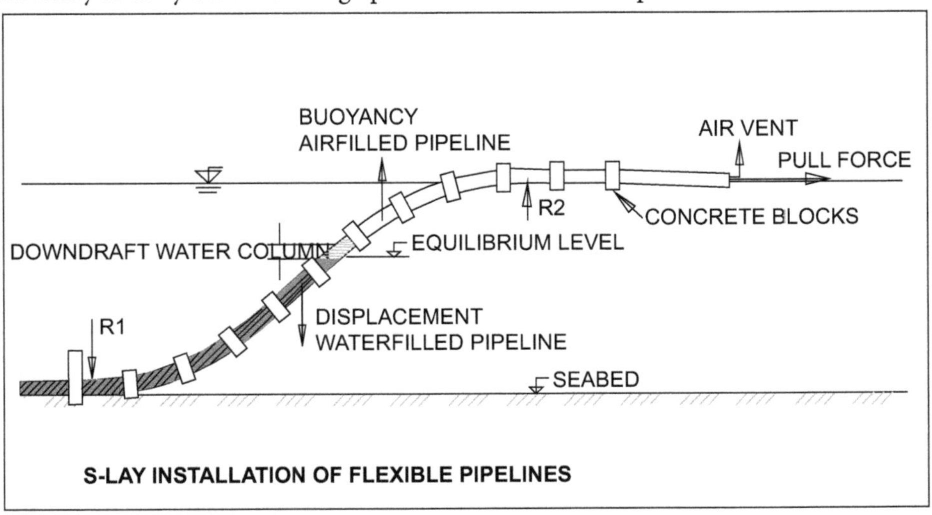

Figure 4.7. Installation of flexible pipelines.

Ballast weights will vary from 15 to 30 tons and distance will not exceed 10 times the outfall diameter. Concrete will be resistant to the chemical aggressivness of seawater. Metallic connecting elements must be made of galvanized carbon steel with sacrifice anodes.

Figure 4.8. Sea outfall. Concrete blocks. Photograph Courtesy of TML&CEVTAS.

For larger diameters, GRP is used. (Connection bell every 12 meters).

When pipe jacking method is used for the installation of underground pipelines, steel reinforced concrete pipes are usually chosen. Concrete pipes can be pushed forward immediately after the soil is excavated. The coupling consists of a stainless steel ring, so it can be potentially corroded.

Below, it is shown the P&ID of an open intake with screening, pumps.

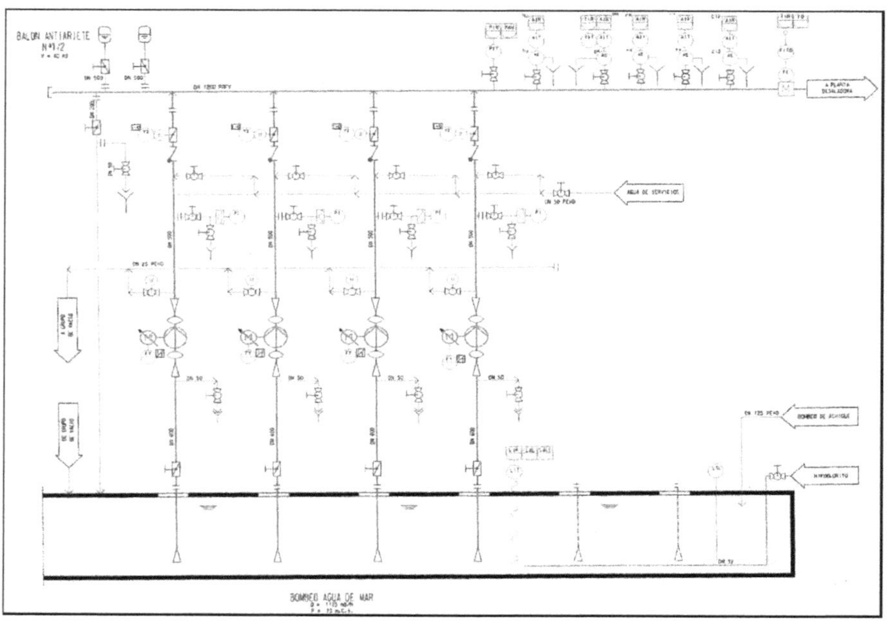

Figure 4.9. P&ID of water Intake. Courtesy of AQUASWISS.

Shore well

The Intake structures are usually constructed of concrete and are arranged to gather the water into a localized area for pickup, serving to support the pumps, as a screen chamber and as a source water pump suction well (see figures below) including isolation penstocks for maintenance. The depth of the shore shaft must be adequate enough to allow for drawdown when the intake operates at maximum capacity and the source water elevation is at its minimum. In addition, the well must provide an ample submergence allowance for the source water pumps. The well must also be capable of resisting and dissipating surges that occur during a power failure. Some shore wells are constructed with a shaft large enough to readily discharge the surge back to the supply source. Usually the depth is about four (4) meters.

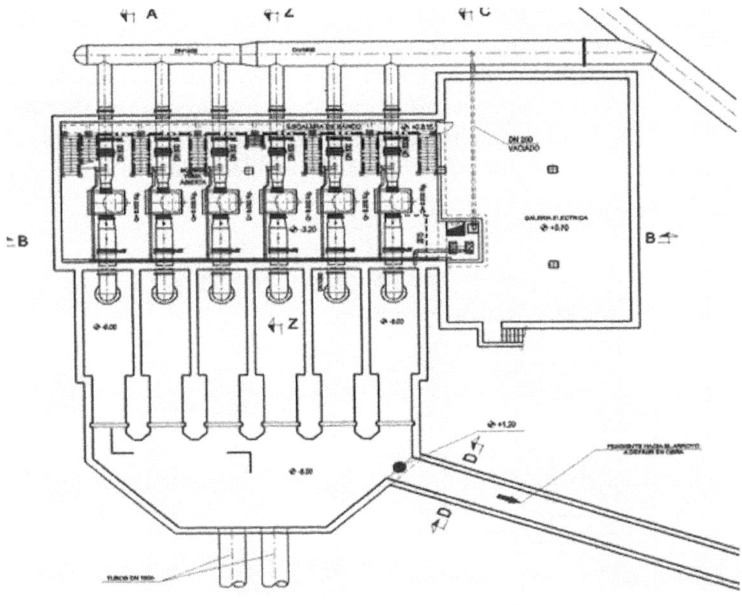

Figure 4.10. Seawater open intake layout.

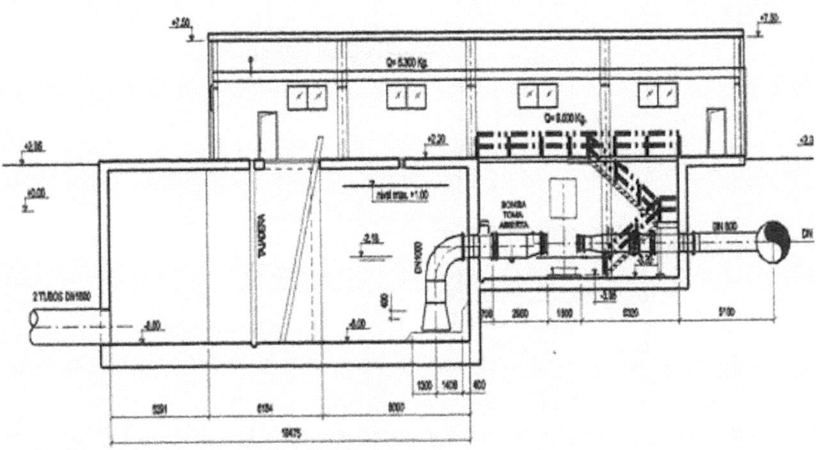

Figure 4.11. Seawater open intake section.

The optimum design will bring relatively clear water directly into the pumps suction area at a low velocity.

It is recommended that the submergence of the pumps and the dimensions of the suction pit in the immediate vicinity of the pumps suction inlet is set up as suggested by the pump manufacturer. Preliminary intake drawings, usually need to be prepared for making studies and estimates and for writing specifications for equipment. During this preliminary stage of intake design for vertical wet pit pumps (or for dry-pit pumps having vertical suction pipe with a bell-mouth entrance), the recommendations found in the American National Standard for Pump Intake Design, ANSI/HI 9.9-1998 should be followed. Figures 4.10 and 4.11 show the basic layout of the pumps and intake structure.

Geometry is generally defined in terms of the pump inlet bell diameter, as shown.

Once the number and size of pumps required is determined, a pumps inlet bell diameter can be estimated. At this point, the bell diameter can be estimated based on an inlet pipe velocity of between 3 and 8 ft/s (0.9 and 2.4 m/sec). The resulting pipe diameter can then be converted to a corresponding bell diameter approximately 1.5-2.0 times the size of the inside pipe diameter. With the bell diameter selected, the proportions of the inlet structure can be estimated from Figure 4.12.

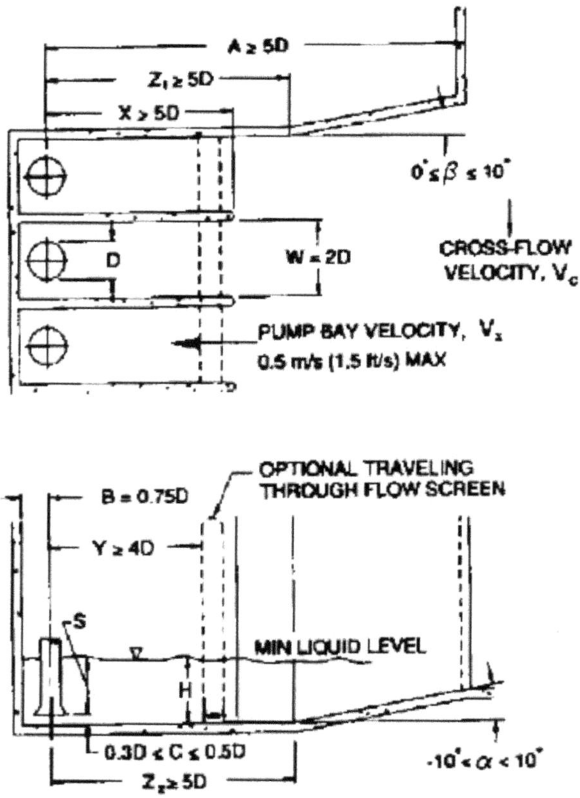

Figure 4.12. Proportions of the inlet structure ANSI/HI 9.9-1998.

For establishing velocities, the minimum submergence over the suction bell for vertical wet-pit pumps can be estimated based on maximum expected flow rate.

Once the intake pump manufacturer has been selected, final intake dimensions can be established on the basis of actual pump inlet pipe dimensions and the equipment suppliers recommendations. Should there be an appreciable variation in any of the dimensions from various sources, a model test of the intake structure is justified and recommended.

For further guidance, ANSI/HI 9.8-1998 provides intake design recommendations for both suction pipes and all types of wet pits that are a result of the combined efforts of sump designers, hydraulic researchers, pump manufacturers, and end users. It is intended to provide designers, owners, and users of pumping facilities a foundation upon which to develop functional and economical pumping facility designs.

Depending on the probable source of debris, the protection usually includes only a bar trashrack with a spacing of 20 mm, but when the source can contain algae or other marine elements, an additional automatic screen with a spacing of 2-3 mm will be installed. Usually three (3) or more wide screens are installed (maximum of 2 meters width).

Penstocks (usually manual) are installed to isolate screens.

If sediment deposit (lime) is likely to happen (sea bottom with sand, current), lime and sand removal channels (including extracting pumps) could be installed. Channel velocity must be very low with a retention time of approximately 15 minutes. The number of channels must be the same as number of screens.

Split casing vertical (with motor in vertical position) pumps are usually installed in large scale desalination plants to reduce space requirements. Split casing pumps are less costly than vertical pumps.

Figure 4.13. Photograph of Split casing vertical (motor in vertical position).

Centrifugal pumps must be submerged deeply enough to provide a pressure sufficient to prevent cavitation in first-stage impeller pumps, referred to as *NPSH* (it is assumed that the proper pump has been selected to perform satisfactorily with available *NPSH*). Usually pumps with a low pH, between 4 and 5, are required.

Calculation of $NPSH_d$ (available)

$$NPSH_d = \frac{P_a}{d} - P_v - \frac{H_g}{d} - \frac{H_l}{d}$$

d = water density.

P_a = Atmospheric pressure.

P_v = Vapor pressure (function of water temperature).

H_g = Head loss.

H_l = Water elevation – Minimum tide level – head loss (intake + pipe + screen)

A security margin of approximately 0.5 meters must be considered between the upper part of pit and maximum tide level.

Priming systems are fully automated vacuum units designed for level-controlled evacuation of the pumps. The design usually includes a surface mounted liquid tank which allows for self supply of operating liquid to the vacuum pumps, two (2) liquid ring vacuum pumps, a level control and service liquid solenoid. Level monitors for start up and dry running protection of the pumps are normally installed.

Pumps must be made of super-duplex (A890 GR5A or 1.4469) or duplex. If duplex is used (1.4517 or 1.4468), cathodic protection is required to prevent the piping from crevice corrosion in flanges or areas where there is not water circulation.

Pumps discharge to a common collector which drives water to the pretreatment system. GRP is the material used for collector and design pressure must be the maximum operation pressure.

A pressure transmitter (PT) and a flow transmitter (FIT) will be installed in the collector. FIT will be used to adjust dosing rate proportional to the flow.

EXAMPLE

Determine sump levels and dimensions.

Desalination plant production (initial): *2 m^3/s*

Desalination plant production (future): *2.1 m^3/s*

Head loss intake pipe (initial): *2.4 m*

Head loss intake pipe (future): *2.7 m*

Head loss at intake at intake (trash racks fully blocked): *0.42 m*

Considering the following extreme tidal levels

LAT = -0.966 m (Lowest astronomical tide to MSL)

For determining the maximum water level of the sump:

HAT = +1.004 m (Highest astronomical tide to MSL)

Pipe diameter: *1.2 m*

Velocity in pipe *1.8 m/s ; 1.9 m/s*

Calculation of minimum water of sump.

The minimum water level of the sump should ensure that the required feed water flow rate can be extracted during the lowest tidal levels, taking into account the maximum friction and local losses at the intake headworks and pipes. At both the ocean and sump level, the pressure is atmospheric and at the ocean, the velocity is negligible. Therefore, using Bernoulli's energy equation:

$$Z = vpipe^2/2g + zsump + hf(pipe) + hL(pipe) + hf(intake)$$

where: vpipe = average pipe velocity (m/s).

z = difference between sea surface and sump level (m).

zsump = set sump level to zero (m).

hf = frictional head loss (m).

hL = local head loss (m).

$$= 1.9^2 / (2 \times 9.81) + 0 + 2.7 + 0.42 = 3.3m$$

The lowest required water level of the sump should be 3.3 meters below lowest tide in order to supply the required extraction rate.

Therefore: ***zsump (min) = -0.966 – 3.3 = -4.27 m to MSL***

Calculating the maximum water level of the sump

The maximum required water level of the sump while the plant is in operation will depend on the highest tidal level, taking into account the minimum friction and local losses at the intake headworks and pipes. Using Bernoulli's energy equation, it was determined as follows:

$$= 1.8^2 / (2 \times 9.81) + 0 + 2.4 + 0.4 = 2.97m$$

Taking into account the sea water level during the highest astronomical tide (HAT), the highest required water level of the sump should be 2.97 meters below HAT in order to supply the required extraction rate.

Therefore: ***z sump (max) = +1.004 – 2.97 = -1.97 m to MSL***

Dimensions of the screening sump

The dimensions of the channels for the screening sump of the feedwater must be based on the same requirements for screening chambers of domestic wastewater works.

- *The average approach velocity should vary between 0.45 and 0.6 m/s;*
- *The maximum approach velocity should not exceed 0.9 m/s;*
- *The minimum approach velocity should not fall below 0.3 ms/s;*
- *The minimum and maximum width of the approach channels is 0.6 and 1.2 m;*
- *A channel length of more or equal to 7.6 m is required in front of the screens.*

Since the pipes themselves have a diameter of 1.2 m, approach channels of 1.5 m will be considered instead of the specified 1.2 m. Therefore, to achieve a maximum velocity of 0.9 m/s for the lowest water level in the sump, a channel width of 1.5 m and a flow rate of 2.1 m³/s, the water height should be 1.55 m. Subsequently, the invert level of the sump should be 1.55 m below the lowest water level in the sump (-4.27 m MSL), which means that the base level of the sump should be at a level of -5.82 m MSL.

During operation the highest water level in the sump is -1.97 m MSL therefore the water height, taking into account the base of the sump level is -5.82 m MSL is 3.85 m. Therefore, calculating the cross-sectional area, together with a flow rate of 2.0 m³/s, the velocity would be 0.35 m/s, which is more than the specified minimum velocity of 0.3 m/s.

Taking into account the above calculations and requirements, the screening sump's dimensions are schematically illustrated in Figure 7.1

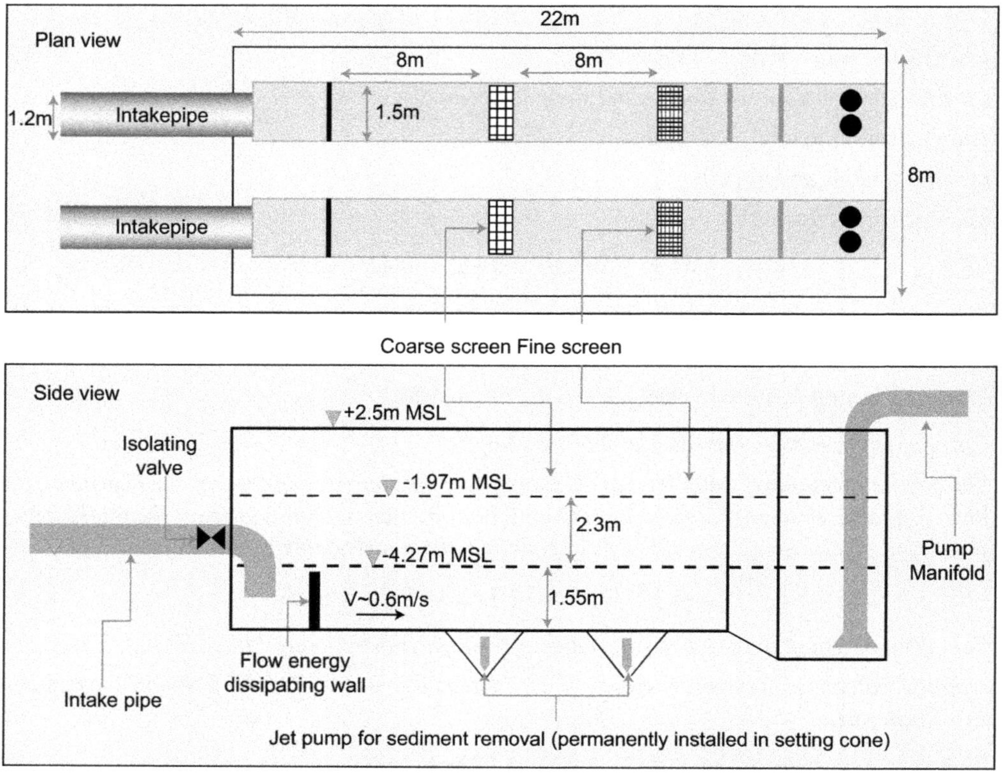

Figure 4.14. Screening sump dimensions.

Pump calculation (Flow and pressure)

Desalination plant production: *20.000 m³/day*

Desalination plant production (future): *27 000 m³/day*

Reverse osmosis recovery: *45%*

0.7% of production for internal services

2% of raw water for filters backwashing

Required raw water flow will be:

$$\frac{20000\ m^3/day \times 1.007}{0.45 \times 0.98} = 45660\ m^3/day \approx 1903\ m^3/h$$

$$\frac{27000\ m^3/day \times 1.007}{0.45 \times 0.98} = 61653\ m^3/day \approx 2569\ m^3/h$$

Four (4) Split case single stage pumps will be installed with a design flow of 636 m³/h.

Pump manometric head will be calculated from:

$$H = Hg + \Delta PT + \Delta PFA + \Delta PFC + Hasp$$

Where

H = Required manometric head.

Hg = Geometric elevation between tank and ground level.

ΔP_T = *Piping head loss.*

ΔP_{FA} = *Sand filters head loss.*

ΔP_{FC} = *Cartridge filters head loss.*

H_{asp} = *High pressure pump suction side pressure.*

4.2 Subsurface Intake. Beach well.

The collected flow from a subsurface intake usually contains low level of solids, silts, oil and grease, natural organic containments and aquatic. Consequently, CAPEX (simple pretreatment) and OPEX (less chemicals consumption, less cleaning of membranes) are lower compared with an open intake.

4.2.1 Concepts.

1. Lithologic column: Composition and characteristics of soil.
2. Piezometric: Pipe of ND80 or ND100 to measure liquid level. It is useful to calculate the cone of depression.
3. Monitoring test: Test to evaluate the permeability of the land at several depts., influence cone.
4. Cone of depression: Directly related to the drawdown in the well. As the pump draws down the water level, a portion of the aquifer surrounding the well is drained in water. A cone shaped depression is formed in the water table around the well. The shape of the cone will vary depending on the type of formation in which the well is located. A fine sand formation will usually create a steep cone of depression, while a shallow cone is usually found in coarse sand and gravel formations. To minimize the overlap of the pumping depressions once the new well is in operation, it must be analyzed in detail the depth of boreholes.
5. Interphase: Separation line with salty water.
6. Borehole "development": Cleaning time with water and air to extract "detritus" and remove bentonite.
7. Bentonite: Material with a high content of Montmorillonite (57% SiO_2 + 20% Al_2O_3 + Lime oxide + magnesium).
8. Casing Well: casing is a steel or plastic pipe installed to keep borehole wall from collapsing and houses the submersible pump and drop pipe.
9. Gravel pack: A gravel pack around the well column increases the effective radius of the well and filters out formational materials such as sand and silt from entering the well through the screens. The typical thickness of the gravel pack is 4 to 6 inches. Well rounded, graded gravel high in silica should be used in the gravel pack. Gravel size is determined based upon the grain size distribution in the strata targeted for screening. The gradation of the gravel pack will also influence the size of the slot openings in the well screen.

4.2.2 Types of drilling.

There are different types of drilling depending on the characteristics of soil. Direct/Reverse rotary circulation systems are mostly used in seawater desalination.

- **Percussion.**

Percussion drilling is a drilling technique in which a drill bit attached to rope or cable is repeatedly raised and lowered, impacting soil and rock, and making the hole deeper. Equipment is easy to build, transport, and simple to use. These drills introduce less contamination than conventional hand drilling methods, and this technique can drill a narrower and deeper hole than hand drilling through many different types of soil and rock.

If the substance being drilled through is sturdy enough, drilling can continue until water is reached. If it occurs in loose soil or sand, a pipe may need to be inserted to keep the walls from collapsing. After the well is deep enough, permanent casing is installed.

- **Rotary-percussion.**

The rotary-percussion drill combines both the rotary (grinding) and percussive (chipping and hammering) action. A hole is formed when energy is transmitted pneumatically through the drill rod to the drill bit, prompting it to thrust into the formation using a repeated hammering motion. The impact of the percussion component is enough to break and dislodge the rock, which subsequently removes debris and cuttings with compressed air or water flow.

The rotational component of the drill occurs as the drill is producing percussion strikes against the formation. Rotary-percussion drilling is usually used for depths that do not exceed 492 feet (150 meters).

The rotary-percussion drill usually requires compressed air to drive the drill bit into the rock. The compressed air has a dual function-it also flushes out the cuttings produced from drilling when the power is transmitted, the drill bit rotates at 10–30 revolutions per minute on a down-the-hole hammer drill for a straight, circular hole.

The rotary-percussion drill is capable of drilling through the hardest consolidated formations, including gravel, very quickly. The diameter limits are between 250 mm and 300 mm.

- **Direct/Reverse rotary circulation systems**

Water, air and sludge rotation drilling.

There are two systems depending on the direction of movement of the drilling fluid:

- **Direct:** The drilling fluid is circulated from the drill stem and then flows up the annulus between the outside of the drill stem and borehole wall. The drilling fluid that carries the drill cuttings flows to the surface and the subsequent settlement pits. Pumps are then employed to lift the cuttings and cycle the free fluid back to the drill stem.
- **Reverse:** The direction of flow of drilling flow is opposite to that of direct circulation drilling. Drilling fluid flows from the annulus between the drill stem and hole wall to the drill stem. The drilling fluid is pumped to a nearby sump pits where cuttings are dropped and settled.

This system uses bentonite to support borehole except in the upper part of borehole, where carbon steel will be used.

Diameter of borehole is usually 500 mm with the initial 10–20 mm of 800 mm; sometimes it is necessary to install a support pipe.

4.2.3 Recommendations before drilling.

Before definitive borehole, it is necessary to collect soil samples, water samples or rock cores, to advance in situ sampling equipment, or to install monitoring wells or piezometers.

1. Well log.

A well log provides information to make well design decisions. It is a written record describing the gravel, sand, silt, and clay in the aquifer formations found during the drilling process. The well log helps identify the depth and thickness of strata that will be the most productive and most effectively transmit water into the well.

2. Monitoring well.

A pipe bored into a groundwater source that collects water for testing purposes.

4.2.4 Soil type.

Find below the classification of the different soil types

1. Fractured limestone gives good quality water.
2. Compact limestone gives very little water.
3. Sand and gravel. Less flow and worse quality of water due to the entrainment of silt.
4. Clay - No water production.

4.2.5 Well elements.

1. Flow.

For the construction of a good quality well, it is essential if the soil is permeable or impermeable. Permeability is a measure of the ability of a soil type to transmit water through it. Flow will depend on permeability of soil, varying from 60 to 150 l/sec.

2. Definitive well casing.

Water well casing may be constructed of steel or plastic. However, permanent steel casing is not used in desalination as it may corrode in certain ground waters and also encourage the development of biofilm slimes.

The driller should certify that the casing is suitable for the specific borehole. Factors that affect casing selection include: cost, diameter, drilling method, water quality, strength requirements, and corrosion resistance. The joints between lengths of casing must be watertight.

Plastic casing may be of several types: Polyvinyl chloride (PVC), Acrylonitrile butadiene styrene (ABS), rubber-modified polystyrene (SR), High Density Polyethylene (HDPE), polyolefin, polypropylene, and Glass-reinforced plastic (GRP or 'glass fiber'). However, in seawater desalination, the most employed system is made of threaded joints PVC pipe. The most important manufacturers are Preussag Engineering, Boodeand Pumpenboese (GWE).

Plastic casing must have sufficient strength to withstand the pressure and setting temperature of the cement grout that will be placed in the annulus outside the casing.

Below the casing, and in line with the aquifer, is another 'casing' with fine slots. This is the well screen, where the slots allow the water to enter the well. It holds back sand and larger particles trying to enter. Slots pass is 1.5, 2 or 3 mm. Three (3) mm is the most common slot. Pass will depend on flow, soil characteristics and it will be calculated for a velocity of 3 cm/sec.

To improve the filtering function, the borehole typically features a diameter that is 2–3" larger than the casing. A fine sand gravel pack filter is placed between the casing and the aquifer when the aquifer comprises sand to fine gravel grain sizes. Some casings come with a pre-made gravel pack filter. Made correctly, this filtering method prevents sand and silt from entering the well.

3. Pump riser pipe.

The pump is supported by a Steel riser pipe.

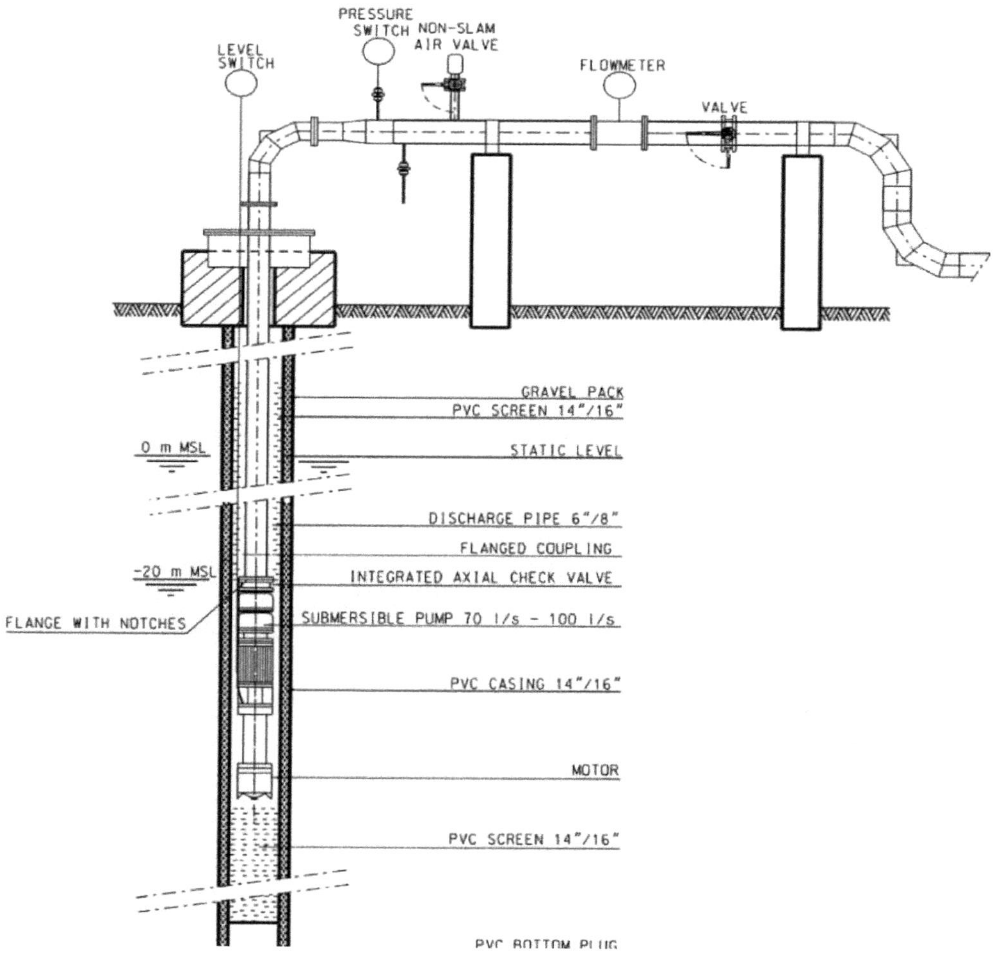

Figure 4.15. Well elements.

To prevent corrosion, the materials used are duplex + cathodic and anodic protection or super-duplex for the elements most prone to corrosion such as bolts and flanges and duplex for pipe. Riser pipes are made with flanged connections every 6 meters (flanges are special). Piping schedule is 10S.

Also of note are rubber coated riser pipes. The results obtained in seawater applications are good. They are supplied by the manufacturer HAGUSTA of Preussag group.

4. Civil.

Concrete sump of approximate dimensions 4x1.5x2 meters. Drainage must be included.

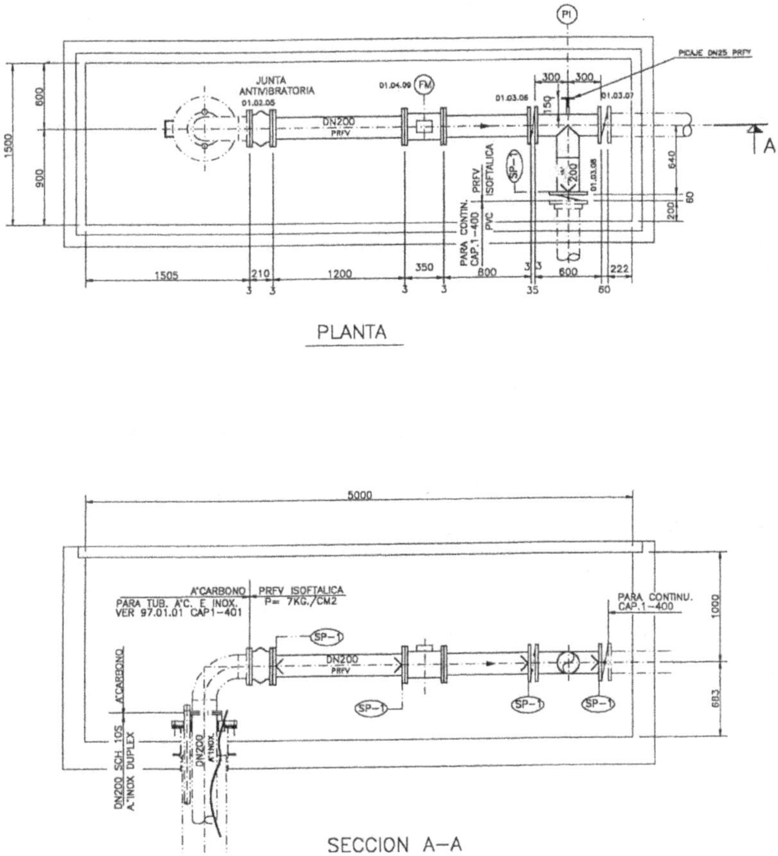

Figure 4.16. Detail of well sump. Courtesy of Pridesa. S.A.

5. Submersible Pump.

A submersible pump consists of a centrifugal pump coupled directly with an electrical motor in a common housing. As the name indicates, the assembly is designed for submerged operation within the tube well. The pump assembly has four major parts:

- Motor unit.
- Suction strainer.
- Bowl type centrifugal pump assembly.
- Column or riser pipes.

The electric motor is a fixed speed squirrel cage induction motor, normally two-pole, with water cooled winding, operating between 1500–3000 rpm depending on water characteristics. If sand is not present, 3000 rpm will work well. If sand content is present in water, 1500 rpm will be the best option. The rotor shaft is normally connected to the pump by a mechanical coupling located within the strainer unit.

The suction strainer is located above the motor and below the pump. Its function is to permit the entry of the water into the first stage of the pump, while at the same time preventing any sizeable matter from entering and damaging the impellers.

The pump unit is made up of a series of pump bowls and impellers connected in series on a single shaft. The head (pressure) developed is dependent on the number of stages in the pump. The impeller diameter and design govern the quantity of water pumped. The outlet of the pump unit is connected to the column pipes or rising main which in turn is connected to the surface pipework. The rising main is usually made of steel or cast iron with flanged or screwed connections.

Stainless Steel 904L, duplex with cathodic protection or super-duplex should be the materials used for seawater desalination plants. Experience has proved that AISI 316 is not suitable due to corrosion problems.

The motor must have a thermal protection PT-100 or PTC's. The position of the motor is important so that the circulating seawater cools the motor (The motor must be installed above the water inlet). Flow past the motor should be set to a minimum of 0.15 m/sec. in order to secure proper cooling of the motor.

If the minimum flow past the motor cannot be obtained the natural way, Grundfos offers a range of cooling sleeves that ensure correct flow and cooling, and are easy to work with. Flow sleeves are typically used when the pump is installed in a reservoir or tank, or in a well, where the water flows to the pump from above, and therefore does not cool the motor. There must be reasonable spacing between the casing and the outer diameter to limit the pressure drop.

Pumps suppliers are: Flowserve, EMU, Grundfos.

Usually the size of pumps is ND 300.

6. **Piping, isolation valves and instrumentation.**

- ND 60 pipe to install the level transmitter. Polyester material will be used for pressure higher than 5 bars or PVC for lower pressure. PE could be an alternative to PVC.
- A by-pass pipe to drain the water when cleaning of well is required should be installed.
- Butterfly valves protected with halar or nylon are the standard type of valves used. The actuator can be either pneumatic or motorized, except the actuator of by-pass valve which is manual (with special type of soils it should be automatic).
- A flow transmitter to control flow (neoprene covered).

4.2.6 Horizontal wells.

Often referred to as radial or "Ranney" collector wells, after their inventor, this horizontal well consists of a reinforced concrete caisson sunk down into the coastal aquifer and well screens that extend out laterally into the formation from inside the caisson (Figure 4.17). The laterals can project radically or in a pattern oriented towards a surface water source if the intended water supply will come from induced infiltration or some preferential direction, such as in a coastal well application. The primary benefits of using a horizontal well (as opposed to a vertical well) are: 1.- It can be drilled from a central location with a long lateral reach (particularly advantageous for sites with limited above-ground access), and 2.- The borehole is exposed to a greater surface area within the geologic formation, so capacity is typically enhanced.

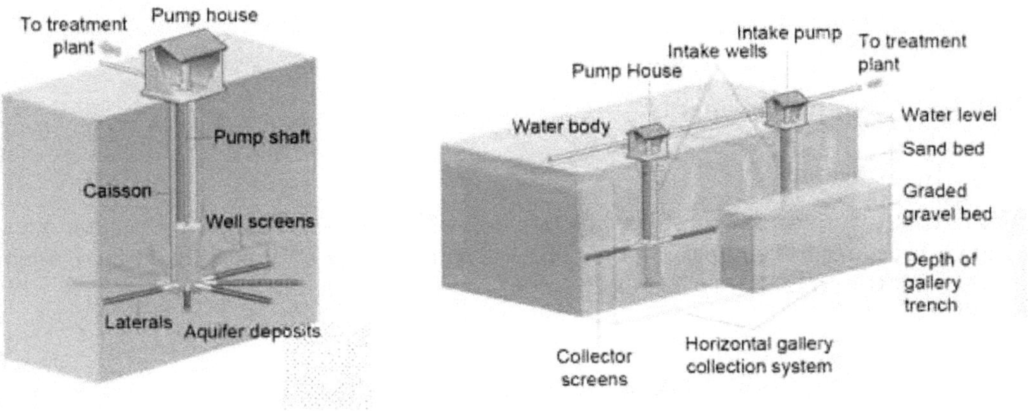

Figure 4.17. Detail of horizontal wells.

4.3 Horizontal directional drilling.

Consists of directional drillings set in geological structure under the sea bed, and fitted with HDPE drains. In beach well intakes, water quality is good (filtrated). Moreover, this system has following advantages:

- No impact on seawater bed, all works being made from the coast, unlike direct intakes.
- Poseidonia fields (protected species in Mediterranean Sea) can be then protected;
- Low overall dimensions for onshore development, inducing low visual impact.
- No impact of fresh water aquifer that may modify seawater salinity and add pollution.
- Constant recharge of seawater aquifer.

Maximum diameter of drains is 700 mm, with a maximum length of 600 m. The quantity of drains depends on soil porosity. Geological studies may be realized to define porosity and pilot drilling may be achieved to confirm the results.

Seawater velocity in the ground must not exceed 2.5 mm/sec. and velocity in drains should be around 0.6 m/sec. Consequently, a 700 mm diameter pipe should have a maximum flow rate of 230 liters/sec.

For seawater applications, the drilling should follow a designed profile below ground (5–10 m depth) and flow out under the seabed to exit on the sea floor, typically in a large-radius arc.

Depending on the soil conditions, a mixture of water and bentonite or other additives can be used for hydraulic excavation. Recent advances in horizontal directional drilling have focused on developing ways to drill using water in lieu of drilling fluids to enable more efficient well screens to be installed and to avoid problems such as smearing of formation soils during screen installation that would clog screen slot openings.

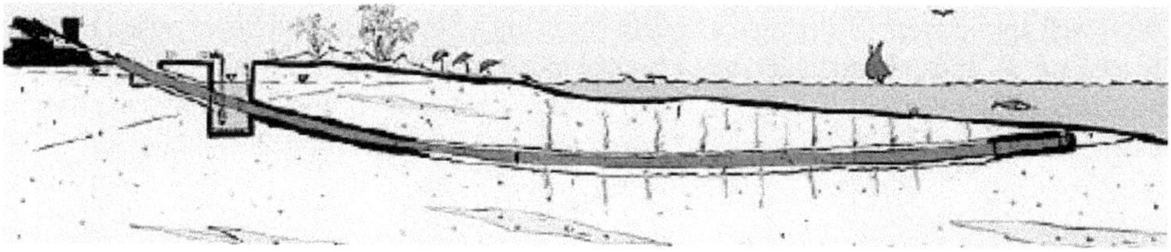

Figure 4.18. Horizontal directional drilling.

4.3.1 Required elements.

1. **Drill machine.**

This equipment typically consists of a rotary drill with a custom-designed drilling fluid program to return drill cuttings, cool the drill bit, and maintain hole integrity during boring, especially in unconsolidated deposits.

2. **Pipe.**

Polyethylene of different sizes depending on flow is the material used.

3. **Drilling fluids.**

Depending on the soil conditions, a mixture of water and bentonite or other additives can be used for hydraulic excavation. This supports the bore hole and reduces frictional forces, while allowing the excavated material to be transported to a separation plant on the surface.

4. **Infrastructure.**

Construction will require some activities on or near beaches or other public-use areas. In many cases, there will be limitations on when construction activities can occur based on seasonal uses for the location, environmental restrictions during certain breeding times, and other limited-access situations. It is also necessary to evaluate how much shoreline frontage is (potentially) available to the project to determine if sufficient property exists to fit in the requisite well system with appropriate spacing. The construction area for horizontal wells and slant wells is similar.

The physical dimensions for the area required for installation of machinery deposits is approximately. 500 m^2.

Area required in sea will be 3-4 times the area occupied by the drains.

5. **Isolation valves.**

Drains must be backwashed periodically. Backwashing frequency will depend on soil conditions type. Backwashing will be automatic (similar to filters). Two (2) automatic butterfly valves must be installed. Valves will isolate pumping pit drain and backwashing system. Valves must have a pneumatic or electric actuator.

6. **Backwashing system.**

Backwashing must normally be carried out with water (filtrated water from drains) for good results. However, it should be considered the option of air backwashing. An independent backwashing pump must be foreseen. The service stand-by pump could be used as an alternative with a flow 3-4 times the average flow of drains and a pressure of 1–1.5 bars.

7. **Desanding unit.**

In order to prevent the pumps from sand accumulation and erosion, a desanding system is installed.

8. **Pumps.**

Pumps installed in HDD wells that do not intersect a wet well or caisson will often need to be submersible so that they can be installed within the angled well casing.

Pumps should be manufactured in duplex material with cathodic protection or super-duplex.

Pit should be sized as a wastewater pit. Distance between pumps and walls will be determined by the manufacturer.

If vertical wet pit pumps (or dry-pit pumps having vertical suction pipe with a bell-mouth entrance) are used, the recommendations found in the American National Standard for Pump Intake Design, ANSI/HI 9.9-1998 should be followed.

9. **Instrumentation.**

The installation of a flow transmitter for each is recommended.

4.4 Natural Pit.

Natural pits include a large stone pitching to protect the intake. The retention time in the pit must be more than six (6) hours to guarantee that small size particles (sand) will settle.

Split casing vertical pumps are employed with an automatic priming system and are usually located far from the sea water intake. Plastic screens will prevent fish and other aquatic objects from being suctioned by the pumps.

Biological activity is very important (surface intake). This aspect should be considered by the design of pretreatment. If sand filtration is used, two stages will be required. For the design of pumps, it will be necessary to study NPSH and tides.

The raw water quality is worse when compared to the wells and it must be pretreated prior to entering the membrane desalination system; this adds significantly to both capital and ongoing O&M costs.

4.5 Collector.

Materials of collector are normally GRP. Design pressure will be the shut-off pressure of pumps. Collector supports, when operating at pressures above 3 bars must comply (special care at the end of collector) with manufacturer recommendations.

Depending on the length and elevation, it could be necessary to install an anti water hammer. It should have a membrane deposit in order to prevent seawater from coming in contact with the steel material.

A flow transmitter (FIT) in the common collector should be used to control reactive dosage. FIT in the collector can be substituted by the instrument installed in the inlet side of pressure sand filters (if installed) .

If a reactive (coagulant and/or acid) is dosed, a static mixer (as close as possible to the pumps discharge) should be installed in the collector.

4.6 Intermediate Deposit.

Installed between the intake pumping station and the pretreatment when intake is more than 1 km from plant.

Dry Split casing vertical (motor in vertical position) pumps with automatic priming system or centrifugal pumps installed in a dry chamber are the pumping alternatives. Dry well stations are the most expensive and complex option and should be avoided, if possible. Speed converters are normally installed so it is important to indicate to the supplier the maximum and minimum operation pressure.

A level transmitter will be used to control level of deposit.

5

SEA WATER DOSING SYSTEMS.

5.1 Reactives.

The following type of reactives are required in a conventional SWRO.

1. Disinfectants.
2. Oxidants.
3. Reactives to adjust pH.
4. Coagulants.
5. Antiscalants.
6. Dechlorination products.

5.1.1 Disinfectants.

Disinfection for RO/NF pretreatment has been applied usually where biological fouling prevention is required (typically for surface waters). Chlorine is added continuously at the intake, and a reaction time of 20–30 minutes should be allowed. A free residual chlorine concentration of 0.5–1 mg/L should be maintained through the whole pretreatment line.

Instead of continuous chlorination, chlorine is preferably applied off-line to the pretreatment section periodically. During off-line chlorination, the feedwater has to be sent to drain prior to reaching the membranes. Before the system goes into operation again, all chlorine containing feed water has to be rinsed out carefully, and the absence of chlorine must be verified (by monitoring of the oxidation-redox potential (ORP)).

1. Sodium hypochlorite & Calcium hypochlorite.

The reaction of sodium hypochlorite is:

$$ClONa + H2O \rightarrow ClOH + NaOH$$

The reaction of calcium hypochlorite is:

$$Ca(OCl)_2 + 2H2O \rightarrow 2ClOH + Ca(OH)_2$$

Hypochlorous acid dissociates in water to hydrogen ions and hypochlorite ions. The sum of Cl_2, NaOCl, HOCl, OCl^- is referred as free available chlorine. Chloramines are formed from the reaction of chlorine with ammonia compounds present in the water. These chlorine-ammonia compounds are referred to as combined available or combined residual chlorine. The sum of free and combined available/residual chlorine is called the total residual chlorine.

The germicidal efficiency of free residual chlorine is directly related to the concentration of undissociated HOCl. Hypochlorous acid is 100 times more effective that the hypochlorite ion OCl^-. The fraction of undissociated HOCl increases with decreasing pH.

Hypochlorite is more effective at pH between 4.5–6.5. At pH above 6.5, there is a loss of efficiency. At pH 7.5, (25°C, 40 mg/L), only 50% of free residual chlorine is present as HOCl but 90% is present at pH 6.5.

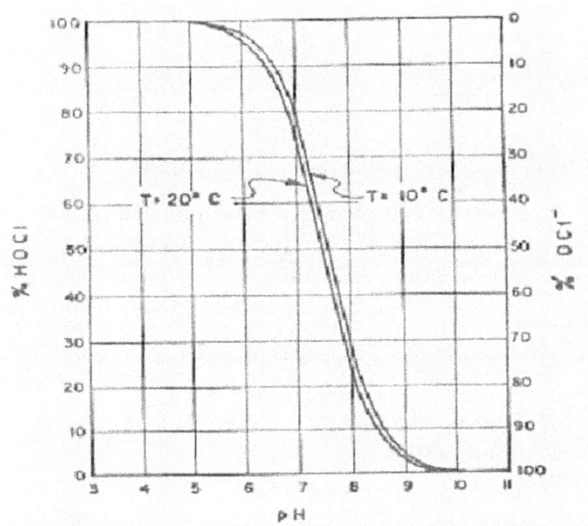

Figure 5.1. Relation pH/Chlorine efficiency.

To determine the optimum chlorine dosage and contact time to prevent biofouling, ASTM D1291 should be applied to a representative water sample.

EXAMPLE

SHOCK DOSAGE

F: Flow m^3/h 2.569

Dr: Dosage, ppm.. 20 (as Cl_2)

Cvr: Concentration of product, %.......................... 14 (as Cl_2)

r: Density, Kg/l 1.22

Dosing flow will be obtained (Fr en l/h) from

$$Fr = \frac{F \times Dr}{\frac{Cvr}{100} \times 1000 \times r}$$

Giving a value of 301 l/h

Mass flow will be obtained from (Mr en Kg/h)

$$Mr = \frac{F \times Dr}{\frac{Cvr}{100} \times 1000}$$

Giving a value of 367 Kg/h

A dosing pump with a capacity of 400 l/h will be installed

CONTINUOUS DOSAGE

F: Treatment flow m^3/h 2,569

Dr: Dosage, ppm...2 (as Cl_2)

Cvr: Concentration of product, %.....................14 (as Cl_2)

r: Density, Kg/l 1.22

Dosing flow will be obtained (Fr en l/h) from the same equation as above giving a value of 30.1 l/h.

Mass flow will be obtained from (Mr en Kg/h) giving a value of 36.7 Kg/h.

Two pumps with a design flow of 44 l/h will be installed.

<u>Chlorine dioxide.</u>

Chlorine dioxide is a greyish-yellow gas, highly soluble in water. It is generated on site. It is dosed diluted in water. It is more effective that chlorine at all the pH. Reaction with water is below indicated:

$$ClO_2 + 4H^+ + 5e^- \rightarrow Cl^- + 2H_2O$$

In a pH range of 6–8.5, chlorine dioxide remains in solution as dissolved gas. The limitations of the chlorine dioxide observed are the chlorites production which can be 30% of ClO_2 concentrations. Due to the fact that the chlorites are not removed during the pretreatment, their impact on the reverse osmosis membranes through an eventual oxidation is possible. The second limitation is often due to the operating cost because ClO_2 is more expensive that sodium hypochlorite.

5.1.2 Potassium permanganate.

Reactive is used to oxidize water and reduce organic matter content. The optimum dosage is usually in the range of 1–2 mg/L, but should be determined case by case. The minimum retention time should be between 20–30 minutes.

5.1.3 Reactives to adjust pH.

The pH adjustment step of pre-treatment must result in the optimal pH level for the desalination system. After coagulants have been added, the pH is often changed significantly. In most cases, the pH must be restrained to a neutral or a slightly acid level.

1. Sulfuric Acid.

Used to reduce seawater pH which usually is about 8 and the tendency is to work in pH between 6.5 to 7. By adding H^+ as acid, three objectives can be achieved:

- At a low pH the bacteriological growth will be lower.
- To get an optimum pH for the performance of membranes.
- By adding H^+ as acid, the equilibrium can be shifted to the left side to keep calcium carbonate dissolved. To control calcium carbonate scaling by acid addition alone, the LSI or S&DSI in the concentrate stream must be negative. Acid addition is useful to control carbonate scale only.

Sulfuric acid reacts with bicarbonates dissolved in water as can be seen from the following equation:

$$SO_4H_2 + 2CO_3H \rightarrow SO_4^{--} + 2CO_2 + 2H_2O$$

Sulfuric acid increases the sulfate ion level in the RO feed, which directly increases the potential for sulfate-based scaling. This aspect must be taken into consideration when adding an antiscalant. Hydrochloric acid is preferred when calcium sulfate, barium sulfate, or strontium sulfate scaling is a concern.

2. Metabisulfite.

In some countries, where it is not possible to have access to acid, it is usual to dose metabisulfite in well waters. This product will be used as disinfectant and will reduce the pH.

3. Caustics.

Caustics, for a few process applications, can be injected to increase the RO feed pH. Typically the only caustic injected is sodium hydroxide [NaOH] based on its cost, its availability, and its solubility in water.

Most times the quality of the NaOH can be technical grade (and free of any additives). If the technical grade NaOH is not clear in color (e.g. has a blackish tint), then a higher grade of NaOH may be desirable such as rayon or nylon grade. NaOH is commercially available as 100% solid flake or as 20% or 50% solution. Caution is required in storing 50% caustic since it can freeze at temperatures just below 60ºF (15.5ºC), with a special concern for cold air drafts.

Caution is required in raising the feed pH as it can decrease the solubility of iron, manganese, and calcium carbonate as measured by an increasing LSI.

The most common process application that uses a caustic feed involves two pass RO systems.

In a two pass RO, the permeate from a 1st pass RO becomes the feed to the 2nd pass. The 2nd pass RO "polishes" the 1st pass permeate to quality levels as good as 4 megohm-cm. Caustic is introduced into the 2nd pass feed for 4 reasons:

- At a pH of 8.2 and higher, all carbon dioxide gas is converted into the bicarbonate ion. The bicarbonate ion will be rejected by the RO. Carbon dioxide, being a gas, would pass through the RO into the permeate stream unimpeded. The carbon dioxide would become an undesirable load to downstream polishing mixed beds.
- Certain TOC constituents are better rejected at a higher pH.
- Silica rejection and solubility are higher at elevated pH levels (particularly above pH 9).
- Boron rejection is higher at elevated pH levels (particularly above pH 9).

5.1.4 Coagulants.

Chemicals added in water and wastewater applications to cause destabilization of suspended particles and subsequent formation of flocs that adsorb, entrap, or otherwise bring together suspended matter that is so fine, it is defined as colloidal. Compounds of iron and aluminum are generally used to form flocs to allow removal of turbidity, bacteria, color and other finely divided matter from water and wastewater.

1. Ferric Chloride.

The most employed coagulant in SWRO is ferric chloride. The efficiency is high and the price is low. Ferric chloride has lower tendency of forming deposits on membranes than aluminum. The reaction is the below indicated:

$$Cl_3Fe + CO_3H \rightarrow Fe(OH)_3 + Cl + 3CO_2$$

The hydrolyzed form consumes alkalinity (see reaction), therefore the raw water pH is reduced (0.1–0.3 units) in proportion to coagulant dosing and feed water alkalinity. Conversion of metal coagulants is necessary to control $Fe(OH)_3$ levels which can precipitate in the reject.

EXAMPLE

Flow (F) .... *2569 m³/h*
Dose (D_r) .... *20 ppm*
C_{vr} (concentration weight %)........................ *42*
Density (r) .... *1.45 kg/l*

Dosage flow (Fr l/h) will be calculated from equation

$$Fr = \frac{F \times Dr}{\frac{Cvr}{100} \times 1000 \times r}$$

Giving a value of 84 l/h

Mass flow (Mr in kg/h) will be calculated from:

$$Mr = \frac{F \times Dr}{\frac{Cvr}{100} \times 1000}$$

Giving a value of 122 Kg/h

Two dosing pumps (one spare) with a capacity of 114 l/h will be installed

Storage capacity will be calculated (V in m³) from:

V= Fr x 24 x t/1000

2. Coagulant aids.

In RO, pretreatment systems polymers are usually used as additives to enhance the effectiveness of metal based coagulants by binding floc particles together. They are applied at low dosages, directly injected to the feed downstream of the dosing point of the coagulant.

5.1.5 Antiscalants.

Scale inhibitors (antiscalants) can be used to control carbonate scaling, sulfate scaling, and calcium fluoride scaling. The formation of inorganic scales, metal oxides and hydroxides, sometimes encountered in the tail elements of BWRO plants, does not present a problem with most seawater feeds, where the precipitation of sparingly soluble salts is less likely to occur due to the relatively low recovery rate of SWRO plants, which is usually limited to 50%, the high ionic strength of seawater, and the low concentration of bicarbonate ions in seawater.

Nevertheless, most SWRO plants use antiscalant, acid or a combination of both to avoid the risk of scale formation. Some plants are also known to operate without any antiscalant on the Cayman Islands or in the Mediterranean.

If a second stage is necessary for boron removal (which requires that the pH is raised to about 10), antiscalants are needed to prevent the formation of magnesium hydroxide ($Mg(OH)_2$). SWRO plants with a single stage often use acid only, while plants with a second RO stage often use polymer antiscalant to maintain a higher pH for boron removal.

There are generally three different types of scale inhibitors: sodium hexametaphosphate (SHMP), organophosphonates and polyacrylates.

SHMP is inexpensive but unstable compared to polymeric organic scale inhibitors. Minor amounts adsorb on the surface of microcrystals preventing further growth and precipitation of the crystals. Food-grade quality SHMP should be used. Care must be taken to avoid hydrolysis of SHMP in the dosing feed tank. Hydrolysis will not only decrease the scale inhibition efficiency, but also create a calcium phosphate scaling risk. Therefore, SHMP is generally not recommended.

Organophosphonates are more effective and stable than SHMP. They act as antifoulants for insoluble aluminum and iron, keeping them in solution. Polyacrylates (high molecular weight) are generally known for reducing silica scale formation via a dispersion mechanism.

Polymeric organic scale inhibitors are also more effective than SHMP. Precipitation reactions may occur, however, with negatively charged scale inhibitors and cationic polyelectrolytes or multivalent cations (for example; aluminum or iron). The resulting gum-like products are very difficult to remove from the membrane elements. For dosage rates on all antiscalants, please contact the antiscalant manufacturers. Overdosing should be avoided. Make certain that no significant amounts of cationic polymers are present when adding an anionic scale inhibitor.

EXAMPLE

Flow (F) ....2569 m^3/h

Dose ((D_r) *..5 ppm*

C_{vr} *(concentration weight %)..................10*

Density (r) ....1 kg/l

Dosage flow (Fr l/h) will be 128 l/h and Mass flow (Mr in kg/h) will be 128 Kg/h.

Two dosing pumps (one spare) with a capacity of 150 l/h will be installed.

Daily consumption will be of 3.1 m^3*, so two deposits with a unit capacity of 3* m^3 *will be installed.*

5.1.6 Dechlorination.

The feed must be dechlorinated to prevent oxidation of the membrane. Eventual degradation may occur after approximately 200–1000 hours of exposure to 1 mg/L of free chlorine (200–1000 ppm-h tolerance). The rate of chlorine attack depends on various feed water characteristics. Under alkaline pH conditions, chlorine attack is faster than at neutral or acidic pH. An acidic pH is preferred for better biocidal effect during chlorination. Chlorine attack is also faster at higher temperatures and higher concentration of heavy metals that can catalyze membrane degradation.

Sodium metabisulfite (SMBS) is commonly used for removal of free chlorine and as a biostatic. The reaction is:

$$S_2O_5Na_2 + 2ClOH + 4CO_3H \rightarrow 2SO_4^{--} + 2Na + 2Cl + 4CO_2 + 3H_2O$$

When dissolved in water, sodium bisulfite is formed from sodium metabisulphite.

$$S_2O_5Na_2 + 2H_2O + \rightarrow 2NaHSO_3$$

In theory, 1.34 mg/L of sodium metabisulphite will remove 1 mg/L of free chlorine. In practice, however, 3 mg/L of sodium metabisulphite is normally used to remove 1 mg/L of chlorine.

EXAMPLE

Flow (F) .... 2569 m³/h
Dose (D_r) .... 100 ppm (shock)/10 ppm (continuous)
Cvr (concentration weight %) 15
Density (r) .... 1.1 kg/l

Shock dosing flow will be 1557 l/h. One dosing pump with a design flow of 1560 l/h will be installed.

Continuous flow will be 155 l/h. Two (1+1) dosing pumps will be installed.

Two dosing pumps (one spare) with a capacity of 150 l/h will be installed.

Daily consumption will be of 3.72 m³/day, so two deposits with a unit capacity of 4 m³ will be installed.

5.2 Dosing points.

5.2.1 Disinfectants.

It is usually added at the intake and diluted to guarantee a proper distribution.

Chlorine dosage will be of 5–10 ppm when Cl_2 is applied off-line (non-continuous). Shock chlorination will be applied once or twice a day during a period of 5 to 15 minutes. In the "shock dosage", the chlorine dose must satisfy the "chlorine feed water demand" at the forecast contact time and a chlorine residual of 0.1 mg/L should be present. There is not a fixed rule for the number and periods of time for shock chlorination, and it must be determined with experience in plant.

If ClO_2 is used, the dosage will be of 0.3–0.5 ppm, to keep the residual chlorine below 0.05 ppm in the plant. Chloride dioxide will be generated in the plant with the use of sodium chloride and HCl with the following reaction:

$$5\,NaClO_2 + 4HCl \rightarrow 5NaCl + 2H_2O$$

Monitoring of dechlorination can be performed by the use of an available Free Chlorine monitor, monitoring for a residual bisulfite concentration, or by an ORP monitor. SWRO systems are equipped with an Oxidation-Reduction Potential (ORP) meter. The feed water can then be continuously monitored for the presence of oxidant. Except in wastewater applications where chloramines are used, the ORP meter reading should always be below 300 mV. If it exceeds 300 mV, the plant operator should receive a warning that a dangerous level of oxidant is getting to the membranes and should take action, such as adding or increasing the dose of SBS, to reduce the oxidant concentration.

If the ORP value reaches 350 mV, the plant should be shut down until the oxidant concentration can be reduced to a safe value (ORP less than 300 mV).

ClO_2 is controlled with a ClO_2 analyzer.

It must be added at the intake and at the discharge of seawater pumps. If sodium hypochlorite is used as disinfectant, both chemicals must be dosed at the same point at the intake. If ClO_2 is used as disinfectant, it is not required.

The reason is that HClO is much more effective at pH between 4.5–6.5. The activity of ClO_2 is constant at a pH between 6–9.

5.2.2 pH adjustment reactives.

Acid is dosed at the discharge of pumps to keep the pH between 6 and 7, as LSI and S&D values can be lowered by reducing pH and also the biological activity will be lower.

Acid dosage must be diluted for a quicker mixing. Mixing is an exothermic and very corrosive reaction, so a static mixer must be used.

5.2.3 Coagulants.

Ferric chloride is used. The optimum dosage is usually in the range of 5–10 mg/L, but should be determined case by case. It is dosed diluted. Rapid dispersion and mixing of the coagulant is extremely important. This is usually achieved by incorporating an in-line static mixer downstream of the coagulant injection point. It is also used to guarantee acid is properly mixed.

5.2.4 Coagulant aids.

Coagulant aids are added to water during coagulation (polyelectrolyte). Generally, they are a very high-molecular-weight. Anionic polyacrylamides are the most effective coagulant aids. The dose will be approx. 0.5–1 ppm. It is directly injected into the feed water downstream of the dosing point of the metal coagulant, at a location where hydrolyzed metal floc has already been formed. If polymers are used at a high dosing rate, and a carryover from filters occurs, cationic polymers may react with anionic scale inhibitors and form a fouling layer on the membrane surface.

A static mixer must be installed close to the pumping station.

5.2.5 Antiscalants.

The antiscalant will be dosed in diluted form, normally added before cartridge filters. However, a line to give the option of dosing antiscalant after the cartridge filters is can be installed in order to have more flexibility.

5.2.6 Dechlorination.

Liquid sodium bisulfite (SBS) diluted at 38% and sodium metabisulfite (MSBS) in solid form are the typical chlorine reducing agents of choice for larger RO systems. The application point will be after cartridge filtration.

Solutions of 10% to 15% by weight should be used. The SBS solution is not stable to air and reacts with oxygen as well as chlorine, within 7 to 14 days.

Designers have been known to use a dosing rate of 3.0 ppm of SBS per 1.0 ppm chlorine. SBS needs to be added far enough upstream of the RO elements to insure at least 20 seconds of reaction time. Proper in-line mixing is required which preferably includes a static mixer.

5.3 Storage of reactives.

Liquid products are stored in deposits with a minimum capacity of 15–30 days.

1. Sodium hypochlorite, richness 14% of Cl_2.
2. Coagulant, richness of 40%.
3. Sulfuric acid, richness of 96%–98%.
4. Sodium bisulphite, richness of 38%.

There are other liquid products which will supplied in IBC (antiscalant and coagulant aid) and are stored in deposits of less capacity (approximately one day).

There are other products which will be supplied as powder (Calcium hypochlorite, meta bi-sulfite and calcium hydroxide) and can be stored in bags and silos.

To store the reactives, the following materials will be used:

1. Sodium hypochlorite — GRP or PVC.
2. Coagulant — GRP or HDPE.
3. Sulfuric Acid — Carbon Steel.
4. Sodium Bisulphite — GRP or HDPE.
5. Antiscalant — GRP or HDPE.
6. Calcium hypochlorite — GRP.
7. Calcium Hydroxide — Carbon Steel.

The following aspects must be taken into consideration:

1. Concrete basin, following design recommendation from the applicable code (capacity of 115% volume of bigger storage tank if double wall tanks are not installed).

2. Level local indicator.

3. Filling system (with pump or without pump) plus protection cabinet with collection system. Centrifugal magnetic driven pumps manufactured in plastic materials are recommended.

4. Sodium hypochlorite. Richness of 13-14% as chloride.

The sizing of the storage tank must be balanced with the fact that the solution loses strength over time. The rate of degradation increases with heat and sunlight. The recommended storage limit is 28 days, with max storage life of 60 to 90 days. As indicated, it degrades at high temperatures, and cooling systems used to keep temperature at 10°C are employed if required (cooling system + heat exchanger + recirculation pump).

Due to off-gassing (the degradation of the NaOCl produces a gas which is mostly oxygen), tanks must be properly vented out of all structures to the atmosphere and a- hydraulic seal must be provided at the overflow.

5. Sulfuric Acid. Silica gel absorbers must be installed in the vents to keep moisture out of the concentrated sulfuric acid. A hydraulic seal must be installed in the overflow with a check valve to prevent from acid being in contact with absorber, in case overflow.

Figure 5.2. Sulfuric acid storage tank. Courtesy of SADYT.

6. Sodium bisulfite deposit. Bisulfite produces SO_2 which is toxic. An air extraction system must be installed in the area and vent must be directed outdoors.

If sodium metabisulfite is used, when mixing with water SO_2 is generated, so each dilution tank must have a vent. If more than one (1) dilution tank is installed, they will be interconnected and extracted to a safe location.

Two deposits in parallel with their corresponding agitators will be installed. They will work in batch mode. The liquid or powdered polymer is added manually to one of the tanks, mixed by the stirrer and the solution is prepared. The other tank where the solution has already been prepared in advance will be used. The typical solution life for a concentration of 15% weight will be of 1 week, so it would be recommended not to have a storage capacity of dilution tanks larger than 5–6 days for material.

Figure 5.3. Sodium bisulfite tanks. Courtesy of Plavisa.

7. **Calcium hypochlorite dilution system**. Calcium hypochlorite is supplied in solid form with a concentration of 65%. It will be diluted at 20%, to obtain calcium hypochlorite at 12% as Cl_2. The agitator must be ebonite lined.

5.4 Reactives dosage.

Number and type of pumps will depend on flow and number of racks.

Liquid metering pumps are usually of the positive displacement type using either plungers or diaphragms. In desalination plants, we will use diaphragm pumps. Plunger pumps will be used for big flows.

Diaphragm pumps can be pneumatically and air operated or mechanically operated.

Positive-displacement pumps can be set over a wide range (10:1) by adjusting the pump stroke length (electronic regulation). In addition, the feed range can be extended by adjusting the pump speed (frequency converter). In this case, the range will go from 30% to 100%, so if motor speed is 1500 rpm, it is not recommended that the motor speed be below 500 rpm.

Figure 5.4. Diaphragm dosing pumps.

If the chemical used is sulfuric acid, it is compulsory to protect the system, so the water dilution system must be operative before the sulfuric acid dosing pump starts. Rotameter must have a low flow limit switch installed which will prevent the pump from starting in case of low flow.

Automatic feed systems can be designed to control the feed flow based on a process variable, such as influent flow, pH or redox. Usually, automatic control is based on seawater flow, so a flow transmitter will be necessary.

The chemical dosage calculation can be obtained from the following equation:

$$\frac{f \times d}{(q \times r \times 1000)}$$

where:

f = Maximum flow (l/h).

d = dosage (mg/L).

q = Maximum dosing pump flow (l/h).

r = richness (kg/l).

For instance, (for pH control and bisulfite) control schemes pace the pump speed based on water flow and use the pH and redox values to trim the pump stroke respectively.

The dosing systems should include pressure relief valves to prevent line failures if all discharge valves or pump isolation valves are closed. Pulsation dampeners are typically located in the discharge piping as closed to the pump discharge connection as possible. They should be isolated with a valve. Back pressure valves are also required to ensure proper pressure differential between the suction and discharge valves and to provide sufficient pack pressure to seat the pump discharge check valve.

Chemical dosing systems can be supplied as skids with two (2) or three (3) pumps. These skids are compulsory in all plants installed within the USA and UK.

Materials used in chemical dosing systems are:

1. **Sulfuric Acid.**
    - **Piping**. Carbon Steel or PVDF (small size pipe) for concentrated acid. When the acid is diluted, PP or PVDF will be used. In the mixing zones, the material will be carbon steel + Teflon.
    - **Valves**. Gate valves of carbon Steel material in concentrated acid line. Valves installed in mixing tee (retention and automatic) will be of membrane type with body manufactured of cast iron + halar and membrane of Viton. Dilution water automatic valves will be membrane type valves with the membrane manufactured in polyetilene. Manual valves will be ball valves.

2. **Hipochloride.**
    - Piping. Material: PVC. Polyethylene could be used but with Teflon or Viton joints for long conductions.
    - Valves. Material: PVC. Type: Ball.

3. **Ferric chloride.**
    - Piping. Material: PVC or polypropylene.
    - Valves. Material: polypropylene. Type: ball.

4. **Bisulfite and antiscalant.**
    - Piping. Material: PVC or polypropylene.
    - Valves. Material polypropylene. Type: Ball valves

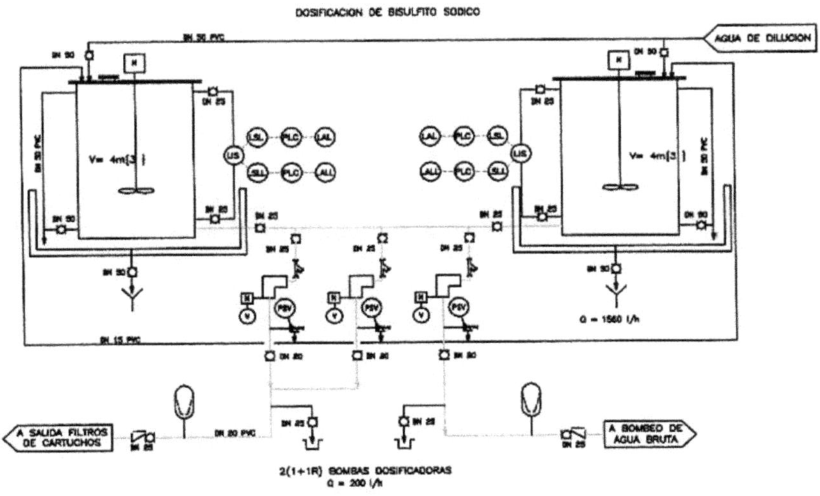

Figure 5.5. P&ID Sodium bisulfite and antiscalant.

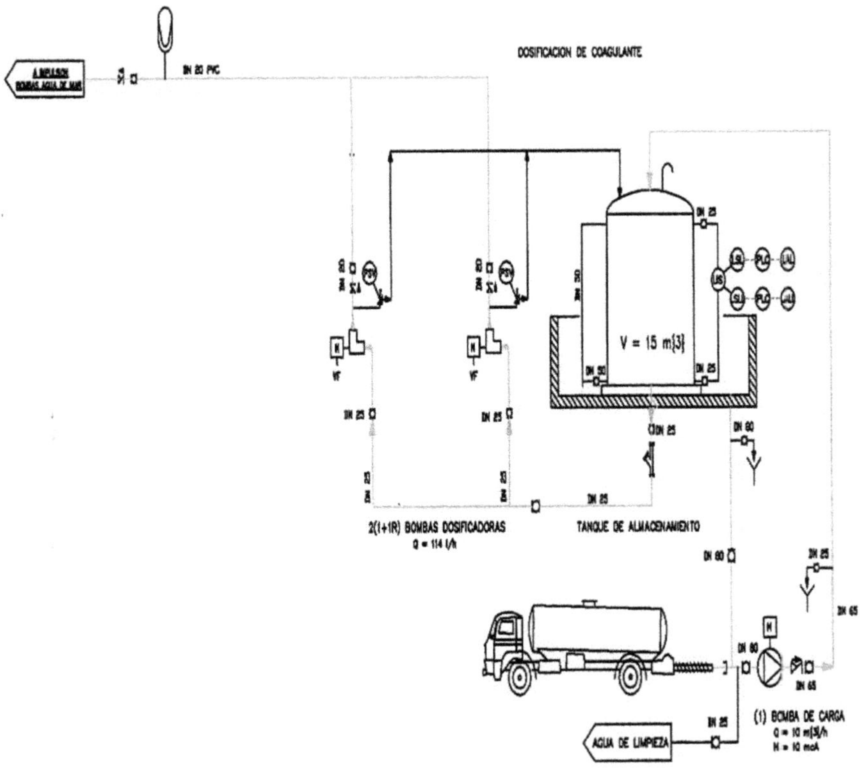

Figure 5.6. P&ID coagulant dosing system.

6
REVERSE OSMOSIS PRETREATMENT. CONVENTIONAL PRETREATMENT.

Seawater pretreatment is a key component of every membrane desalination plant. The main purpose of the pretreatment system is to remove particulate, colloidal, organic, mineral and microbiological contaminants contained in the source seawater and to prevent their accumulation on the downstream seawater reverse osmosis (SWRO) membranes (to protect the membranes from fouling). The content and nature of foulants contained in the source seawater depends on the type and location of the desalination plant intake.

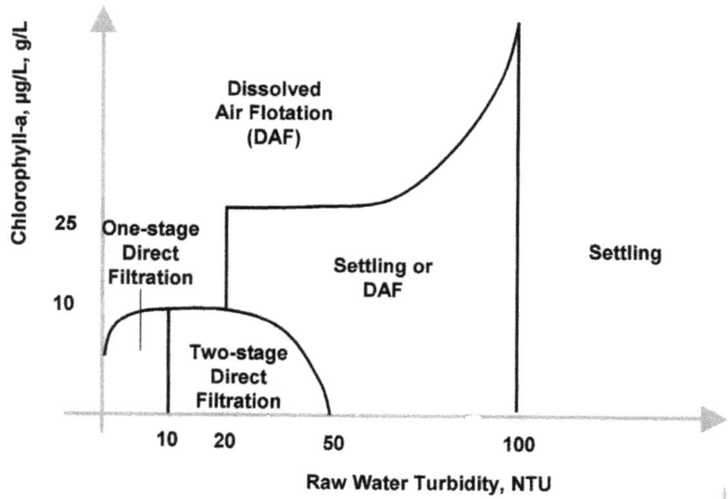

Figure 6.1. Graphic. Pretreatment recommendations.

The main operational concerns in SWRO plants which need to be resolved by the pretreatment are particulate fouling by suspended particles, bio fouling by microorganisms caused by nutrients in the feed water, organic fouling by dissolved organic matter, scaling by sparingly soluble inorganic compounds, and oxidation and halogenation by residual chlorine which is added during the pretreatment. The type and amount of pretreatment depends on the intake water quality and the desalination process. As surface intakes have to cope with more variable water quality due to seasonal weather conditions and algae blooms, pretreatment is generally more complex and extensive than for sub-surface intakes, which can achieve a silt density index (SDI) of less than 3 with single stage sand filters (without coagulant) or with cartridge filters only.

For SWRO, plants operating with an open intake, primary and secondary pretreatment systems are installed to ensure acceptable RO feed water quality and stable operation during algal blooms. Primary pretreatment typically includes microstraining/screening to remove large suspended materials (more than 50 μm), followed by coagulation and sedimentation or DAF (dissolved air flotation). Secondary pretreatment typically comprises granular (dual) media filtration and/or UF.

An overview on the main pretreatment alternatives for SWRO plants is given in figure 6.2.

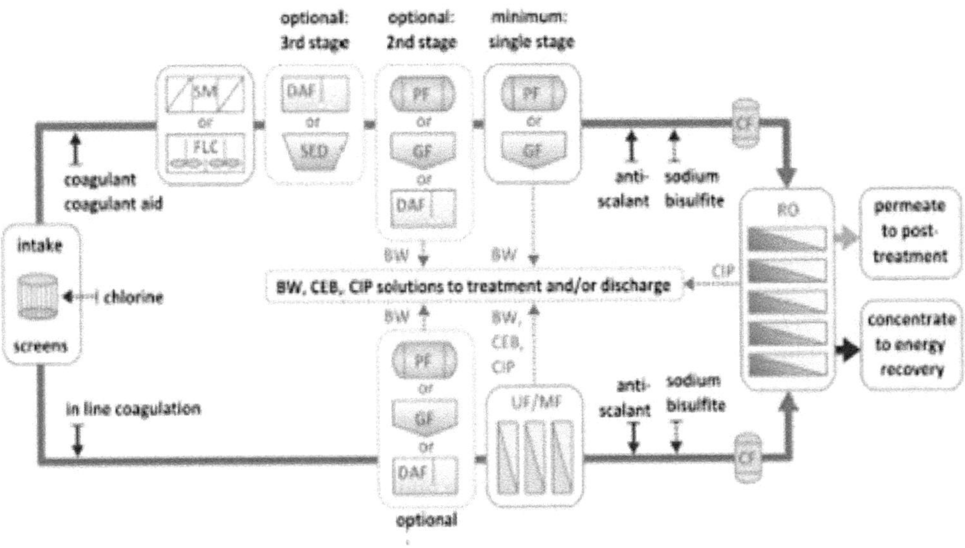

Figure 6.2. shows simplified flow-scheme of a SWRO plant with conventional pretreatment (top) or UF/MF pretreatment (bottom). Abbreviations: SM: static mixer, FLC: flocculation chambers, DAF: dissolved air flotation, SED: sedimentation, PF: pressurized media filter, GF: gravity media filter, CF: cartridge filter, BW: backwash, CEB: chemically enhanced backwash, CIP: cleaning in place. Dotted lines show intermittent flows/doses.

6.1 Settling.

6.1.1 Objective.

Sedimentation is typically used upstream of granular media and membrane filters. It is implemented if the plants source water has a daily average turbidity higher than 30 NTU *or* experiences turbidity spikes of 50 NTU or more which continue for a period of over one hour.

If sedimentation basins are not provided, large turbidity spikes may cause the pretreatment filters to exceed their solids holding capacity (especially if pressure-driven granular media filters are used), which in turn may impact filter pretreatment capacity. If the high solids load continues, the pretreatment filters would enter a condition of continuous backwash, which in turn would render them out of service.

Sedimentation basins for seawater pretreatment are typically designed to produce settled source water of less than 2.0 NTU and SDI15 below 6. To achieve this level of turbidity removal, sedimentation basins are often equipped with both coagulants (most frequently iron salt) and flocculants (polymer) feed systems. The needed coagulant and flocculent dosages should be established based on jar and/or pilot testing.

If the source water turbidity exceeds 100 NTU, then conventional sedimentation basins are often inadequate to produce turbidity of the desired level of less than 2 NTU for low silt and algal content. Under these conditions, sedimentation basins should be designed for enhanced solids removal by installing lamella plate modules or using sedimentation technologies that combine lamella and fine granular media for enhanced solids removal.

Typically, enhanced sedimentation technologies are used for treating source water from open ocean intakes which are under a strong influence of river water or wastewater discharges of elevated turbidity.

6.1.2 Clarification alternatives.

High rate clarifiers are a basic type of clarifier used as pretreatment in desalination plants. The advantages of these systems are:

- Compact design.
- Clarified water of good quality.
- Good response to water characteristics variations.
- High sludge concentration.

Four basic types of processes are used:

- Dense-sludge process.

The dense sludge process system uses coagulation and flocculation followed by lamella clarification to achieve high TSS and BOD_5 removal efficiencies in a compact unit. A coagulant (for example ferric chloride or alum) is added to the raw wastewater as it enters the air mixing chamber (1), where flash mixing and grit removal occur. Polymer is added as a flocculating agent as the wastewater flows to the second mixing chamber (2), which is equipped with an axial flow turbine. Thickened sludge produced in another part of the system is re-circulated to the second mixing chamber to serve as a ballasting agent and nucleus for floc growth. The third chamber (3) is a plug-flow reactor where flocculation continues and scum and grease separate and floats to the surface (4). The water and flocculated sludge enter an initial clarification zone (5) where most of the solids settle. The clarifier contains a lamella settling zone (6) where most of the remaining solids are removed. The settled sludge is thickened (7) and a portion of the thickened solids are re-circulated back to the second mixing chamber (2) with the remainder of the solids wasted.

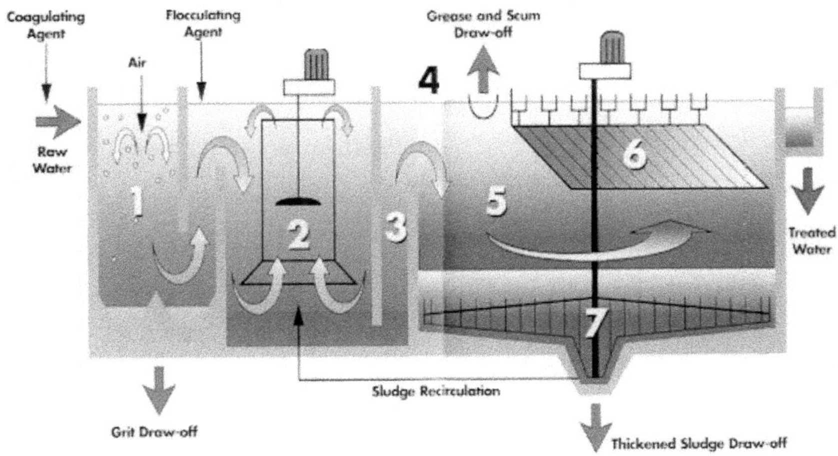

Figure 6.3. Lamella plate clarification.

Hydraulic load ranges between 35-75 $m^3/m^2/h$ with average values for design purposes of 50 $m^3/m^2/h$. Suspended solids removal percentages are 60-70% with a percentage of organic matter removal higher than 30-40%. Materials in contact with water must be suitable for salted water.

- **Ballasted floculation.**

Ballasting refers to the use of materials added during flocculation, that increase the density of flocs resulting in faster settlement. The ballasted flocculation process is a high rate settling process that combines the advantages of ballasted flocculation and lamella clarification. The use of micro-sand gives the advantage of treating a wide range of raw water. Density and shape of micro-sand particles increase the flocculation and settling efficiencies. As a consequence it is easier and faster to create a strong floc, even with lighter solids coming in. Once the floc is formed and attached to the micro-sand, the floc settling velocity is high enough to allow for high rise rates. Ferric chloride is injected into the coagulation chamber at a dosing rate of 10 ppm. Rapid mixing in this basin starts the coagulation process. Coagulated raw water then enters the injection tank where micro-sand and polymer are added. Polymer is injected at a dosing rate of 0.2 ppm. Dynamic mixing induces a high probability of contact between coagulated solids, polymer and micro-sand. Flocculated water enters the maturation chamber where a slow mixing process enables floc maturation and the increase of floc size. The micro-sand then becomes the nucleus of the newly formed floc. Rapid settling of ballasted floc is achieved in the lamella clarifier section. Micro-sand allows for high rise rate design of the lamellar clarifier. Clarified water is then collected at the surface. Sludge is extracted from the bottom of the clarifier by means of an abrasion-resistant centrifugal pump. Using the centrifugal vortex principle, a hydrocyclone then separates the sludge from the reusable micro-sand. These micro sand particles are discharged through the underflow of the hydrocyclone and recycled in the injection tank. The fine and lighter sludge particles move upwards with the major fraction of the water out through the vortex overflow.

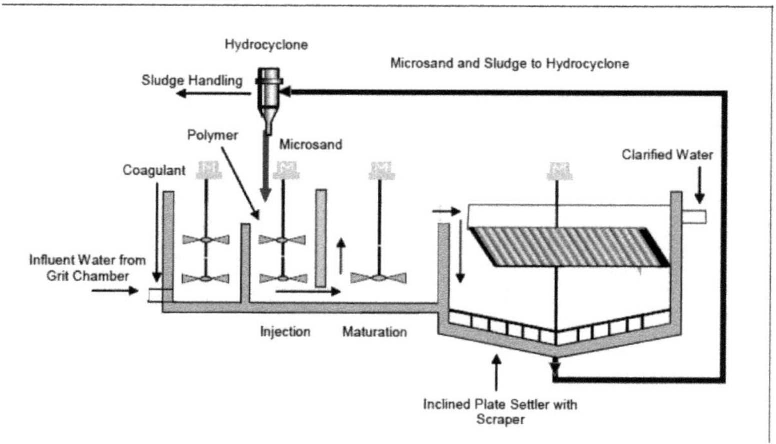

Figure 6.4. Ballasted floculation.

Hydraulic load ranges between 50-150 $m^3/m^2/h$ with average values for design purposes of 100 $m^3/m^2/h$. Suspended solids removal percentages are 60-70% with a percentage of organic matter removal higher than 30-40%.

- **Lamella plate clarification.**

Lamella plate clarification uses chemical addition followed by three stage floculation and a lamella plate. Coagulant and polymer are injected into the influent wastewater. When the chemically conditioned waste water passes through each of the flocculation zones, the mixing energy gradient is decreased as the wastewater proceeds from one stage to the next. The chemically conditioned/flocculated wastewater then passes to the lamella clarifier for solids separation. A portion of the clarifier underflow can be recycled to the influent of the process to enhance settling, or the entire underflow can be sent to a thickening tank and the solid-processing facilities

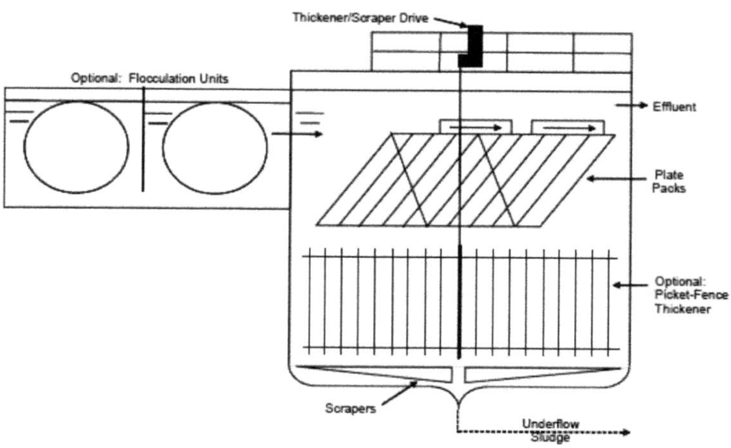

Figure 6.5. Lamella plate clarification. Courtesy of Parkson.

Hydraulic load ranges between 35-75 $m^3/m^2/h$ with average values for design purposes of 50 $m^3/m^2/h$. Suspended solids removal percentages is 60-70% with a percentage of organic matter removal higher than 30-40%.

- **Solid–contact clarifiers.**

Most upflow designs are called either "sludge blanket" or "solids-contact" clarifiers. After coagulation and/or flocculation in the sludge blanket units, the incoming water passes through the suspended layer of previously formed floc.

Because the centerwell in these units is often shaped like an inverted cone, the rise rate of the water decreases as it rises through the steadily enlarging cross section.

Sludge blanket efficiency depends on the filtering action as the freshly coagulated or flocculated water passes through the suspended floc. Higher sludge levels increase the filtration efficiency. In practice, the top sludge interface is carried at the highest safe level to prevent upsets that might result in large amounts of floc carryover into the overflow. Excessive sludge withdrawal or blowdown should also be avoided. The sludge blanket level is often highly sensitive to changes in throughput, coagulant addition, and changes in raw water chemistry and temperature.

"Solids-contact" refers to units in which sludge are circulated internally. The term also describes the sludge blanket unit and simply means that prior to and during sedimentation the chemically treated water contacts previously coagulated solids. Solids-contact, slurry pool units do not rely on filtration as in sludge blanket designs.

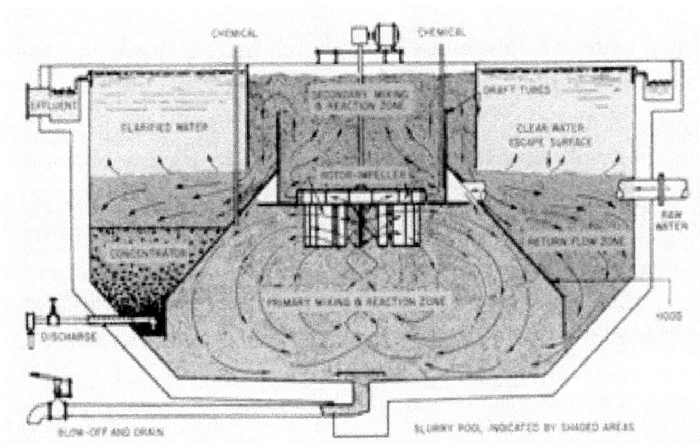

Figure 6.6. Solid contact clarifier. Courtesy of Degremont.

6.1.3 Commonly monitored and controlled elements in the clarification and sedimentation processes.

- Monitoring: Level, flow, influent turbidity, chlorine residual, pH, effluent turbidity, sludge level.
- Control: Chemical feed, flocculation control, sediment Basin sludge, sludge removal valves/pumps.

6.2 Dissolved Air Flotation.

6.2.1 General characteristics.

As opposed to settling, flotation is a solid–liquid separation technique that is applied to particles whose density is lower or has been made lower than the liquid they are in (algae, oils, grease..) and cannot be easily removed with conventional clarification processes. The Dissolved Air Flotation (DAF) process is proving to be a very efficient and cost effective pretreatment option. Several suppliers have selected the DAF as the preferred pre-treatment provider for a large RO desalination plant.

Dissolved air flotation (DAF) utilizes the property of micro-bubble adherence to suspended solids, increasing the tendency of the particles to float. The flocculated water meets a water flow with supersaturated air (85% to 95%) which is supplied through nozzles. Due to the pressure drop of the supersaturated water at the nozzles, small air bubbles are formed. These micro-bubbles attach themselves to discreet floc particles created in the flocculation process. The rising velocity of the air bubbles is higher than the water velocity and the air bubbles will thus collide with the flocs in water. The density of the aggregates decreases until values below the water density. As a consequence, the aggregates will float on the water surface. As they rise to the surface, the buoyant flocs form a stable sludge layer above the waters surface and mechanical scrapers skim the solids from the surface into a collecting bin. A sludge of dry solids content in excess 2-3% may be produced. Air bubbles of 20 to 50 μm are considered the best for the recovery of fats. The air to solids ratio has a major effect on the performance of a DAF unit. Depending on the raw water quality and the efficiency of mixing of the recycle stream with the flocculated water, the amount of recycle required typically lies somewhere between 8% and 12% of the influent flow.

The DAF is best suited to remove oil from seawater. If oil were to get through the flocculation zones as free oil it will almost certainly be trapped by the micro bubbles and be floated off into the sludge blanket for removal along with the floc. The percentage of removal is around 90%.

Algae are difficult to remove by conventional treatment such as sedimentation, as they are naturally less dense than water and do not settle well. Poor removal of algae (diatoms, green algae, flagellates, blue green algae) can lead to clogging of granular media filters and short filter runs. 99–99.9% removal can be obtained with DAF compared with 90–99% by sedimentation. The removal of algae with flotation is more effective than sedimentation (60–90%) depending of the inlet concentration algae. DAF systems can achieve values of NTU lower than 0.2 and can be combined with filtration or ultrafiltration.

The severe red tide event in the middle east/Persian gulf have redirected the attention of the desalination industry to DAF as part of SWRO pretreatment schemes. DAF is now being regularly used in new SWRO plants in the Persian Gulf upstream of granular media filtration or UF.

6.2.2 Types of DAF.

There are various manufacturers of DAF equipment, each with their own proprietary technologies. However, proprietary systems do not present significant design differences and major EFC without proprietary DAF technologies carry out their own designs. We will focus on the high rate DAF with the larger presence in the sector.

- **AQUADAF™**

AquaDAF is a high rate dissolved air flotation system designed by Infilco Degremont, Inc. which can operate at loading rates of 30-40 m/h. Clarification rates for this system are based on flow per square foot area of the false floor at the bottom of the DAF cell. A clarification rate defined in this manner does not account for the increased footprint due to remainder of the system's components.

AquaDAF utilizes dual flocculation chambers where polymer may be added followed by a DAF Cell. An internal angled baffle wall and a false floor create a recirculation effect (8-12% of clarified or filtered water) throughout the DAF Cell which increases bubble density and efficient flotation for removal of the "float". The clarified water exits the system through a patented hole-patterned piping on the false floor and out an effluent channel while the sludge blanket hydraulically flows over into a collection channel of a saturated air/water recycle stream.

Advantages

- It can operate efficiently with peaks up to 200 ppm.
- It is efficient for cold water clarification (0°C)
- The retention time employed in the flocculation zone is between 8 to 10 minutes.

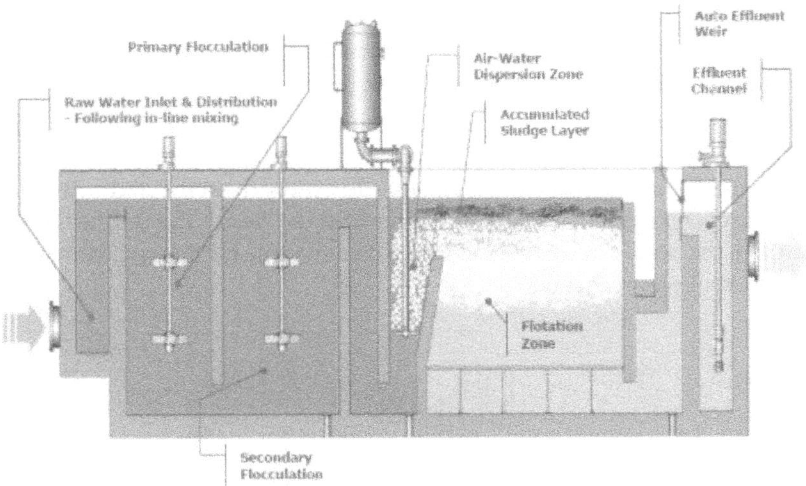

Figure 6.7. Sectional drawing of AquadaF. Courtesy of Degremont.

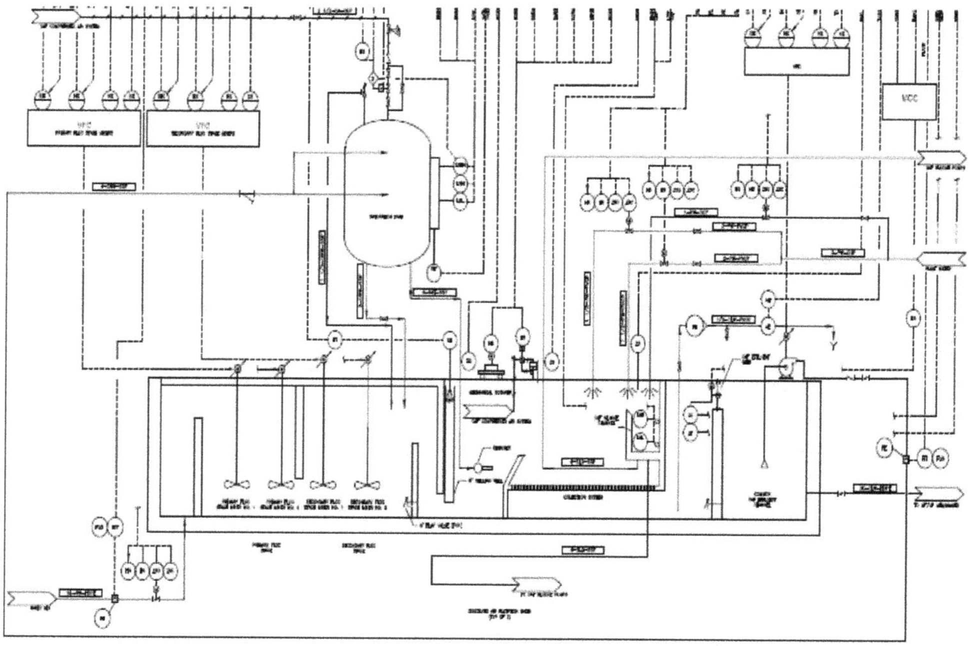

Figure 6.8. P&ID Aquadaf

- **CLARIDAF LEOPOLD™**

ClariDAF is a high rate dissolved air flotation system designed by Leopold. ClariDAF can be operated at loading rates up 20-25 m/h. Clarification rates for this system are based on flow per m^2 area occupied by the effluent laterals located at the bottom of the DAF cell.

The ClariDAF system, as depicted in the schematic in Figure 6.9 utilizes a rapid mix zone where polymer may be added followed by one or more flocculation basins and a DAF cell. An internal angled baffle wall creates a recirculation effect throughout the DAF cell which increases bubble density and efficient flotation for removal of the "float". The clarified water exits the system through the laterals and into an effluent channel while the float solids are mechanically skimmed into a collection channel.

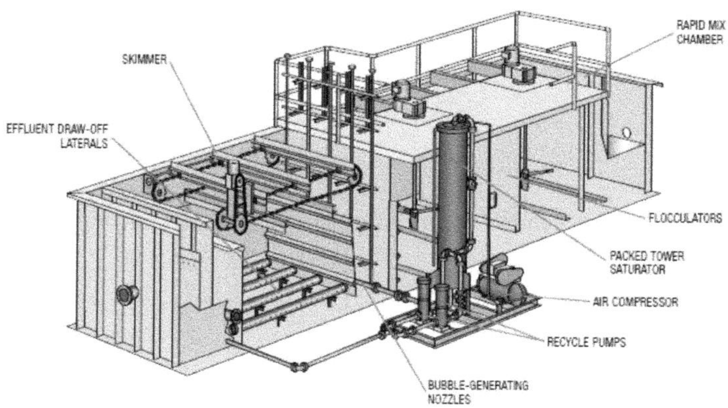

Figure 6.9. ClariDAF schematic .

6.3 Sand filtration.

6.3.1 Objective.

To remove suspended solids from raw water. There are materials which cannot be removed due to their small size. In line floculation will be used to agglomerate the small particles in filterable particles (flocs).

Filter media selection will determine the effectiveness of the filtration as pretreatment system. The materials used in filtration are :

1. **Gravel.**
2. **Sand.** Silica > 98%. Loss of HCl <5mg/100gr. Density 1.5 to 1.7 compacted and 2.5 of supplied in grain .
3. **Anthracite.** Silica <0.02%, density 1.4 in particle and 0.7 apparent, Loss of HCl < 1%.
4. **Volcanic stone.** Density similar to anthracite.

6.3.2 Concepts.

- **Effective size.**

The effective size is such that approximately 10% of the total grains by weight are smaller and 90% are larger. Therefore, the effective size is the minimum size of most of the particles.

- **Sand.**

Sand is used with an effective size of 0.2-0.3 mm for 2nd stage of filtration and 0.4 to 0.65 mm in the 1st stage of filtration.

When only one stage of filtration is used as pretreatment, the effective size ranges from 0.8 to 1 mm.

- **Anthracite.**

The effective size is usually from 1.2 to 1.5 mm for 1st stage and from 0.7 to 0.9 mm for 2nd stage.

- **Gravel.**

Gravel will be used to cover nozzles. The effective size ranges from 2 to 3 mm.

- **Uniformity coefficient.**

Uniformity is measured by comparison of effective size at which 60% of the grains by weight are smaller and 40% are larger. The larger size, divided by the effective size, is called the uniformity coefficient. The smaller the uniformity coefficient, the more uniform the media particle sizes. The maximum uniformity coefficient will range from 1.4 to 1.6. For sand a coefficient from 1.45 to 1.5 is required and for anthracite from 1.5 to 1.6 .

- Filtration rate.

It is measured in m/h and ranges from 7 m/h to 18 m/h depending on the turbidity of raw water.

Filtration rate depends on the following factors:

- Raw water turbidity.
- In line floculation.
- Double filtration.

To classify filtration rate, we shall analyze the origin of water:

Well water.

In water with very low turbidity (from 1 to 2 NTU) very small sizes of sand or silt may be present. Filtration rates range from 14 to 18 m/h.

Only one filtration stage is required, unless there are special circumstances, and it has been demonstrated that single–medium bed (sand) filters can operate with successful results. Washing normally only includes upflow water flushing, but when silt is present it will require an auxiliary scour.

It is not required a maturation period.

Water from horizontal drains.

This water has a low turbidity (between 2 and 5 NTU) and it carries water and silt. Filtration rates range from 10-12 m/h with only one filtration stage. The backwashing is carried out with water and air scour and it is necessary a rinse period of 15-30 minutes.

Open intake.

These waters have a high turbidity. In periods of storm the turbidity can reach values up to 20 NTU. A double filtration with sand and anthracite is required. Raw water should be previously flocculated and a rinse period of 15-30 minutes is required.

The 1st stage filter can be a dual media with 1.2-1.6 m of total effective media depth. The first layer is anthracite (or pumice) with a large diameter in size give larger void spaces to the media bed, with greater solids holding capacity. The second layer is sand with a smaller diameter which act as a final barrier for the fine particles responsible for fouling. The 1st stage filters operate as roughing filters and are sized for a filtration velocity around 10-12 m/h. The 2nd stage filter is a standard-bed sand filter with 1 m of effective sand media depth. The 2nd stage filter uses a smaller diameter sand size (0.2-0.3 mm ES) which maximizes the sand media surface area and subsequently the possibility for the attachment of finer particles present in the water. The 2nd stage acts as a polishing filter to remove the majority of the remaining particles producing an effluent with a low Silt Density Index and are sized at a greater hydraulic loading rate (14 – 18 m/h). The combination of increased sand media surface area and a higher hydraulic loading rate enables one to meet the expected performances. Experience shows that the 2nd stage filter can remove 0.3 – 0.4 SDI.

It is highly recommended one carry out pilot testing when the origin of water is from an open intake in order to optimize pretreatment design and minimize operation risks due to a flawed design.

When sedimentation or flotation is installed upstream of granular media, the design criteria of filtration is similar to the above indicated with filtration rates for the first stage filters not higher than 10-12 m/h.

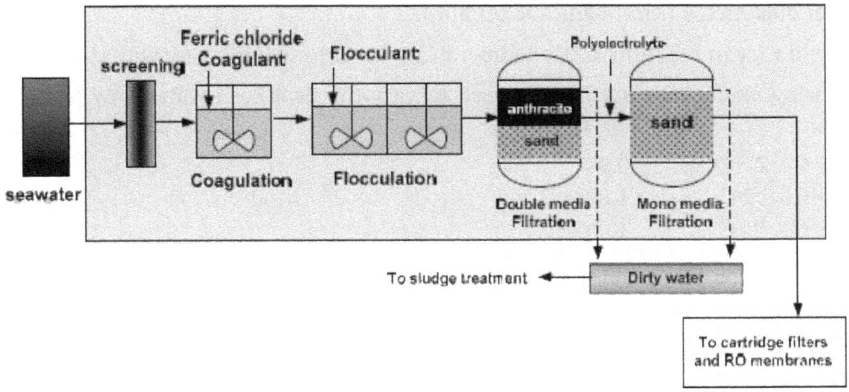

Figure 6.10. Schematic of two stages filtration.

Other options to be considered.

A single stage with a velocity lower than 7 m/h with an enhanced three (3) stage filtration media of anthracite, sand and garnet. Tri-media filters have 0.45 to 0.60 m of anthracite as the top layer, 0.20 to 0.30 m of sand as a middle layer and 0.10 to 0.15 m of garnet as the bottom layer. These filters are used if the source seawater contains a large amount of fine silt or the seawater intake experiences algal blooms dominated by pico-plankton and usually have open filters.

Two stages filtration with only sand and filtration media and with different size at each stage. This option is only recommended at low velocities 7 m/h at first stage and 10 m/h at second stage

The design of filters will be different if one or more filtration media are used. Expansion height and the size of the backwash water collector increases.

6.3.3 Design characteristics.

1. **Backwashing fluid.**

As the amount of solids retained in the filter media increases, bed porosity decreases. At the same time, head loss through the bed and shear on floc captured increase. Before the head loss builds to an unacceptable level or turbidity break-through occurs, washing is required.

Wash water source options are different. However, brine (reject) at upflow velocities between 30 to 50 m/h and air at velocities which range between 50 to 70 m/h are the most usual alternatives.

Backwash velocities vary depending on the type of bed. Single-media (sand) filters need less water than dual media (sand +anthracite) filters.

Velocity will vary in function due to temperature. The flow will increase with the temperature.

The following coefficients are usually used to vary the backwash water flow in function of temperature.

- 15ºC - 20ºC 0.94
- 20ºC - 25ºC 1.00
- 25ºC - 30ºC 1.06

Media loss must be prevented through wash water collection troughs, so backwash water velocity and time must be carefully controlled. High-rate backwash is recommended but as indicated is limited by the potential media loss.

Experience shows that air scour essentially eliminates mudball formation. Air scour usually is employed before water backwash.

If air and water are used simultaneously the cleaning may be more effective and the water consumption reduced. However, air-flow must be stopped before the wash water overflows into the wash water collection troughs in order to prevent media loss.

Below recommended backwash reference values are indicated:

– *Single media filters.*

- Backwashing with water and air:	7-10 m/h.
- Backwashing with water:	14- 20 m/h (between 5 to 10 minutes).
- Maturation:	7 to 10 m/h (10 to 15 minutes).

– *Multi media filters.*

- Backwashing with water:	20-45 m/h (between 5 to 10 minutes).
- Maturation:	3 – 4 times filtration bed (1st stage).
	7-8 times filtration bed (2nd stage).

2. **Bed expansion.**

Bed expansion ranges from 30 to 50% of media height.

If sand is used, 30% of bed expansion is recommended.

When anthracite is the media, 50% of media is used.

If sand and anthracite it is recommended a bed expansion between 30 to 40%

Media may be lost during backwash if expansion is excessive.

3. **Head loss and length of filtration run.**

The head loss in a filter bed depends on the following values:

- Water flow per nozzle (m^3/h/unit).
- Depth and type of bed (see figure 6.11).
- Water temperature.

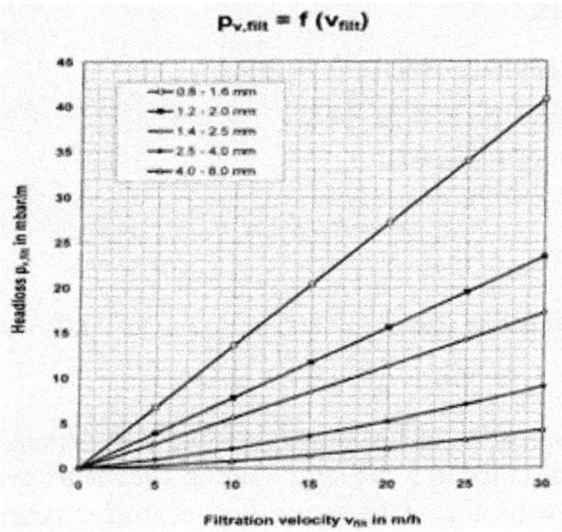

Figure 6.11. Head loss of filters.

In normal operation, the head loss is approximately 0.3 bars when the filter is clean, and 1 bar when the filter is clogged. Filter should be backwashed when terminal head loss is reached; otherwise turbidity break-through may occur.

Filter performance is often judged by the length of filter runs, but long filter runs may not be good for operation. Long filter runs make washing more difficult because of particle matter compaction. In addition, long filter runs indicate that filters are not working at their most cost-effective capacity. Filter runs will depend on the filtration rate, amount of solids removed and the terminal head loss.

Rapid sand filters are generally operated with run lengths between 12 or 72 hours, typically with 24 hour runs. Pressure filters may have somewhat longer filter runs than gravity filters if they can be operated at higher head loss without turbidity break through.

An approximate formula to calculate a filter run:

$$Hours = 100 \times \frac{Headloss\ increase(ft)}{\left(Turbidity(NTU) \times Velocity(gpm/sq/ft)\right)}$$

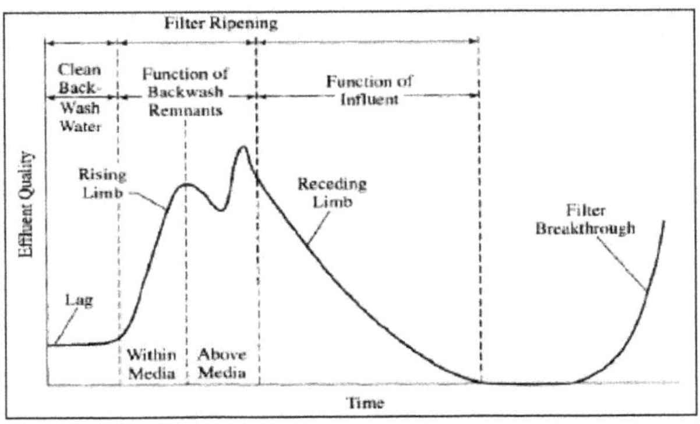

Figure 6.12. Characteristic filter run.

6.3.4 Type of filters.

There are three (3) type of filters which can be used

1. Pressure filters: mono, dual, three-bed.
2. Gravity filters: Mono, dual-bed.
3. Continuous back washing.

6.3.4.1 Pressure filters.

1. **Parts of a pressure filter:**

- *Vessel.*

Vessels are usually manufactured in carbon steel+ebonite. However, GRP has started to be an alternative and is being installed in some large SWRO. The main advantage of GRP is that any potential corrosion risk is eliminated. However, it is necessary to check and verify during long operation periods of time its behavior especially the mechanical behavior.

Standardized sizes are the following ones:

- 2800 mm of diameter x 10 m of cylindrical length. Surface 27 m^2.
- 3600 mm de diameter x 12.5 m of cylindrical length. Surface 44 m^2.
- 4000 mm of diameter x 12.5 m of cylindrical length. Surface 50 m^2 (be careful with transportation)

Dual media bed filters with a media depth 1000 mm must have a diameter of 4000 mm.

Filters in large sea water desalination plants are usually horizontal with two (2) supports. When concrete supports are used, 4 to 6 supports must be used.

Design pressure is the maximum operation pressure, defined as the maximum pressure of sea water pumps minus geometric height. To be conservative, the shut-off pressure of pumps (zero flow) can be used for design of vessels.

Safety valves (valid for sea water) made of rupture disc will be installed to avoid an over pressure according to ASME.

If the over pressure is not reduced by the safety valves, two pressure switches (working in parallel) can be installed in order to stop the sea water pumps, by wiring them directly to the MCC of the pumps (see EUROPEAN DIRECTIVE or CODAP CODE).

Figure 6.13. Pressure sand filters manufactured in GRP.

Shell thickness (E_{cv}) calculation will be based on ASME VIII:

$$E_{cv} = \frac{P_d \times R_V}{(S_v \times E_1) + (0{,}4 \times P_d)}$$

P_d = Design pressure.
R_v = Shell radius.
S_v = Work load of the material.
E_1 = Welding efficiency in shell joint.
If there is no coating an over-thickness per shell corrosion has to be added.
Select the minimum commercial thickness (more than calculated thickness) = E_{rv}
The maximum pressure (P_{mv}) will be calculated with the following formula

$$P_{mv} = \frac{S_v \times E_1 \times E_{rv}}{R_v - (0.4 \times E_{RV})}\left({}^{kg}/_{cm^2}\right)$$

Torospherical thickness calculation will be based on ASME VIII:

$$E_{cf} = \frac{P_d \times R_f \times M}{(2S_F \times E_2) + (M - 0.2) \times P_d}(mm)$$

P_d = Design pressure.
R_f = Head radius.
M = Asme coefficient (depending on type of head).
S_F = Work load of head material.
E_2 = Welding efficiency in head joint (normally 1)
Consider the over-thickness per corrosion (If there is not rubber liner) Select the minimum commercial thickness (more than calculated) = E_{RF}
Maximum pressure :

$$P_{mf} = \frac{2 \times S_f \times E_2 \times E_{rf}}{(M \times R_F) - E_{rf}(M - 0.2)} \, {}^{Kg}/_{cm^2}$$

False bottom will be calculated as:

- Material ⇒ AE-275 B
- S_p ⇒ Work load=1.050 Kg/cm^2
- P_d ⇒ Design pressure=0.9 Kg/cm^2
- E_p ⇒ Plate thickness (Code ASME VIII.Div 1/Timocherko)

B_1 = distance between supports.
B_2 = distance between hole shafts.
A_1 = hole diameter.

$$E_p = \sqrt{\frac{6 \times 0.0833 \times P_d \times B_1^2 \times B_2}{S_p \times (B_2 - B_1)}}$$

- *Inlet distributor.*

Distributes the raw water along the filter and collects backwash water.

It can be manufactured of the following materials:

- Carbon steel+ebonite section trapezoidal and very expensive.
- Same shape but in polyester.
- Pipe made in PVC with a slot for water passage. It must be well leveled for it must work as not only a discharge overflow but as a wash water collection.

The latter is the employed in sand filters. Typically, PVC collectors of 500 mm and 600 mm diameter are used.

The more critical the filtration system is, the better the collector should be sized and designed.

For filters of 4000 mm of diameter, two (2) collectors must be used, as backwash water flow will be 2000 m^3/h .

Figure 6.14. Water distribution collector

- *Filtration nozzles.*

Have three functions: To collect water that passes through the filter media, avoid media loss through filtration and to distribute wash water (and air if used) uniformly across the filter bed. The density of nozzles will be 45 a 55 units per m^2.

If filtration velocity is low, the density of nozzles will be 50 units per m^2.

If filtration velocity is high, the density of nozzles will be 55 units per m^2.

A metal plate is used to support the nozzles and sand. This plate is supported on round metal pillars fixed to the plate. The nozzle tubes are typically designed for a differential pressure of 1.5 bar.

If nozzles are too short, the air bed will not be uniform, and consequently the air distribution will hamper backwash efficiency.

Figure 6.15. Filtration Nozzles. Courtesy of Acciona Agua.

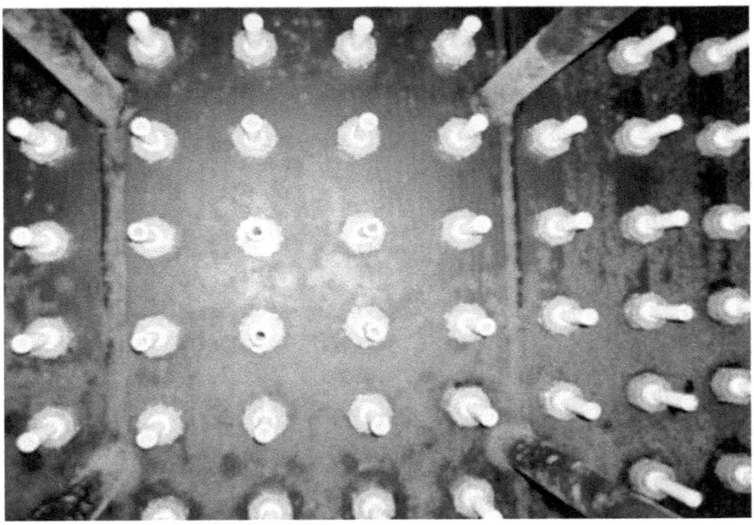

Figure 6.16. Filtration Nozzles. Courtesy of Acciona Agua.

- *Valves installed in a filter.*

Below list of valves used and symbology.

- Inlet of filter. A
- Outlet of filter. E
- Backwash inlet. W
- Backwash water outlet. C
- Air inlet. M

- Maturation outlet S
- Vent. V
- Drainage and rinse outlet. J
- Collector of treated water and back wash water pipe.
- Control valve (butterfly+positioner) to adjust backwash water flow.
- Common valves for drainage of air piping ND 80.

Figure 6.17. Pressure filter piping.

- **Filter ND 3600.**
  - Inlet and outlet of filter. ND 250. Inlet valve with positioner.
  - Backwash inlet and outlet. ND 400 and DN 500.
  - Air inlet. ND 250.
  - Maturation outlet. ND 250.
  - Vent. ND 150.
  - Drainage and rinse outlet. ND 200.
  - Control valve (butterfly+positioner) to adjust backwash water flow if back washing is carried out with seawater.
  - Common valves for drainage of air piping ND 80.
  - Control valve with positioner in maturation outlet +plate orifice.
- **Filter ND 4000.**
  - Inlet and outlet of filter. ND 300, Inlet valve with positioner
  - Backwash water inlet and outlet. ND500 and ND 600.
  - Air inlet. ND 250.
  - Maturation outlet. DN 250.
  - Vent. ND 150.

  - Drainage and rinse outlet. ND 200.
  - Control valve (butterfly+positioner) to adjust backwash water flow.
  - Common valves for drainage of air piping ND 80
  - Control valve with positioner in maturation outlet +plate orifice.
- **Second stage filters.**

Filters are the same size as the filters of the 1st stage. However, they operate at higher velocities so the diameter of service valves is bigger and the number of filters is lower.

2. Operation of pressure filters.

Normally filters work at a constant flow and variable head loss. The inlet valve with positioner regulates this flow. The filters can also work at variable flow and at a constant headloss. This option creates more disturbances when a clean filter is put into operation.

Number of filters installed must be selected analyzing the scenario of one of the filters being backwashed and the occasional need for a filter to be out of service for maintenance. It is recommended to clean all the filters following a sequence: when the dirtiest filter is washed, the following washing will start.

There is a formula as a guide to determine the required number of filters:

$$N = 1.2 \times Q^{0.5}\ (Q = gallons\ per\ day)$$

3. Backwashing periods.

- Emptying.
- Agitation with air.
- Some times agitation with water and air (as commented previously it can be dangerous when dual media with anthracite is used).
- Rest period.
- Water washing.
- Filling.
- Brine rinse.
- Maturation.

- **Maturation water recycling.**

If the designer wants to reuse maturation water, it will be necessary to install a drainage valve in the wash water main collector to drain the water during maturation. Thus, with only one (1) valve for the entire battery of filters, it will be possible to carry out the operation.

If it is intended for both backwash and maturation to occur simultaneously, it will be necessary to install one (1) additional drainage valve per filter.

In maturation, it is necessary to install a butterfly valve with positioner to regulate flow with maximum head loss of (2) bars. The rest of loss will be through an orifice plate at the inlet control valve (5) bars.

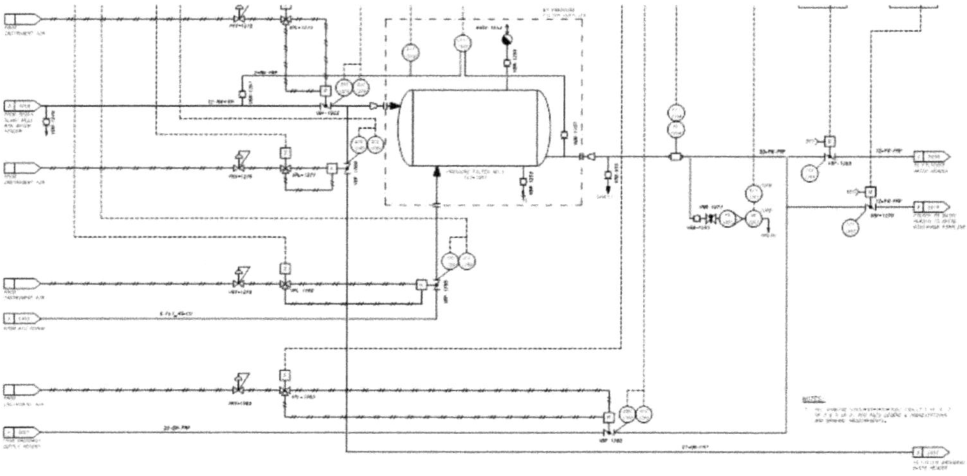

Figure 6.18. P&ID filtration (1 filter)

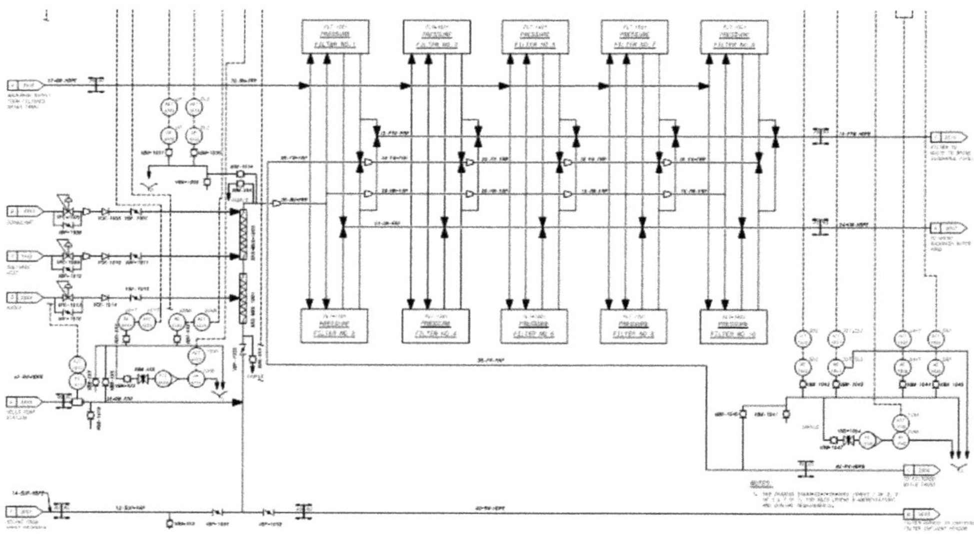

Figure 6.19. P&ID pressure filtration.

- **Washing system.**
- *Water.*

Backwashing is carried out with brine. Two pumps plus one additional stand-by pump or (1+1) will be the typical arrangement. The selection of the number of pumps will depend on the cost. Design pressure for pumps will be 2-2.5 bars.

Flow regulation will be done with a butterfly valve with positioner. It must be taken into consideration that backwash flow varies with water temperature. A thermometer plus a flow transmitter will be installed in common pipe. Pump isolation valves must be manual.

The entrance of air into the pumps must be prevented by a two baffle arrangement. The first baffle will impart an upward velocity to the entering bubbles, thus assisting them in the escape from the liquid. Any remaining bubbles can then be removed by redirecting the flow by means of an overhanging baffle. This arrangement is especially important when brine is used.

- *Air.*

Positive displacement blowers are used for the agitation of sand or anthracite bed with a flow of 50-70 m/h.

Air pipe is manufactured of glass reinforced plastic. To prevent water backflow to the blowers, a siphon is required at the entrance of filters. A drain valve will be installed between siphon and filters.

Check and butterfly valves will be installed in discharge pipe of blowers. One blower and one additional stand-by blower are the typical installation.

- **Back wash manhole.**

Backwash flow is important so the depth of individual manholes must be (1.5 m). Pipe diameter must also be big enough to prevent overflow. Take into consideration that discharge is from 3-4 meters.

6.3.4.2 Spruce filters™

The Spruce Filter is currently being used as a pretreatment for RO membranes in the Middle-East. It is a multi media deep bed filter with:

- 4 layers of natural, long lasting, inert media.
- An increasing density, decreasing particle size and specific shape factor.
- A natural positive charge attracting bacteria such as Legionella and Cryptosporidium.
- Independently proven to achieve submicron filtration.

Feed water is delivered to the top of the Spruce Filter™ and passes through the bed as in the standard method of feeding granular media filters.

The filter works in two ways:

- Mechanical Retention of the suspended solids.
- Adsorption of negatively charged pathogens and microorganisms removing 99.954% of solids down to 0.2 microns.

The specific shape factor of the media contributes significantly to the fluidization on backwashing the system, typical backwash water volume being 0.1 to 1% of the filtered water.

The Spruce Filter™ is capable of running at much higher velocities than dual media and at a water temperature of 15°C and above will run at 60 m^3/m^2/hour plus. It requires the addition of less chemicals to achieve the required SDI. This delivers vast CAPEX and OPEX savings.

6.3.4.3 Gravity filters.

1. Elements of a gravity filter.

Vessel.

Vessels are manufactured in concrete with a unit surface with range from 28 to 100 m^2. Sixty (60) m^2 is a standard value in desalination plants with dimensions of 4 m x15 m.

The bigger the unit filtration area is, the bigger the backwash equipment should be. Consequently, the designer must find the equilibrium between the economic impact of increasing the number of filters and the increase of the backwash equipment if the number of filters is too low.

The influent water distribution channel is used to collect backwash water.

Usually, these filters are covered as sunshade protection.

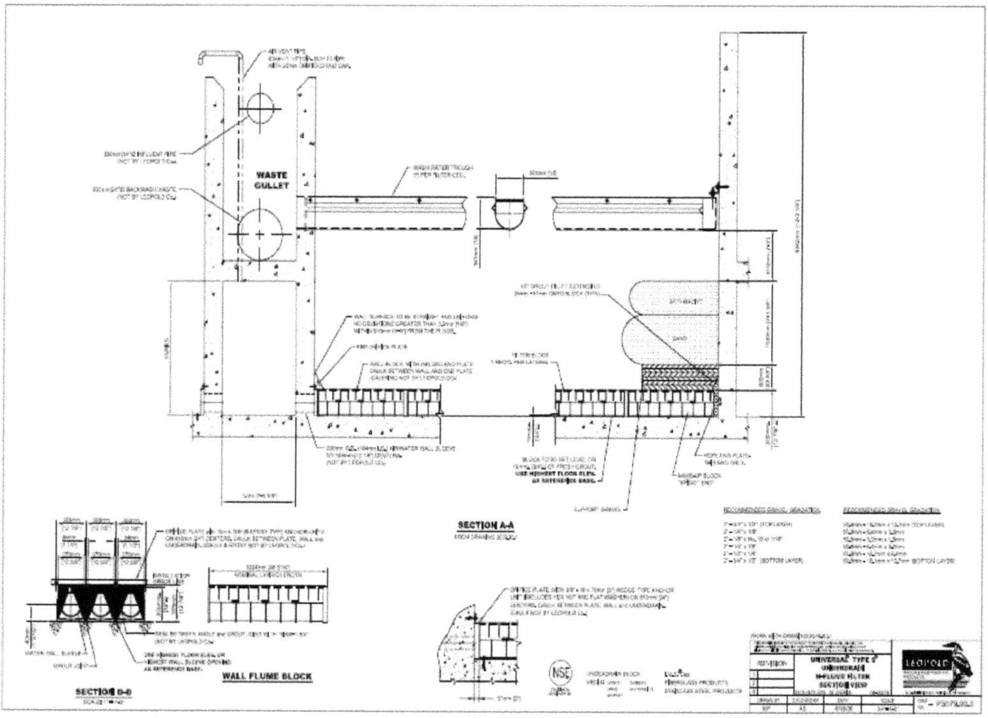

Figure 6.20. Gravity Sand filter. Courtesy of Leopold.

Media.

The same media as in the pressurized filters is used but with an approximate height of 1350 mm (compared to 1-1.1 m for the others). The combination of the different elements are: 600mm of anthracite, 600 mm of sand and 150 mm of gravel.

If all three materials are used, the usual distribution will be 400 mm of anthracite, 400 mm of sand and 300 mm of garnet.

Underdrain systems.

Three basic types of underdrain systems that are common:

- False bottoms.
- Leopold porous blocks.
- Pipe laterals.

The philosophy of false bottoms with nozzles is basically the same as pressure filters. The only difference is that the air inlet is longer.

Leopold porous blocks are easier to install and the risk of sand leakage is lower

Leopold porous blocks.

The dual parallel lateral was developed to solve flow distribution problems of other systems. As shown in Figure 6.21 the system consists of a primary lateral and two secondary compensating laterals which are parallel to each other. Control orifices open from the primary lateral directly into the secondary laterals. Backwash water is introduced into and flows through the primary lateral, rises and discharges from the control orifices into the secondary compensating lateral. Any unbalanced flow from the primary lateral creates an opposing flow, forming a directly compensating velocity pressure gradient which puts the secondary lateral into a uniform hydraulic pressure condition throughout its length. This uniform pressure services the top deck dispersion orifices that discharge from the compensating lateral into the filter box.

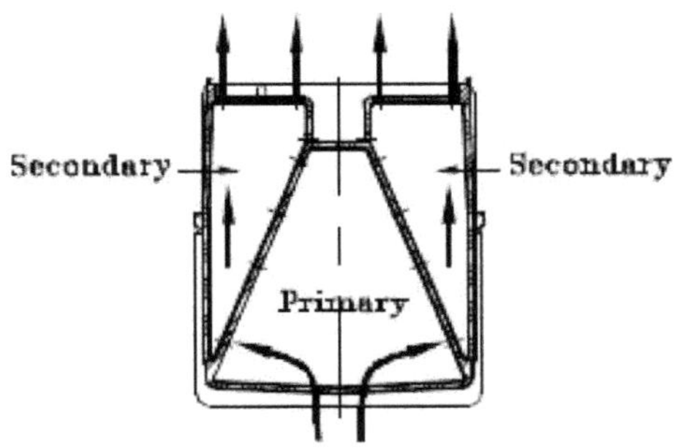

Figure 6.21. Detail of Leopold Porous blocks. Courtesy of Leopold.

The function of the secondary/compensating laterals is to provide a compensation or balance to these variances. This system provides an automatic equalization of pressure prior to discharging the backwash water from the underdrain, resulting in complete uniform discharge along the entire length of the lateral. Dual parallel laterals can be sized up to 15 meters (50 feet) in length with a backwash lateral mal-distribution of less than ±2%. Maldistribution is the percent difference in flow from one part of a lateral or underdrain system to another. Usually the highest and lowest points of flow or pressure are used to determine mal-distribution.

Widely spaced nozzles release large volumes of air creating disruptive "explosions" propelling media to the surface and providing zones of low or no turbulence.

The triangular shape of the primary lateral collects incoming air uniformly along its length. Small dispersion orifices as shown in Figure 6.21, then distribute and meter the air uniformly to the top deck discharge orifices. The top deck orifices are closely spaced at 5 cm (2 inch) centers to provide a more uniform coverage of the filter floor. A baffle in the bell end of the secondary laterals ensures that the air and water that are metered into each 1.2 m (4 feet) section are discharged from that section. Without the baffle, the backwash water tends to push the air up and down the lateral causing pulsations.

The filter compartment is constructed like a box with a rough-finished, plain flat floor. The design does not require any special sub-structure such as supporting piers or beams, nor does it require added filter depth for a plenum or false floor. The 1.2 m (4') dual parallel lateral blocks are set in a base grout directly on the flat filter floor.

The dual parallel lateral blocks are placed end-to-end in rows so the laterals are continuous through the length of each row. Rows of blocks are placed on ± 30.5 cm (12 inch) centers across the width of filter.

Figure 6.22 shows the typical installation of the dual parallel lateral in a filter with a front flume. The backwash water enters the flume, travels into the dual parallel laterals, is dispersed up into the media, and finally is collected by fiberglass washwater troughs.

Note the air header piping in the flume that uses "j-risers" to distribute the air to each dual parallel lateral. The air riser "j" shape is used to flush any water that accumulates in the bottom of the air header piping as the air fills the header.

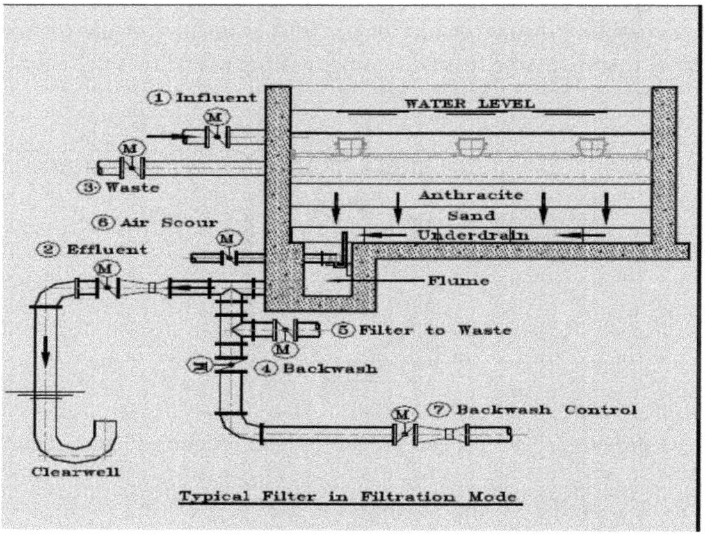

Figure 6.22. Typical installation of dual parallel lateral in filter.

Pipe laterals.

Pipe lateral underdrain systems generally consist of centrally located manifold pipe to which smaller, equally spaced laterals are attached. Lateral pipes usually have one or two rows of 6 to 19 mm perforations on their bottom sides. The lateral pipes may be fitted with nozzles as illustrated in figure. Guidelines for lateral design include the following ratios:

- Total area of orifices (surface area of bed): 0.0015 to 0.0015:1
- Cross sectional area of lateral (total area orifices): 2 to 4:1
- Cross sectional area of manifold (total area laterals):1 :5 to 3:1

Orifices are normally spaced at 8 to 30 cm and laterals at roughly the same spacing as the orifices. Approximately 45 cm of support gravel, is required to cover lateral network. Three to five graded layers are usually involved. With sizes varying from 38 to 3mm. The bottom layer should extend 10 cm above the highest wash water outlet.

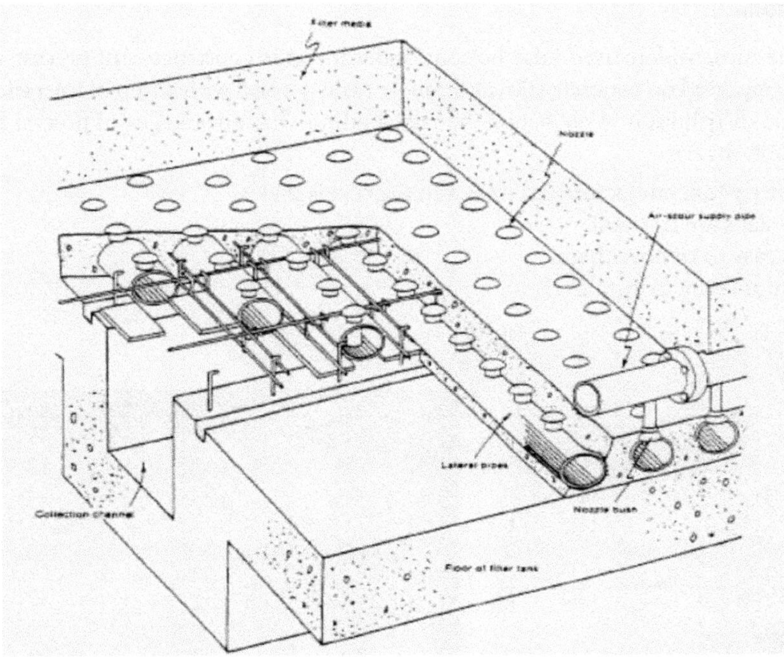

Figure 6.23. Detail of pipe laterals underdrain.

Figure 6.24. Photography of pipe laterals underdrain. Courtesy of HidroambIente.

False bottoms.

One of the most widely used false bottom underdrains is constructed of precast reinforced concrete supported on concrete sills, and can be easily leveled with 2 mm of tolerance to guarantee equal distribution of air below the false bottom. The advantages of precast reinforced concrete bottom are:

1. It is easy to manufacture the slabs and the cost is low.
2. The slabs are resistant.
3. It is easy to remove them.
4. They are easy to install.

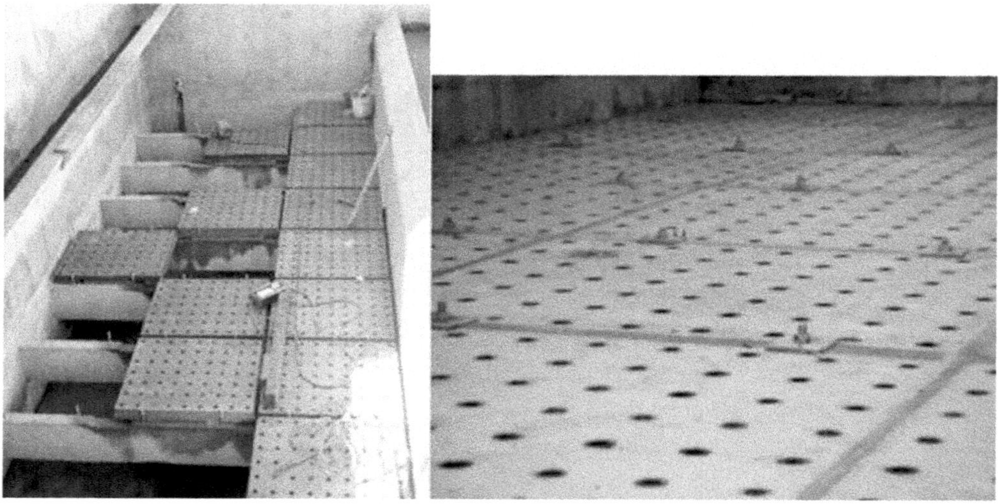

Figure 6.25. Photography of pre cast concrete underdrain.

Monolithic concrete underfloor.

One particular bottom system is the The Monoflor HD filter underdrain system. It is constructed using one-piece molded formed of high-impact polystyrene. The forms include preformed openings spaced on 6" and 5" centers, for receiving adapters and nozzles. The forms become a permanent part of the finished false floor once the concrete has been poured. The polystyrene forms are supported on 150 mm diameter PVC pipe columns, which are filled with concrete when the floor is poured. A chisel-pointed stainless steel all-thread rod is inserted through each pipe column and chemically anchored into the concrete floor of the filter cell. To tie the Monoflor HD underdrain into the filter walls, reinforcing rods are doweled into the walls. The result, after the concrete pour, is a filter floor that is firmly anchored, both at the periphery and into the filter floor. The six different standard structural designs will ensure that any foreseeable applied load, either during backwashing or filtration, can be met.

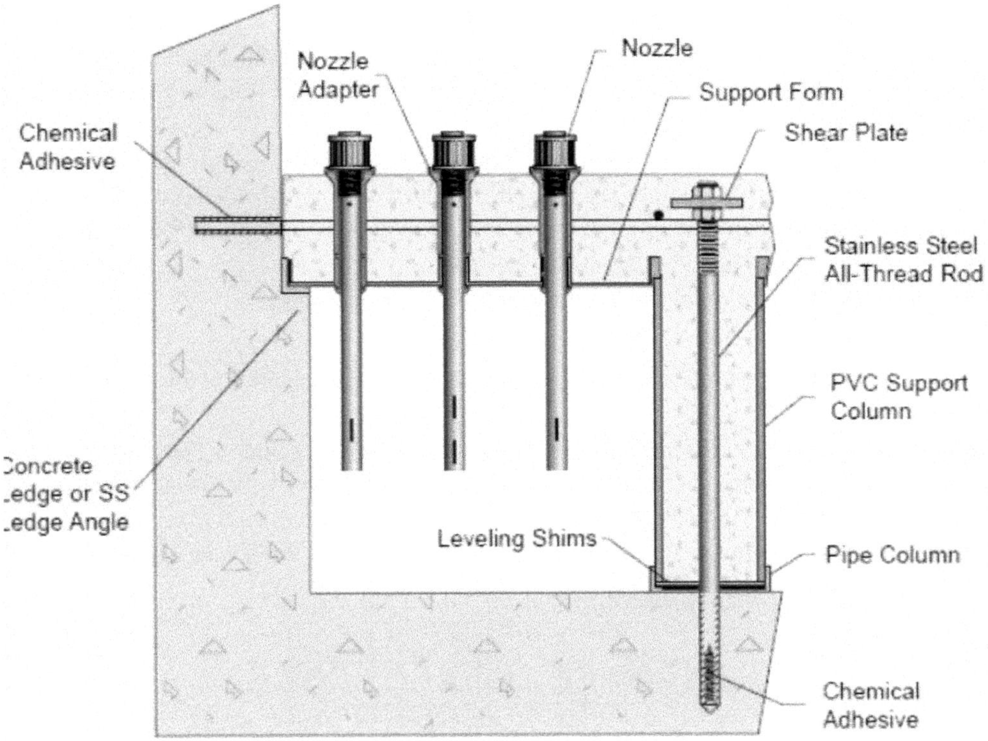

Figure 6.26. Constructive detail of system MONOFOR HD. Courtesy of Degremont.

Figure 6.27. Constructive detail of system MONOFOR HD. Courtesy of Degremont.

- *Washwater troughs.*

The wash water troughs are suspended at even spacing above gravity filter beds to provide uniform removal of wash water during backwashing. The channels are transversal to the central concrete channel.

If troughs are placed too close to the surface of the unexpanded bed, media may be lost during backwash. The design elevation of the weir edge of the trough may be determined by adding the depth required for maximum bed expansion (usually 50%) and the overall depth of the trough, plus a small margin of safety (0.15 to 0.3 m).

The criteria for the location of the troughs based on trough spacing is:

$$H = 0.34 \times S$$

Where H=height of the top edge of the trough above the fluidized bed and S=center to center spacing of the troughs

Troughs are usually made of fiberglass-reinforced plastic (FRP), stainless Steel, or reinforced concrete. Troughs made of FRP and stainless steel usually have semicircular bottoms, and concrete troughs have v-shaped bottoms.

The required cross sectional area of the trough for a given wash water flow can be estimated from Figure 6.28.

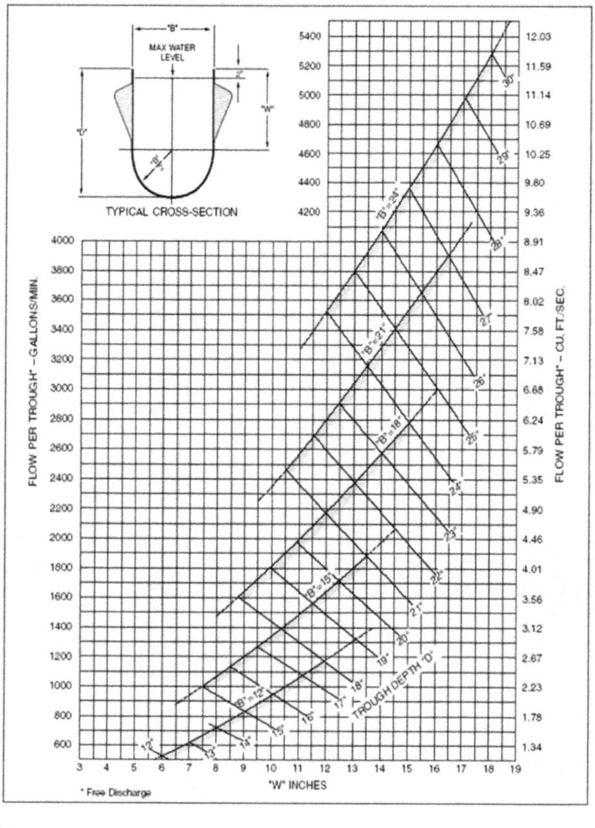

Figure 6.28. Cross sectional calculation of washwater troughs.

2. Gravity filter headloss.

Filter performance is often judged by the length of filter run, but too long a filter run may not be good for filter operation. Long filter runs make washing a filter much more difficult because of particle matter compaction in the filter media. In addition, a long filter run indicates that the filter is not working at its most cost-effective capacity.

Filters should be washed when terminal head loss is reached; otherwise turbidity breakthrough may occur. Gravity filters typically have a terminal head loss of 0.4-0.5 bars when ferric chloride+coagulant aid is used. These values are slightly lower than terminal head loss of pressure filters which is approx. 0.7 bars.

If reactives are not used, terminal head loss will range from 2 to 3 meters. These values are similar to the values used in drinking water treatment plants.

Head loss is normally monitored by differential pressure measuring the water pressure above and below the filter media. Aside from head loss developed by particle retention within the filter media, operating head loss depends on the filtration rate, the clean bed head loss through the filter media and head loss through the filter under drain system and the effluent rate controller. Terminal head loss is the difference between the static head available between the water elevations in the filter and the filtered water effluent control weir, less the operating head loss through the clean media, underdrain system, effluent piping connections and bends and the effluent rate controller. Typical terminal head loss is 2.5 to 3 meters.

3. Retention time.

Gravity filters with a head loss of 4-5 meters per column of water usually have a retention time of 15 to 20 minutes (for a velocity of 8 m/h). Compared to the retention time of pressure filters with a retention time of 9 minutes, the gravity filters offer higher retention times.

This retention time is used to enhance flocculation before water gets into contact with filter media, so these types of filters can operate without flocculation chambers (only static mixers or mixing chambers are required).

4. Feed and recovery of backwash water system.

It must be taken into consideration that sea water is corrosive, so butterfly valves can be a cheaper alternative than penstocks. Also, piping material in PVC can be a cheaper alternative than concrete.

5. Differences between pressure filter and gravity filter.

- Pressure filters are manufactured in a workshop. Gravity filter are erected on site.
- Pressure filter material is carbon steel+rubber / ebonite (recently GRP is being used for medium sized filters). Gravity filters are manufactured in concrete. Carbon Steel+ebonite is more expensive.
- Maximum surface of pressure filters is 50 m^3. However, gravity filters can have higher filtration surfaces (50-70 m^2).
- Retention time in gravity filters is larger than in pressure filters. As indicated previously, it would improve floculation process.
- Filter run (period between backwashing) is longer in pressure filters than in gravity filters.
- Quality of treated water can be better in pressure filters than in gravity filters.

6. Double stage filtration.

In double stage filtration, pressure filters can be used in the 1st and 2nd stage, with the advantage of only installing one pumping station.

When it is not possible to use only one pumping station, and two (2) stations are required, a pressure filter+gravity filter can be used. An in line flocculation will also be used.

When floculation chambers are used, it will be better to install a gravity filter+ pumping+ pressure filter with floculation in line.

When the range of seawater temperature is high (14-28°C), the third option is the best alternative from an energetic standpoint as the second pass pump can absorb pressure system variations (6 bars) due to the temperature. In some cases, pressure range can be up to12 bars.

7. Filter media.

Bed depth is typically 1350 mm instead of 1000/1100 mm.

A typical dual media bed filter contains 600 mm of anthracite, 600 mm of silica sand and 150 mm of gravel.

Mixed media filters contain 400 mm of anthracite, 400 mm of silica sand and 300 mm of garnet.

8. Filter arrangement.

Filters are normally placed next to one another along one or both sides of pipe gallery. This approach provides the most compact arrangement and also simplifies filter operation and maintenance. If possible, areas for future expansion should be provided at one end of the row (or rows) of filter, and piping in the gallery should be installed with blind flanges at the ends to make future filter additions easier. In large plants, placing filters in rows on opposite sides of a pipe gallery is common practice.

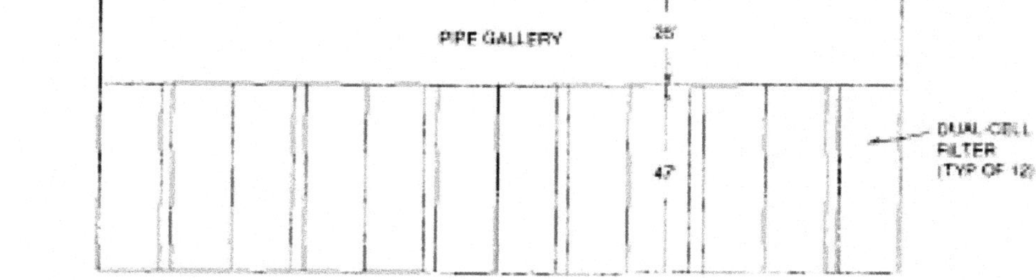

Figure 6.29. Filter arrangement.

9. Filtration by-pass.

It is necessary to install a manual by-pass at the filtration outlet to provide the plant with flexibility enough re-circulate the 50% of filtration flow before it is treated in cartridge or pre-coat filters.

6.3.5 Equipment used in filtration.

1. Conduits.

Typical piping serving a gravity filter is shown. Influent conduits should be sized to limit velocities to 0.6 m/sec. This may result in the use of an influent flume rather than a pipe in large plants. Hydraulic considerations generally result in velocities of 0.9 to 1.8 m/second in wash water and filtered water piping. At higher velocities, head loss often becomes excessive, and undesirable effects such as water hammer are more likely to occur.

Design considerations include avoiding high points and including air release valves in wash water supply piping.

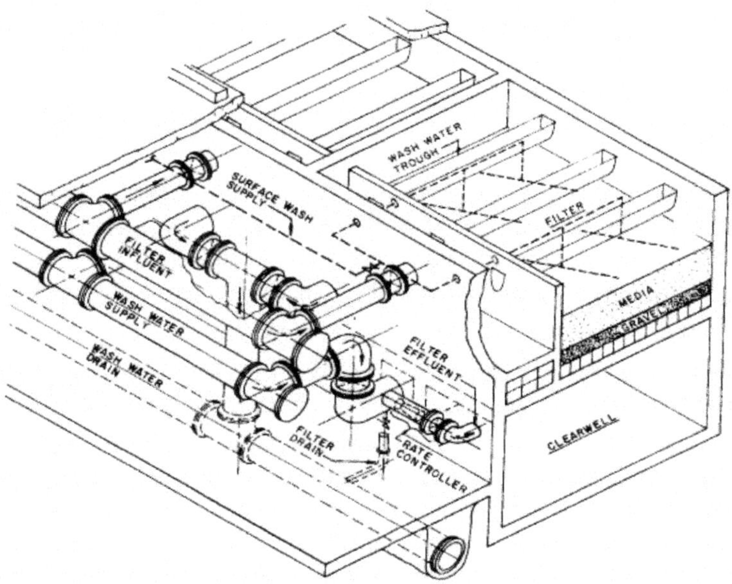

Figure 6.30. Conduits of gravity filters.

2. Backwashing pumps.

Backwashing deposit (brine) is usually buried, so the type of pumps which could be used are:

- Horizontal centrifugal pumps with a positive suction head. A dry chamber is required.
- Horizontal centrifugal pumps without positive suction head+priming system and submergible pumps.

Superduplex is the material used for pumps.

Figure 6.31. Photography of filtration backwash pumps.

3. Blowers.

Positive displacement blowers are usually employed.

4. Butterfly valves.

A typical filter is equipped with at least five valves: influent, filtered water, wash water supply, wash water drain and surface wash or air water supply. A filter-to waste valve may also be included; however waste lines constitute a potential cross-connection and must be equipped with air gap protection against backflow from the drain to filter.

Rubber seated cast iron or superduplex butterfly valves are used.

Typical manufacturers of these valves are LSB (AMVI) with HALAR cover. BRAY uses NYLON 11 (cheaper) and TYCO which manufactures the valves in super duplex.

Valve actuating systems may be pneumatic or electric. Electrical actuators offer greater reliability, but initial cost is higher and in the event of that maintenance is required , electric controls may require greater technical skill to service than pneumatic controls, so pneumatic actuators are used in most filters. Automatic double effect valves are used, but in some cases the wash water drain valve has been installed with a single effect NC actuator to prevent that the filter empties.

Inlet and outlet valves must be control valves with a positioner depending on type of filter. It typically should operate between 0 and 45º open. It should also automatically close during power failure to prevent dewatering the filter media.

Drain and backwash inlet valve should be installed with a mechanical opening limit.

Electro valves boxes (one for each filter) are used for control. Usually, each electrovalve has an installed air flow restrictor.

5. Flow meters.

E+H, KROHNE, SIEMENS (DANFOSS), ABB or similar are the standard manufacturers of flow transmitters.

6. Differential pressure transmitters.

They are used to measure the headloss between the pressure filters collectors.

7. Level transmitters.

They are used to keep the level constant in the gravity filter. Ultrasonic, hydrostatic or differential pressure transmitters can be used.

Example of the design of two stages filtration

SAND FILTERS – OPEN/1ST STAGE

DESIGN PARAMETERS

No. of units.. 6

Operation... 6 filters operating or 5 operating and one washing

Type .... By gravity

Execution... Rectangular open

Design flow .... 1903 m^3/h (20,000 m^3/day)

Filtration velocity .... 7 m/h

Type of washing.. Air-water

Backwashing velocity............................... 25 $m^3/h/m^2$

Air velocity... 54 $m^3/h/m^2$

FILTERS SIZING

- *Minimum surface required for each filter is*

$$\frac{1903\ m^3/h}{6 \times 7 m/h} = 45.31\ m^2$$

Selected filtration Surface will be 43.2 m^2. (4800 mm width and 9000 mm length)

Filtration velocities are the following ones:

Normal conditions - 6 filters operating and a daily production of 20,000 m^3/day

$$\frac{1903\ m^3/h}{6 \times 43.2 m^2} = 7.34\, m/h$$

Maximum flow : 5 filters operating and one stand-by

$$\frac{1903\ m^3/h}{5 \times 43.2 m^2} = 8.8\, m/h$$

- *Media*

The media of each filter consists of:

	Media	*Depth (mm)*	*Size (mm)*	*Volume(m^3)*
(I)	Gravel	150	6.7 - 13.2	6.5
(II)	Gravel	100	2.36 - 4.75	4.3
(III)	Gravel	100	2.18 - 2.8	4.3
(IV)	Sand	550	0,5 – 1	24
(V)	Anthracite	350	1.18 - 2.36	15

2246 nozzles per filter will be installed (52 nozzles/m^2.)

MEDIA BACKWASHING

Filters will be backwashed with water and air

- *Air*

Agitation air will be obtained from two rotary blowers (one spare)

Required flow will be:

43.2 m^2 x 54 $m^3/h/m^2$ = 2333 m^3/h

Effective pressure will be 0.4 bar.

- *Water*

Filtrated water from 1st stage will be used as backwash water

Required flow will be:

With air

43.2 m^2 x 10 $m^3/h/m^2$ = 432 m^3/h

Without air:

43.2 m^2x 25 $m^3/h/m^2$ = 1080 m^3/h

GRAVITY SAND FILTERS – ANTHRACITE (2ND STAGE)

DESIGN PARAMETERS

No. of units..............................4

Operation................................4 filters operating or 3 operating and one washing

Type..Pressurized

Execution................................Closed

Flow ....1903 m^3/h (20,000 m^3/day)

Filtration velocity ....15 m/h

Type of backwashing ....Air-Water

Water velocity..........................25 $m^3/h/m^2$

Air velocity..............................54 $m^3/h/m^2$

FILTERS SIZING

- *Minimum surface required for each filter is*

$$\frac{1903\ m^3/h}{4 \times 15 m/h} = 31.7\ m^2$$

It will be selected an horizontal pressure sand filter with a diameter of 3600 mm
Required length will be:

$$\frac{31.7}{3.486} = 9.1\ m$$

In a first approach, the length of each filter will be 8 m
Filtration surface will be: 3.486 x 8 = 27.9 m^2.
Torispherical heads give an additional length of:

$$2 \times (800 - 3.600 + \sqrt{3.600^2 - 450^2}) = 1.543 mm.$$

Additional surface will be: 1.543 x 3.486 = 5.38 m^2,
Resulting in a total filtration Surface of: 27.9+ 5.38 = 33.28 m^2.
Filtration velocity will be of:
Normal conditions - 4 filters working and a daily production of 20,000 m^3/day

$$\frac{1903\ m^3/h}{4 \times 33.28 m^2} = 14.3\ m/h$$

Maximum flow: 3 working, 1 filter backwashing

$$\frac{1903\ m^3/h}{3 \times 33.28 m^2} = 19.1\ m/h$$

Plant upgrade –If it would be necessary to increase the capacity of plant to 27.000 m^3/day, to keep the same filtration velocity 5 additional filters would be required.
Nominal flow – 5 filters operating

$$\frac{2569\ m^3/h}{5 \times 33.28 m^2} = 15.4\ m/h$$

Maximum flow - 4 filters in service; 1 filter backwashing

$$\frac{2569\ m^3/h}{4 \times 33.28 m^2} = 19.3\ m/h$$

FILTER MEDIA
Filter media will consist of anthracite and sand of he following characteristics.

	Sand	*Anthracite*
Size	*0.4-0.5 mm.*	*1-1.2 mm.*
Uniformity	*1.6*	*1.6*

	Sand	*Anthracite*
Height	*400 mm.*	*600 mm.*
Amount	*10 m^3/unit*	*15 m^3/unit*

BACKWASHING

Air-Water backwashing will be as below indicated

- *Air*

Required flow will be:

33.28 m^2 x 54 $m^3/h/m^2$ = 1797 m^3/h

Effective pressure will be 0.4 bar.

- *Water*

Prefiltrated water from 1st stage will be used:.

Required flow will be::

**With air injection*

33.28 m^2 x 10 $m^3/h/m^2$ = 333 m^3/h

**Without air injection:*

33.28 m^2x 25 $m^3/h/m^2$= 832 m^3/h

6.3.6 Continuous backwashing filters.

These types of filters are based on the counter flow principle. The water to be treated is admitted through a water inlet distributor in the lower suction of filters and is cleaned as it flows upwards through the sand media. The sand containing the entrapped solids is conveyed from the tapered bottom of the section of the unit by means of an air-lift pump to a sand washer at the top. Cleaning of the sands starts at the pump, in which impurities are separated from the sand grains by the turbulent mixing action. The contaminated sand spills from the pump outlet into the washer labyrinth in which it is washed by a small countercurrent flow of clean water. The separated solids are discharged through the wash water outlet while the clean grains of clean sand are returned to the sand bed. As a result, the bed is in slow, constant downward motion through the unit. Thus waste purification and sand back washing take place continuously.

These filters usually work in series of two stages for open intakes.

The first filter has a coarse (0.9- mm) sand media bed with 2 m deep and a velocity of 12 $m^3/m^2/h$. The second filter contains finer (0.5-mm) sand media and its media depth is 1 m. Second stage filters work at approx. 8-9 $m^3/m^2/h$.

The Figure 6.32 depicts this pretreatment system. Both first and second-stage filters have instrumentation for continuous turbidity monitoring and data logging. The second-stage filter is equipped with a particle counter as well.

The source seawater is conditioned with coagulants prior to entering the first filter stage. In addition, polymer is fed to the source water during heavy rain and red tide events to enhance the flocculation of the source water particles prior to filtration.

No chemicals are added to the second stage feed water.

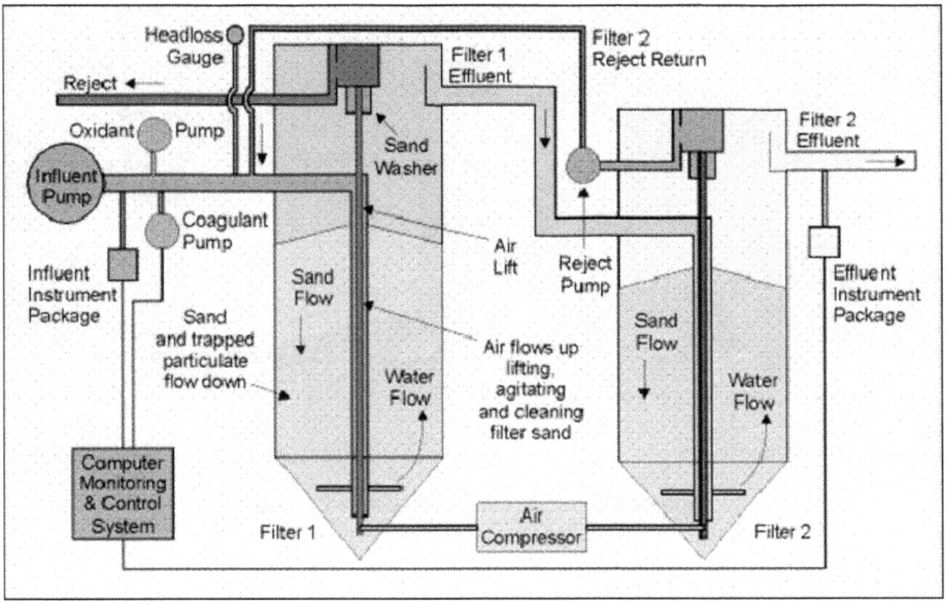

Figure 6.32. Schematic of continuous backwash filters. Two stages.

1. **Advantages.**

No need for back wash pumps, wash water storage tank, wash water collecting tanks and in general all the equipment associated to water backwash. Washing is carried out with filtrated water (loss of 10%). Air backwashing requirements are of approx. 4-6 Nm^3/h

No need for bottom filter nozzles.

2. **Disadvantages.**

It cannot be fully guaranteed the efficiency of system when water is flocculated either in line or in flocculation chambers.

6.3.7 Automatic self-cleaning filters.

Self cleaning filters are used before UF membranes, offer valuable protection by removing biological organisms, organic matter and inorganic compounds.

Acting as a pre-filter in substitution of standard cartridge filters, they increase recovery rates and lessen chemical consumption and disposal for membrane cleaning. However, they are basically used as pretreatment of MF/UF systems

Automatic cleaning screen filters commonly operate by the following procedure. As water flows from the inside to the outside of the screen, suspended material is collected on the inside screen surface. As the silt builds up, this causes the differential pressure to rise. When a pressure of typically seven (7) psi is reached, the cleaning cycle is triggered and an exhaust valve opens removing filter cake from the screen and carrying it out of the system. The cleaning cycle consists of a set of suction nozzles that back flush on less than one square inch of the screen at a time. The nozzles rotate in a spiral pattern and cover the entire screen in a cleaning cycle that's typically in the range of 8-20 seconds depending on the filter used.

Figure 6.33. Self cleaning filters before UF system.

6.4 Precoat Filtration.

Diatomaceous earth media is not recommended for primary filtration because of its characteristically high head loss and short run times. The diatomaceous earth precoat filter technology is a long-term, established method in conventional water and wastewater treatment. Its use in seawater applications is limited and few references exist worldwide and it is used as alternative for second stage filtration.

These consist of special cartridge filters with a precoat deposit on the filter (filtration less than 2 microns).

The elements included in a pressure filter system are shown in figure 6.33. In the filtering mode, source water is pumped into the filter shell, and a baffle plate distributes flow and prevents it from eroding the cake on the septa. A flow-measuring device installed in the filter outlet line activates a valve that maintains a constant flow rate.

In the precoat cycle, the source water inlet and filtrate outlet valves remain closed. If the system includes a precoat tank, as shown in the schematic, first the system is filled with previously filtered water to a depth sufficient to reach the top of the shell. Then the pump recirculates the water as precoat slurry is added. In an alternative arrangement, a recirculation tank is not used. With level controls in the filter shell, the unit is first partially filled with water, then filled completely along with the addition of precoat slurry.

Precoat accumulates on the septa as the slurry is recirculated. Because the flow tends to seek the path of least resistance, recirculation tends to concentrate on those element areas not previously covered, until a reasonably uniform precoat thickness is eventually formed.

After full cover is achieved, the remaining suspended precoat material is intercepted by the filter, and the turbidity of recirculated flow quickly clears up. At this time, in a simultaneous operation, the recirculation valve closes, and the filter inlet and outlet valves open to initiate the filter cycle.

When the maximum available head differential is reached or the quality of filtrated water is low, the filter run is terminated and the filter is cleaned. In the cleaning cycle the inlet valve is closed, the waste valve is opened, and after the unit is fully drained, sluice water is turned on to flush away the filter cake.

Duration of precoat (filtration period) will range from 18 to 48 hours. Filtration velocity is 4 m/h.

After the precoat has been deposited, filtering begins, and sometimes at the same time a small amount of DE material (called *body feed)* is added to the source water to maintain the porosity of the media and increase the water quality.

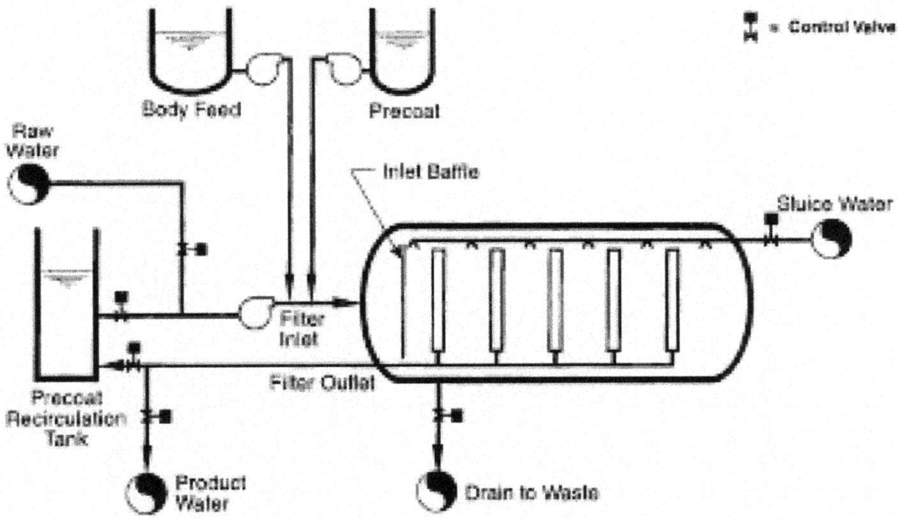

Figure 6.34. Precoat filtration. Pressure filter.

6.4.1 Types of filters.

Filters are always constructed as cylindrical pressure vessels of diameters of 1500, 1800 and 2000 mm. Most units fabricated are manufactured of carbon steel with a coating of ebonite (3-4 mm thickness) to protect from corrosion.

Top flanges are installed to dismantle cartridges. A bridge crane is required for filters maintenance and each filter has two sight glasses to follow precoat formation.

There are two types of pressure vessel closures:

- Top entry bolted cover for standard flanges.
- Special cover clamps. Smaller size of flanges.

Figure 6.35. Precoat filter vessel.

6.4.2 Types of cartridge.

Plastic support core with a length of 1500 mm and 60 mm of diameter. The filtration Surface per filter is 0.28 m^2. There are examples of support core made of duplex.

6.4.3 Precoat.

DE is a fine powder (90 % SiO_2) that has abrasive properties when handled both in dry form and in slurries. Special precautions must be taken to protect plant staff against dust, and equipment and pipelines must be protected against erosion. Because of its abrasive properties.

DE breaks down in size under turbulent operating conditions. DE can be delivered in plastic-lined paper bags or in woven plastic bags

Diatomaceous earth media contains a 90% of SiO_2 with a granulometry of 30-40 microns up to 77 microns.

Dose will be between 400 to 800 gr/m^2. For example, for a 105 m^2 filter the consumption will be between 40 to 80 kg/filter.

1. **Valves (Each filter has the following valves).**

   - **Service.** ND 250 (inlet/outlet). Inlet valve with positioner and a by-pass valve of ND 25 to pressurize the filter.
   - **Recirculation.** ND 200 (Inlet and outlet).
   - **Drain.** ND 50.
   - **Vent (upper part).** ND 200.
   - **Vent (lower part)** ND 80.
   - **Air inlet.** ND 50.

2. **Instrumentation (For each filter).**

   - Electromagnetic flowmeter inlet ND 250.
   - Differential pressure transmitter.

3. **Recirculation system.**

   - **Compensation tank.** It compensates water loss during precoat deposition. It has an approx. volume of 15-20 m^3.
   - **Recirculation pumps.** Two (1+1) pumps are installed with a flow of 75% of service flow 300 m^3/h and a pressure of 2.5 bars.
   - **Automatic valves.** Four (4) ND200 valves and one (1) ND100 for compensation tank filling.

4. **Precoat dosage.**

For DE concentrations of 4% or less, case-hardened centrifugal pumps may be used for slurry transfer. Abrasion-resistant chemical feed metering pumps are also acceptable for these concentrations. Thicker slurries must be conveyed with either peristaltic-type pumps or rubber-lined centrifugal pumps. To minimize breakdown of DE material, pipeline velocities should be limited to no more than 8 ft/s (2.4 m/sec.), but must be maintained at over 3 ft/s (0.9 m/sec.) to prevent settling. Depending on the amount to dose, it will be necessary to install one (1) or two (2) deposits with manual or automatic loading.

If dosing is automatic, dosing pumps directly from Big-bag will be used. SODIMATE and FLEXICOM are the main suppliers.

5. **Cleaning system.**

For cleaning of precoat filters and cartridge filters, a tank with a capacity of 6 m^3 and a pump of 100 m^3 for filling and cleaning solution recirculation is installed.

6.5 Cartridge filtration.

6.5.1 Design basis.

A cartridge filter with an absolute pore size of less than 10 μm is the suggested minimum pretreatment required for every RO system. It is a safety device used to protect the membranes and the high pressure pump from suspended particles. Usually it is the last step of a pretreatment sequence. The better the prefiltration the less RO membrane cleaning required. If there is a risk of fouling with colloidal silica or with metal silicates, cartridge filtration with 1 to 3 μm absolute pore size is recommended. The filter should be sized on a flow rate according to the manufacturer's recommendation and replaced before the pressure drop has increased to the permitted limit, but at least every 3 months.

This filter works as a mechanical filtration, but depending on the characteristics of cartridge electrostatic adsorption is also possible.

Usually, the filtration rate is of 14-20 m/h which are equivalent to 10-15 l/min/10 for string wound filter cartridge or thermally bonded cartridges.

Figure 6.36. Cartridge filters. Courtesy of Cadagua.

A pore size of 5-10 microns nominal is the standard for cartridge filters, though some companies have started an absolute pore size of 20 micron filter in some certain type of filters. This value can cause problems and it is used 40 micron (PALL-Ultipleat HIGH FLOW).

An absolute pore size rating specifies the pore size at which a challenge organism of a particular size will be retained with 100% efficiency under strictly defined test conditions. Among the conditions that must be specified are: test organism (or particle size), challenge pressure, concentration and detection method used to identify the contaminant. A nominal pore size rating describes the ability of the filter media to retain the majority of particulate at (> 98%) the rated pore size. A filter with an absolute pore size of 20 microns is more expensive than nominal pore size of 5 microns.

Head loss in a cartridge filter varies from 0.3 meters of column of water (mcw). to a maximum of 10 to 15 mcw. which is the maximum value.

EXAMPLE

Flow (F) $2569\ m^3/h$

Design flow is calculated for a flow rate of $0.7\ m^3/h$ *for every 10" of cartridge. If the length of the element is 1000 mm, the flow through each filter is:*

$$0,70 \times 4 = 2.8\ m^3/h$$

Then, the number of filtrating elements will be 2569/2.8 = 917 units

If the number of filtrating elements per filter is 156, the number of cartridge filters will be six (6).

6.5.2 Types of cartridge filters.

Cartridge filters can be horizontal or vertical depending on the supplier.

Cartridge filters are manufactured of the indicated materials below:

- Glass reinforced plastic up to 1200 mm of diameter.
- Carbon steel plus ebonite in all the sizes.
- Stainless Steel in all the sizes.

The false bottom is designed for a pressure of 2.5 bars.

There are two models of vertical cartridge filters:

a. **Hanging** with filter elements hanging from the perforated plate. It needs a maintenance crane and the removal of piping.
b. **Supported,** Filter elements are supported in the perforated plate. They require a maintenance platform (AMAZON) and machining holes of the plate, but they have the advantage that it is not necessary neither to dismantle pipes for maintenance nor a crane.

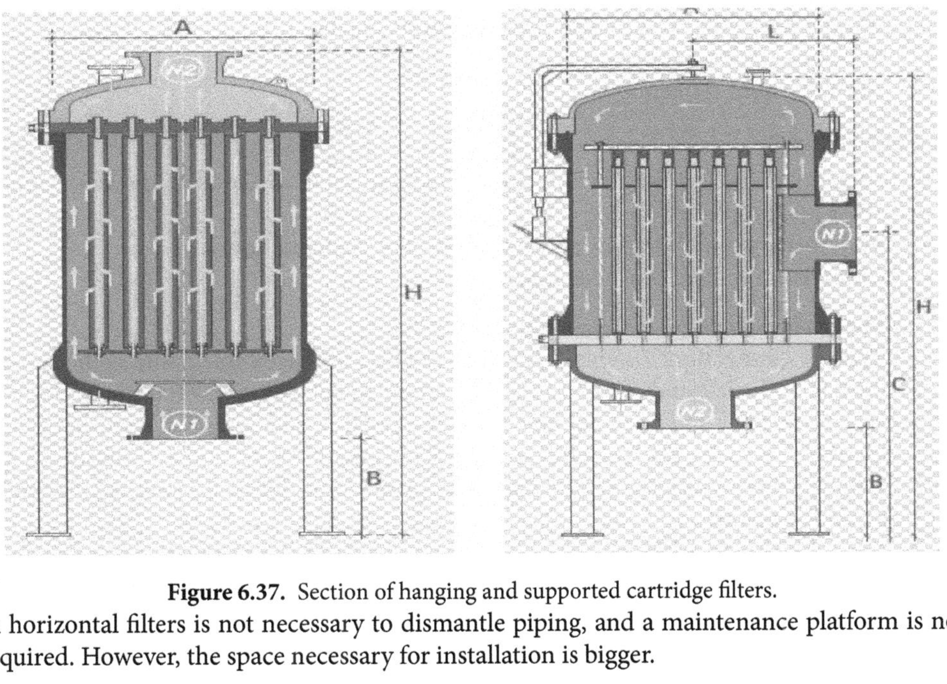

Figure 6.37. Section of hanging and supported cartridge filters.

In horizontal filters is not necessary to dismantle piping, and a maintenance platform is not required. However, the space necessary for installation is bigger.

Figure 6.38. Multiple horizontal filters.

6.5.3 Types of filtration elements.

Different types of cartridge filters.

- **Length**: 40" (1000 mm) and 60" (1500 mm).
- **Type of cartridge:** melt blown media, wound filter and pleated.

<u>Wound filter.</u>

- High experience.
- It is not a selective particle size. This is nominally rated.
- Fit all the standard housings.
- Normal length :1250 mm.

<u>Melt blown media.</u>

- Larger particles is retained in outer sections/removal of smaller particles in the inner.
- 100% polypropilene.
- Velocity between 0.6 to 0.75 mm/h per 250 mm elemental cartridge.
- Clean filter pressure drop. 200 mbar.

- **Materials**: Polypropilene, glass fiber, and cellulose.
- **Type of connection**: Double Open End (DOE) cartridges are, as the name implies, open at both ends and rely on the sealing arrangement within the housing to close one end of the filter and so direct flow through the other end. Double Open End (DOE) cartridges are ideal for non-critical, general purpose filtration and pre-filtration applications, where operating economies are of prime importance. It is not recommended in SWRO plants.

  Code 3 utilizes 2 x 222 'O' rings at the open end and a flat blanking cap at the closed end. This is one of the most common styles used in industrial and pure water applications around the world.

  Code 8 utilizes 2 x 222 'O' rings, but with a spear (or bomb-fin) at the closed end. The function of the spear is to help centralize the cartridge on installation of longer cartridge lengths.

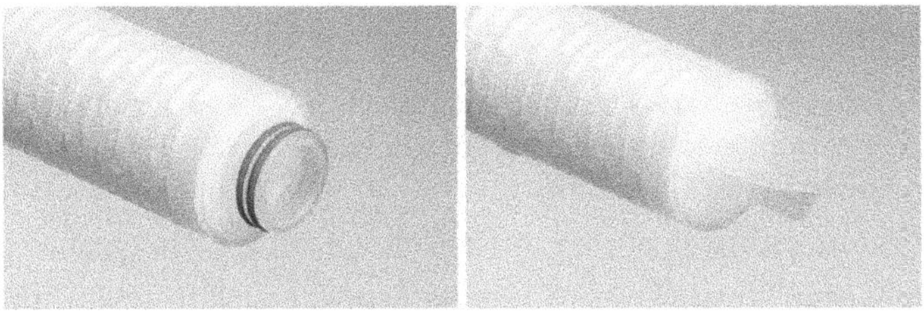

Figure 6.39. Detail of cartridge filters ends.

Cartridge filters that suffer from biological/mineral fouling could be backwashed with detergent or disinfectant. For that reason, it is advisable to install a cartridge filter that can be backwashed.

For example, in a hanging filter, the cartridges are mounted on swords, supported with their heads, screwed into the false bottom plate with a nut threaded on the top to seal the cartridge. There will be a lower plastic grill that provides the necessary rigidity to prevent the cartridge filter from breakdown due to internal vibration.

It is necessary to talk specifically about the Ultipleat® High Flow filter (horizontal installation) supplied by PALL. It is a large diameter, coreless, single open ended, pleated cartridge with an inside-to-outside flow pattern. The filter's crescent-shaped pleat geometry, combined with its large 152.4 mm diameter allows you to use significantly fewer filters and smaller housings for high flow rate applications. However, the initial investment is higher.

The length of these filters is 60" (1500 mm) and can work at velocities of 25-30 l/min/10" of filter.

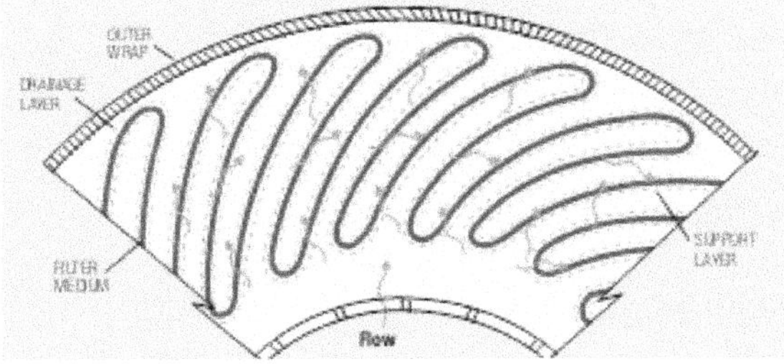

Figure 6.40. Section Ultipleat® High Flow filter.

6.5.4 Installation.

The equipment installed in a cartridge filter consists of:

- Manual valves for raw water inlet and outlet.
- Manual valves for backwashing water inlet, outlet and drainage.
- Instrumentation; It is necessary to install a flow transmitter (electromagnetic or rotameter) at each filter because cartridge filters work at variable flow. A differential pressure transmitter will be installed between collectors. The cartridge filter with less flow will be cleaned when head loss is above the set point (usually 1 bar).

6.5.5 Cleaning.

Cartridge filters need to be cleaned periodically (due to biological or mineral fouling). Precoat cleaning system or membranes cleaning system will be used. Valves will be manual as leaning period will never be less than 3-6 months.

Cleaning products are usually hypochlorite (biological fouling), detergent or water.

Cartridge filters must be replaced every 5 years if the water quality is good. If the water quality is not good, replacement rate will be of approximately one (1) year.

6.5.6 Maintenance.

Vertical filters need a hoist and enough vertical space to extract the filters and move them above cartridge housing. With horizontal cartridge filters, the problem can be the distance between filters.

6.5.7 Pressure vessels design.

Pressure vessels are designed to operate safely at a specific pressure and temperature, technically referred to as the "Design Pressure" and "Design Temperature". A vessel that is inadequately designed to handle a high pressure constitutes a very significant safety hazard. Because of that, the design and certification of pressure vessels is governed by design codes such as the ASME Boiler and Pressure Vessel Code VII div 1 and AD Merkblätter: German standard.

6.6 Pretreated water quality.

6.6.1 Quality control.

Quality of water must be controlled at different points of pretreatment with the instruments listed below:

- **Seawater.**
  Hydrocarbures analyzer (if required).
  Automatic SDI (if required).
  Chlorofile analyzer (if required).
- **Seawater pumping static mixer.**
  pH control.
- **2nd stage sand filters Outlet.**
  Chlorine control.
  Turbidity control.
  SDI control.
- **Precoat filters outlet.**
  Chlorine control.
  Turbidity control.
  SDI control.
  Particles counter.
- **Cartridge filters outlet.**
  Conductivity. (high and low pH alarm).

pH. (high pH alarm).

Redox potential (< 250 mV).

Particles counter.

SDI analyzer. Usually manual though semiautomatic and automatic alternatives are available. It is necessary to have a pressure of 2.1 bars in the pipe.

If analytical parameters (pH, redox, conductivity, SDI, temperature) are not valid for the reverse osmosis system, automatic valves installed in the suction of high pressure pumps will be closed. A valve named (LA) placed before the isolation valve will be installed to divert the water until the desired parameters are achieved. During start-up, once the parameters are acceptable, the automatic valve installed in the suction side of HP pumps will open and LA will close gradually.

Parameters for invalid values before RO:

- SDI > 4.
- 6,5 > pH > 8.5.
- RH (oxidant water).
- Temperature > 40°C.

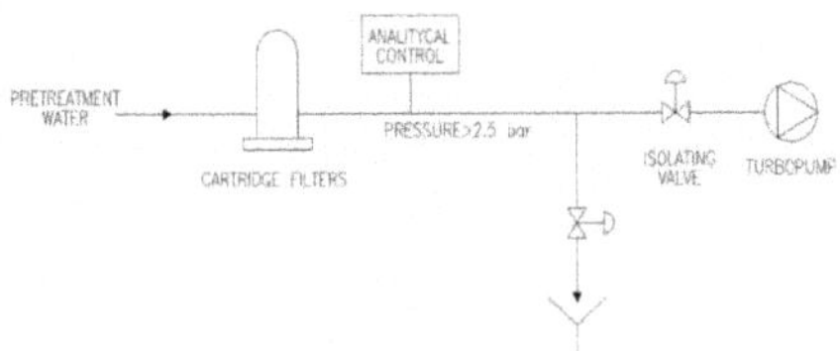

Figure 6.41. LA valve schematic.

If there are several R.O. Racks, the flow to pass through the LA valve will be variable. So, using a regulating valve is recommended. A control loop, in this case, (2.5 bar) of pressure or flow, will be established. The opening and closing of this valve should not imply the depressurization in the suctioning of the operating turbo pump.

6.6.2 Pressure and flow control.

For flow and quality control of a system (sand filters), it is necessary to install a manual by-pass, which should be used during start-up of filters before putting into operation the next step of treatment.

Size of by-pass must be at least 50% of total flow.

After precoat filters, the by-pass must be automatic.

After cartridge filters, the drainage system will have an automatic positioner so that it will be open before starting the turbo-pumps and will open when a pump stops. The drainage will be sized for the flow rate of at least (2) racks. However, it is advisable to have a manual drain for 50% of the rate.

6.7 SWRO plants with conventional pretreatment.

Table 6.1 shows SWRO plants with conventional pretreatment completed after year 2000 with capacities between 20,000-330,000 m^3/dayTable 6.1 SWRO with conventional pretreatment

Plant characteristics	Intake & pretreatment	Chlorine & SBS	Coagulant	Antiscalant & acid
Sydney, Australia commissioning in 2009, (design parameters given) 250,000 m³/d, provides 15% of Sydney's water, two pass SWRO	4 risers 200-300 m off shore, tunnel to plant, intake velocity <0.1 m/s single stage DMF (gravity filters were used in pilot tests; flocculation and second stage filtration was not found to be necessary	intermittently 10 mg/L NaOCl at intake 3 mg/L at plant intake 3-30 mg/L SBS	3 mg/L $FeCl_3$ and 0.2 mg/L poly DADMAC or 6 mg/L $FeCl_3$ without Poly DADMAC,	antiscalant not anticipated (if needed: 1.5 mg/L), 55 mg/L H_2SO_4 (pH 6.9), or pH 6.5 with more H_2SO_4
Tugun, Gold Coast, Australia commissioned in 2009, (design parameters given) 125,000 m³/d, two pass RO, 41% total recovery	twin risers 1 km off shore, 20 m water depth, intake tunnel to plant, intake velocity <0.1 m/s gravity DMF	intermittently 10 mg/l NaOCl at intake screen, dechlorination with 3 mg/L SBS prior to RO	18 mg/l $FeSO_4$ (4-20 mg/L) 0-0.4 mg/L polyelectrolyte,	2 mg/L antiscalant (1st pass) 35mg/L H_2SO_4
Perth, Australia commissioned in 2006, 144,000 m³/d, provides 17% of Perth's water, 45% recovery (1st pass), 95% recovery (2nd pass)	364,000 m³/d intake, one 8 m diameter inlet, 10 m water depth, 200 m off shore, buried glass reinforced pipe (diameter 2.3 m), one pressurized DMF	intermittently in intervals of 1-2 weeks, NaOCl at intake,dechlorination with SBS	$FeCl_3$ and poly DADMAC (organic coagulant),	antiscalant (1st pass) H_2SO_4
Tuas, Singapore commissioned in 2005 136,000 m³/d, 38.5%	Open off shore intake, DAF, gravity DMF, pressurized DMF	no information	$FeCl_3$, BW blended with Cc, occasionally visible plume	H_2SO_4 to pH 6-6.5

Plant characteristics	Intake & pretreatment	Chlorine & SBS	Coagulant	Antiscalant & acid
Fujairah, UAE Gulf of Oman, commissioned 2003, 170,500 m^3/d, recovery: 43% (1st pass) 90% (2nd pass), 41% (total)	380 m off shore, 10 m below sea surface, gravity DMF	Intermittently electro-chlorination at the intake (max. 4 mg/L) and in the coagulation chambers (max. 10 mg/l) for max. 4 hours	up to 5 mg/L $FeCl_3$, up to 1.5 mg/L coagulant aid BW blended with Cc,	up to 25 mg/L H_2SO_4
Tampa, Florida 95,000 m^3/d, providing 10 % of Tampa Bay water Supply	166,000 m^3/d intake from PPCW, coagulation, flocculation, single stage sand filtration, followed by DE filters	chlorine dioxide, dechlorination with SBS	$FeSO_4$	adjustment to pH 6.5-6.8
Barcelona, Spain, commissioned in 2009 200,000 m^3/d, provides 20% of Barcelona's water, recovery: 45% (1st pass), 85% (2nd pass), 44% (total)	Off shore submerged intake (2.2 km, -25 m), 3 stages: 1. flocculation, DAF 2. gravity DMF 3. pressurized DMF	shock chlorination at 3 points (offshore, pumping station, before first pretreatment stage), dechlorination with SBS	$FeCl_3$ and coagulant aid,	Antiscalant H_2SO_4
Alicante II, Spain 65,000 m^3/d, recovery: 43%	mixed open and well intake (Neodren), each providing 50% of feed DMF, CF; SDI 2.2, NTU 0.01, TOC 1 mg/L	none used	none used	1.1 mg/L

Plant characteristics	Intake & pretreatment	Chlorine & SBS	Coagulant	Antiscalant & acid
Hamma, Algeria commissioned in 2006, 200,000 m^3/d, provides 25% of Algier's water 44.5% recovery	1. flocculation, sedimentation 2. gravity DMF	no information	no information	No information
Skikda, Cap Djinet, Beni Saf, Hounaine and Mostaganem, Algeria 100,000-200,000 m^3/d, all commissioning 2008/09	two pressurized DMF	no information	no information	no information
Chennai, India 100,000 m^3/d commissioning in 2009	one gravity DMF, one pressurized DMF	no information	no information	no information

Abbreviations: BW: backwash, DE: diatomaceous earth, SBS: sodium bisulfite ($NaHSO_3$), CIP: cleaning in place, Cc: concentrate, S: salinity, SCC concentrate salinity, DMF: dual media filtration, DAF: dissolved air flotation, PPCW: power plant cooling water, SHMP: sodium hexametaphosphate, PAA: polyacrylic acid, DADMAC: diallyldimethylammonium chloride, $FeCl_3/FeSO_4$: ferric chloride, ferrous sulphate, NaOCl: sodium hypochlorite, Cl_2: chlorine, ClO_2: chlorine dioxide, H_2SO_4: sulfuric acid

7

REVERSE OSMOSIS PRETREATMENT. MICROFILTRATION & ULTRAFILTRATION.

Microfiltration (MF) and Ultrafiltration (UF) are low pressure or vacuum membrane filtration processes that are used for pathogen and suspended solids removal. MF/UF membrane processes have gained wide acceptance in the drinking water industry because of their ability to produce a high-quality and consistent water product. More recently MF/UF membrane filtration has gained acceptance as a pretreatment for nanofiltration (NF) and reverse osmosis (RO).

Numerous manufacturers of MF/UF membrane systems exist, each with their own proprietary technologies. The differences between proprietary systems can be overwhelming and present significantly varying design considerations.

The global recovery of microfiltration & ultrafiltration systems is approx. 94-96% with a water loss of 4-6%. Filtrated water quality is good, but not as good as the quality of a properly designed well. The advantages of MF/UF systems compared to conventional systems are:

- Less space requirements.
- Better quality of pretreated water. This point impacts on operation of reverse osmosis. It is possible to work with higher fluxes reducing the investment costs of RO systems.
- Less operation costs.

The typical values obtained with MF/UF systems are:

	MF/UF	Conventional
Turbidity (NTU)	<0.3	<5
SDI	<3	<5
TSS(mg/L)	<5	<5
TOC	<1	<5
Virus removal	Log 4 remove	No
Bacteria removal	Log 6 remove	No

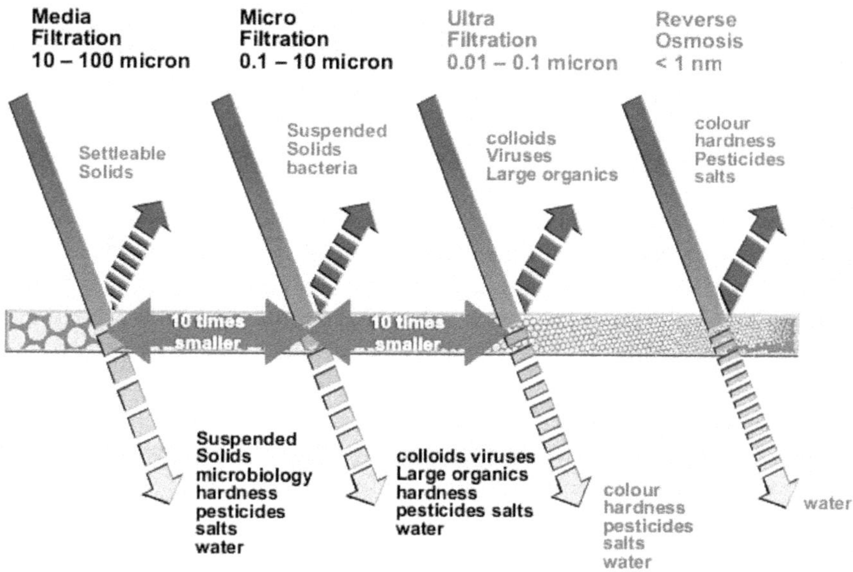

Figure 7.1. Different treatments per size.

7.1 Types of UF systems.

7.1.1 Introduction.

Virtually all of the UF/MF plants being installed today are proprietary systems.

Non-proprietary systems are designed to accept as many different pressure/encased MF or UF modules as are available. Currently several manufacturers are standardizing on 8-inch diameter by 40-inch or 60-inch long hollow fiber modules, similar to spiral wound RO membranes. As more manufacturers standardize, there will be a greater variety of modules available. A non-proprietary system provides the flexibility to change modules if the raw water quality or finished water goals change. These systems also offer the owner the benefit of competitive purchasing. It is likely the development of non-proprietary MF/UF systems will help advance the industry as the technology will then become more cost-effective and widely accepted. Additionally, the pairing of MF/UF technology could be used with reverse osmosis systems.

7.1.2 Submerged membrane systems.

Submerged membrane systems are proprietary systems with Zenon and US Filter being the most prolific. Although the system components differ slightly between the manufacturers, the mode of operation remains similar regardless of the proprietary equipment used.

Submerged membrane systems are those where the membrane fibers are immersed in an open tank and are exposed to the feed water. Groups of membrane fibers are bundled together in racks or modules.

The membrane fibers are slightly longer than the distance between the upper and lower attachment points, allowing the membrane fiber to shake during operation to dislodge accumulated solids. Multiple modules can be submerged in the process tank, depending on the permeate flow required. During normal operation, raw water enters the process tank and completely submerges the membrane modules.

The raw water is drawn through the membrane fibers (outside-in) to the inside of the fiber (lumen) by a vacuum pump. Filtered water (permeate or filtrate) exits through the top of the module to the permeate manifold. Generally multiple modules are connected to one permeate pump through a common manifold. Concentrate is continuously withdrawn from the process tank so that the raw water flow is equal to the sum of both permeate and concentrate flow.

Process tanks for submerged membrane systems are designed for a plug flow mode of operation, and recoveries will vary from 50% at the lead module to 99% at the end module. Suspended solids concentration also increases from the raw water inlet to the concentrate outlet. Since the solids concentration and module recovery increases from the lead module to the end module, flux decreases from the lead to the end module. However system recovery and flux is reported as an average of all modules.

Design flux velocity will range between 40-80 $l/h/m^2$

It is necessary a previous filtration of 50 to 100 microns of feed water.

Concrete cells must be covered with a coating to protect them during chemical washings (pH of 2).

A vacuum system is necessary to prime filtrate pumps, filtrated water backwashing pumps and air blowers

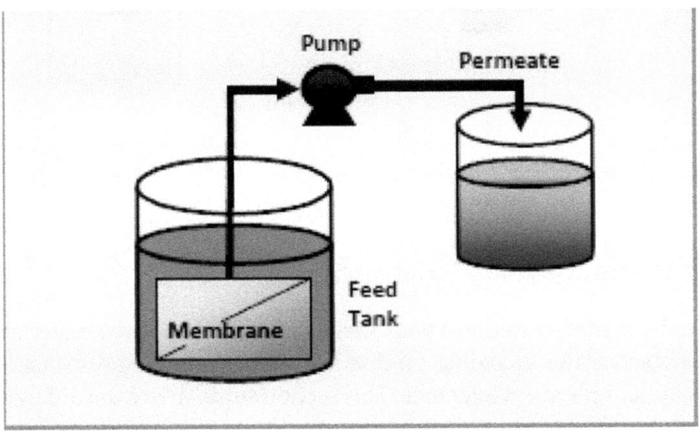

Figure 7.2. Submerged membrane system.

7.1.3 Encased membrane systems.

Encased membrane systems are becoming increasingly non-proprietary in nature. As stated earlier, membrane elements from the various manufacturers are standardizing dimensions to allow for increase flexibility within a particular system. This will allow for relatively simple changes to a system due to alterations in feed water quality, system performance standards, etc. The major manufacturers of UF are Pall, Koch, Norit (X-Flow), Aquasource, and Hydranautics.

A German company called Inge has recently produced a hollow fiber membrane with a multi-bore fiber orientation. The multi-bore design fuses separate capillaries into a single support structure for added stability and reduced fiber damage. The module also incorporates an annular space where permeate is collected. This spacing may optimize backwashing effectiveness. The system is non-proprietary, and can easily be retrofitted into existing systems, as well as vice versa.

Encased membrane systems utilize the same hollow fibers as the submerged systems except that they are packed into a cylindrical casing, usually with a diameter of 8", and a length of 40" or 60". These fiber packages, or elements, are then arranged end-to-end in a pipe, or vessel. Most commonly, in UF applications, the vessels are oriented vertically and each vessel houses one element. For large capacity systems – particularly in Europe – multiple modules can be installed in long vessels in a horizontal configuration similar to reverse osmosis systems. Some manufacturers provide their modules with end structures that are already configured to accept the required piping connections, thus eliminating the need for another housing. This can save a significant amount of capital cost.

An encased system may utilize a cross-flow design, which can allow for a high flux with minimal solids build-up on the membrane surface, or it can operate in "dead-end" mode where all water processed through the filter comes out as filtrate.

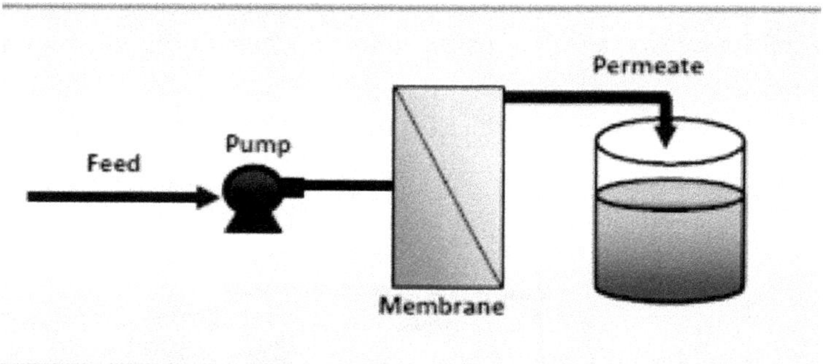

Figure 7.3. Encased membrane system.

7.2 Membrane material properties.

Membranes can be manufactured in a wide variety of materials. These materials differ in their performance characteristics including mechanical strength, fouling resistance, hydrophobicity, hydrophilicity, and chemical tolerance. This section summarizes the differences in material properties.

7.2.1 Hydrophobicity vs. Hydrophilicity.

The degree of hydrophilicity or hydrophobicity influences the wettability and applied pressure requirements for water flow through the membrane. Hydrophilic membranes require less operating pressure than hydrophobic membranes.

Hydrophilic membranes tend to exhibit greater fouling resistance than hydrophobic membranes. Particles that foul in aqueous media tend to be hydrophobic. Hydrophobic particles tend to cluster or group together to form colloidal particles because this lowers the interfacial free energy (surface tension) due to surface area exposure. General tendency will favor particle attachment to any material less hydrophilic than water because less exposure of hydrophobic particles can be achieved by attachment of the particles to the membrane surface. To prevent fouling, a membrane requires a surface chemistry which prefers binding to water over other materials. This implies that the material must be very hydrophilic.

7.2.2 Materials of construction.

Membrane manufacturers offer membranes constructed of a wide variety of materials. These materials vary widely in their chemical and mechanical properties including mechanical strength, burst pressure, oxidant tolerance, VOC tolerance, pH operating range, and so forth. The end user must be aware of the strengths and limitation of each material type and ensure the selected material is compatible with raw water quality, pretreatment requirements, and other operating conditions. Following is a list of commonly available materials and the characteristics of each.

- **Polyethersulfone (PES) Polyvinylpyrrolidone (PVP) Blends** PES / PVP blends are highly oxidant tolerant (higher than 250,000 ppm hours for chlorine, tolerant to permanganate and ozone), are tolerant to wide pH range (between 2 and 12 continuous operation, and lower than 1 for cleaning), exhibit resistance to oils and grease, and are highly hydrophilic. The weakness of PES/PVP blends is that they are susceptible to attack from organic solvents such as benzene
- **Polypropylene (PP)** PP membranes exhibit resistance to a wide operating range of pH (2-14), exhibit good chemical resistance, good mechanical strength (35 psig TMP), but are not oxidant tolerant (lower than 102 ppmh). PP membranes can tolerate limited exposure to chloramines in the feed stream (up to 0.5 ppm).
- **Polysulfone** Polysulfone membranes exhibit a wide pH tolerance (continuous exposure range of pH 1-13), high temperature limit (typically 75° C), good oxidant resistance (chlorine exposure: storage 50 mg/L, short-term sanitation 200 mg/L, >106 ppmh).
- **Polyvinyldeneflouride (PVDF)** PVDF membranes are highly oxidant tolerant (up to 5000 ppm chlorine), have a moderate pH operating range (2-10.5), have moderate temperature limits (40°C), and exhibit good mechanical strength (36 psig TMP, 36 psig inlet).
- **Cellulosic Derivatives (CD)** CD membranes exhibit a narrow operating pH range (4-8.5), low temperature limits (lower than 35°C), and moderate oxidant tolerance (more than 105 ppmh).

Polyacrylnitrile (PAN) PAN membranes have moderate temperature limits (40°C), a moderate pH operating range (2-10), good mechanical strength (44 psig TMP, 44 psig max inlet pressure), moderate oxidant tolerance (up to 1000 ppm during cleaning, <103 ppmh), and moderate resistance to acids and caustic (acid = 1N, caustic = 0.1N)

7.3 Mode of operation.

7.3.1 Pretreatment.

Pretreatment can be employed prior to membrane treatment to control fouling, to provide additional treatment, to meet manufacturer warranty requirements, and to satisfy regulatory requirements. Types of pretreatment employed depend on the nature of the raw water quality and the goal of the pretreatment.

Pretreatment can reduce the solids loading in the membrane feed stream, permitting higher flux rates and in turn reducing initial capital costs by reducing the required membrane area. A reduced solids loading also has the effect of reducing membrane operating costs because of a reduced transmembrane pressure requirement. Reduced solids loading results in benefits including longer runtimes between cleanings, lowered operating costs, and increased membrane longevity.

Pretreatment can also be employed to protect the membranes from constituents in the raw water that could cause damage. Examples of this include large objects and some bacteria that could cause system damage or plug the membrane pores.

Pretreatment may remove raw water contaminants that would not normally be removed by the membrane itself. For example oxidation, coagulation, and sedimentation can be employed to remove dissolved contaminants such as Fe, Mn,

The use of disinfectants, either continuously or intermittently, and the use of inline coagulation, is a common practice to improve the process performance and filtrate quality. The reported coagulant levels in UF/MF full scale plants ranges between 0.3 to 10 mg/L, with values of 0.2 -2 mg/L, assumed to be the norm and 10 mg/L as the exception. Chlorination levels range between 0.3 to 2 mg/L (continuous).

Several coagulation modes may be applied in coagulation prior to MF/UF (Figure 7.4)

- Rapid mixing and no flocculation (in line coagulation).
- Rapid mixing and flocculation.
- Rapid mixing, flocculation and sedimentation.
- Rapid mixing, flocculation and flotation.

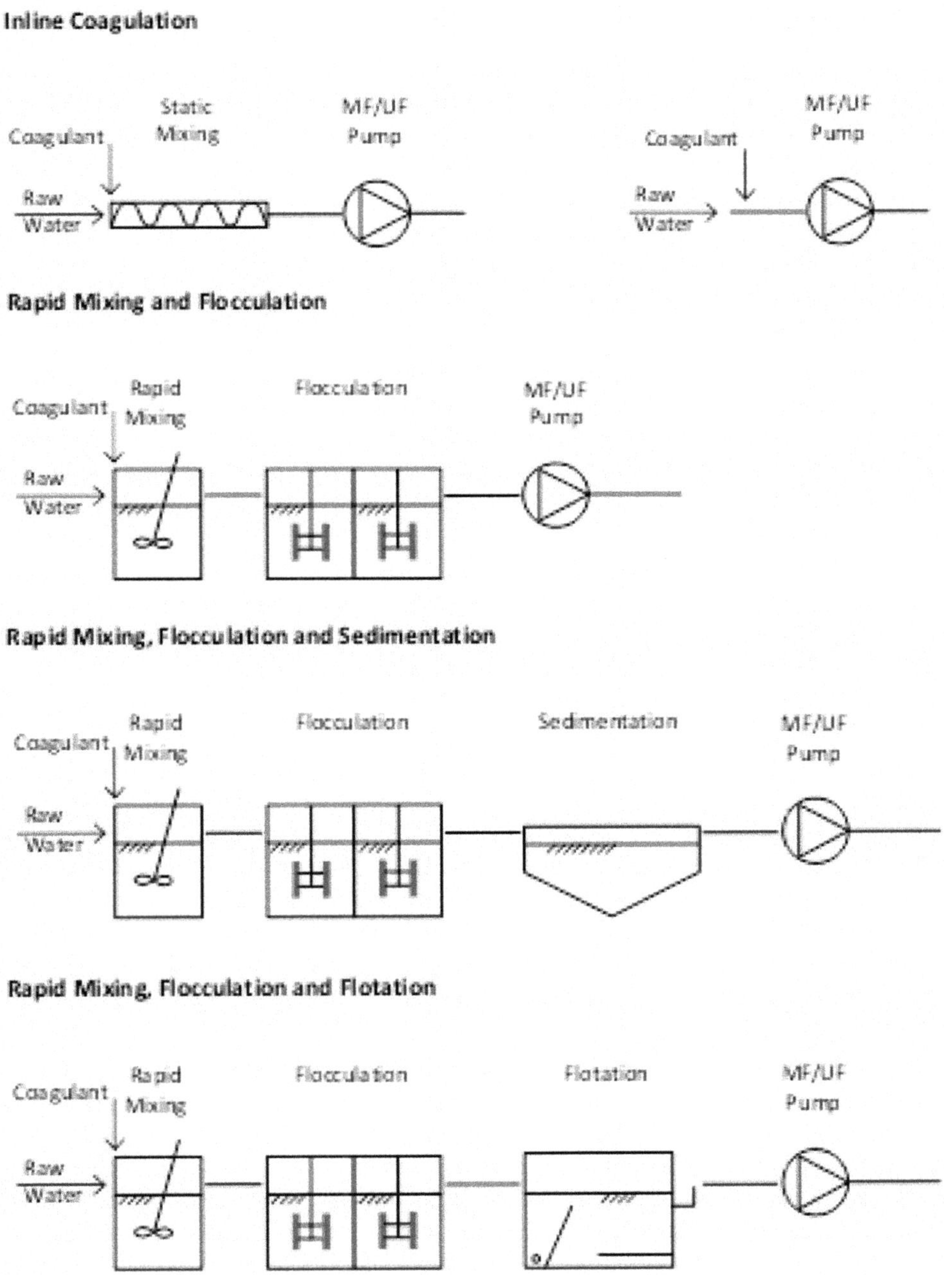

Figure 7.4. Types of coagulation before MF/UF systems.

7.3.2 Cross flow vs. Dead-end filtration.

The direction of feed water flow, in relation to the membrane surface, determines the type of filtration in a membrane system. In a dead-end filtration system the feed water effectively contacts the membrane surface at a perpendicular angle. In contrast, feed water in a crossflow system flows parallel to the membrane surface. Dead-end flow can also be achieved in a cross-flow system by simply closing the discharge valve. Feed water flow direction accounts for a variety of differences between the two systems. Particularly, the two modes of operation may experience differences in fouling rate, flux and recovery, and finished water quality.

Cross flow filtration in MF/UF is the same as in NF/RO: $Q_{feed} = Q_{permeate} + Q_{concentrate}$.

In dead end operation $Q_{feed} = Q_{permeate}$, just like in conventional multimedia filtration. In cross-flow mode, solids are continuously flushed from the system resulting in less frequent back-pulses and backwashes, and possibly longer membrane life. In dead end operational mode, solids build up in the system, thus more frequent backwashes and backpulses will be required. Some systems are capable of operating in cross flow or dead end mode such as Koch, Norit and Hydranautics. The Zenon and US Filter submerged systems should be considered crossflow since a continuous waste stream is produced.

Recovery is a term that describes the amount of water that is treated versus the amount of filtrate that is produced: Recovery = $Q_{filtrate}/Q_{feed}$. Even systems operating in dead-end mode should not be considered as 100% recovery systems. Water that is used up during backwashes and flushes to drain must be considered when calculating system recovery.

7.3.3 Inside-out vs. outside-in flow.

Inside-out flow and outside-in flow refers to the direction of feed water passing through the membrane, as well as the orientation of the feed water in relation to the membrane surface. For example, in an outside-in system, the feed water surrounds the membrane, and the filtrate is collected from inside of the hollow tube fibers (lumen). The outside-in scheme has the advantage of a larger membrane surface area which allows for a slightly higher flow than the inside-out model while still maintaining the same flux rate and solids concentration the membranes. The feed water enters the fibers at one end of the membrane element, and the discharge passes through the element and exits the fibers on the other side. The filtrate is collected inside the element on the outside of the fibers.

Both methods have been shown to operate successfully by different manufacturers. What the user should consider in evaluating the best options for a particular situation are the back-washing implications of the two different configurations. The backwash cycle for virtually all systems consists of introducing the cycle filtrate back into the modules through the filtrate conduit within the module. Thus the inside-out modules backwash from outside-in and vice versa. Depending on the water quality, the configuration of the modules and the water treatment goal, one type of configuration and backwash scheme may be more suited than another. It may only be determined by pilot studying which may be the most effective for a particular application.

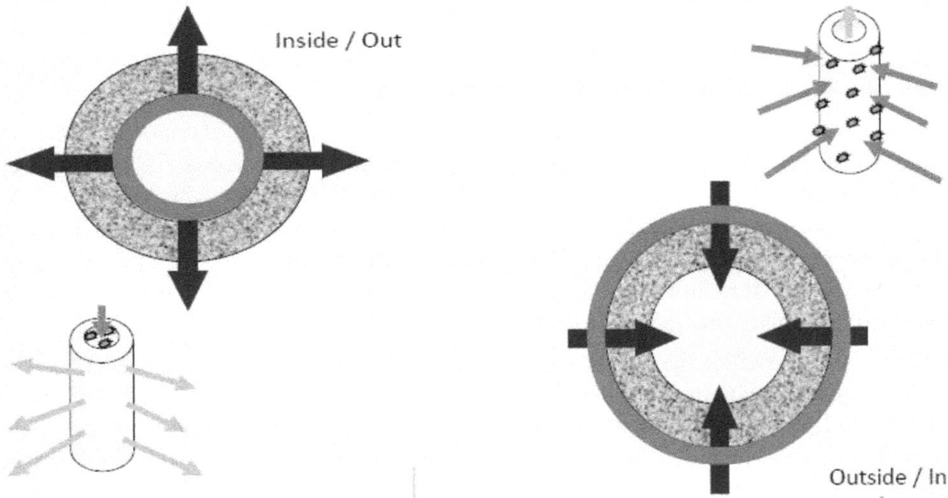

Figure 7.5. Detail of inside/ out and outside/in UF flow configurations.

7.3.4 Cleaning method.

Cleaning MF and UF systems is very similar in nature to RO and NF cleanings. However, the MF/UF systems have the added advantage of backwash and backpulse during the cleaning cycle. MF/UF membranes, depending on the material of construction can be cleaned with caustic solutions to remove organics. Citric acid or other acid solutions are used to remove scalants, and oxidants may be used to remove biological or inorganic foulants, if the membrane material is tolerant of oxidants. The chemicals to be used depend on the type of fouling and the material of membrane construction. Cleaning cycles can be automated and implemented into the system design. A typical cleaning cycle can last for anywhere from 10 minutes to 2 to 3 hours per bank, including soaking and rinsing times. Cleaning frequency for surface water applications can again vary tremendously, from once per day to about 4 to 6 weeks. Cleaning procedures, durations, and frequencies will vary significantly depending on feedwater quality, system design and operation, flux rates, and recovery rates. Cleanings may be automated and performed in place or they may be manual with the modules actually being removed from operation and cleaned in a special cleaning area. Submerged systems have the ability to remove cartridges service and clean them in a special tank. Encased modules are usually cleaned in place, but can be removed to a special cleaning rack.

7.3.5 Back wash procedure.

All UF/MF systems incorporate a backwash cycle, whether operating in dead-end or cross-flow modes.

A backwash cycle commonly operates on a preset timeframe, usually about every 15 to 60 minutes and is initiated by the PLC controlling the system. A typical backwash lasts around one to two minutes. The flow rate can range from lower than the feed flow to as much as ten times as high as the feed flow.

All elements in a given array are backwashed together. The backwash water is typically collected in a backwash tank. The tank ensures the backwash fluid is released to atmospheric pressure. Backwashes can also be initiated when the transmembrane pressure reaches a predetermined limit as recommended by the system manufacturer.

Typically, filtrate water is used for the backwash. The filtrate flows out of a tank and through the filtrate end of the membrane, then to discharge. The backwash releases solids that are attached to the feed side of the membrane surface. Backwashing improves the operational flux of the system to maintain desired performance. It is recommend to include a forward flush to drain to ensure that solids dislodged during backwashing are thoroughly flushed out of the system prior to directing filtrate to the filtrate tank Experience has shown that this will greatly reduce turbidity spikes that may be seen after backwash cycles.

Brine can also be used for the backwash increasing the global recovery of the system (3-4%), but it is recommended to limit the use of brine for non chemically enhanced backwash pulses. For CEB, ultrafiltrated water should be used. Obviously, a forward flush of the system after backwash should be included to remove salinity of membranes. The control of the system would get more complex.

Compressed air, as in the Memcor (US Filter) system, can also be used for backwash purposes. The compressed air, usually less than 7 bars, is introduced on the filtrate side to remove accumulated suspended solids. Many systems also have the ability to have a continuous gas stream flowing through the membrane. This provides added protection against solids build-up.

The membrane systems that employ outside-in flow during normal operation generally use a backpulse feature to dislodge solids from the outside surface of the membrane fibers. During a backpulse, membrane permeate is drawn from a holding tank and is pumped in reverse direction, from the inside of the fiber out effectively dislodging solids from the outside of the membrane.

Some MF and UF systems use *chemically enhanced backwashes* (CEBs) or “mini-cleans” to maintain permeability, control TMP, and extend the cleaning frequencies. CEBs are usually less frequent than typical backwashes and may include a short time for the membranes to soak in the chemical solution. Common CEB chemicals are sodium hypochlorite, citric acid, and caustic soda, and their concentration sand use are dependent on the specific membrane product and the targeted type of foulant.

Backpulses typically are initiated automatically by the PLC controlling the system and occur at 15-30 minute intervals with a duration of 10-30 seconds.

7.3.6 Integrity testing.

Membrane integrity testing ensures the membrane is still intact. These tests check the membrane structure for wear or damage which would result in a pore size that is larger than permitted. It is important that these tests be carried out regularly since finished water quality can be adversely affected by damage to the membrane surface. All tests are non-destructive, and are designed to isolate the problem area, should the need arise. Such tests include bubble point, diffusive flow and pressure hold.

The tests are characterized as direct or indirect. Direct tests challenge the physical structure of the membrane utilizing air or pressure hold tests. Indirect tests monitor filtrate water quality parameters such as turbidity or particle counts to determine membrane integrity.

One of the more popular tests is the "bubble point" test. The bubble point test uses capillary properties, of the membrane, to locate any problems. According to capillary theory, the smaller the pore size, the higher the pressure is required to expel water form the pore. Pressurized air hits the filtrate side of the membrane, while feed water covers the feed side. The feed side is at atmospheric pressure via an open backwash valve. The test is successful when, at a specified air pressure, no air is released into the feed end for a specified time period. Therefore, the membrane passes the test when the feed water remains in the pores of the membrane at a specified air pressure.

7.3.7 Control.

In below figures, the basic controls which are active during inline pressurized UF plant operation are described.

RO feed pressure control.

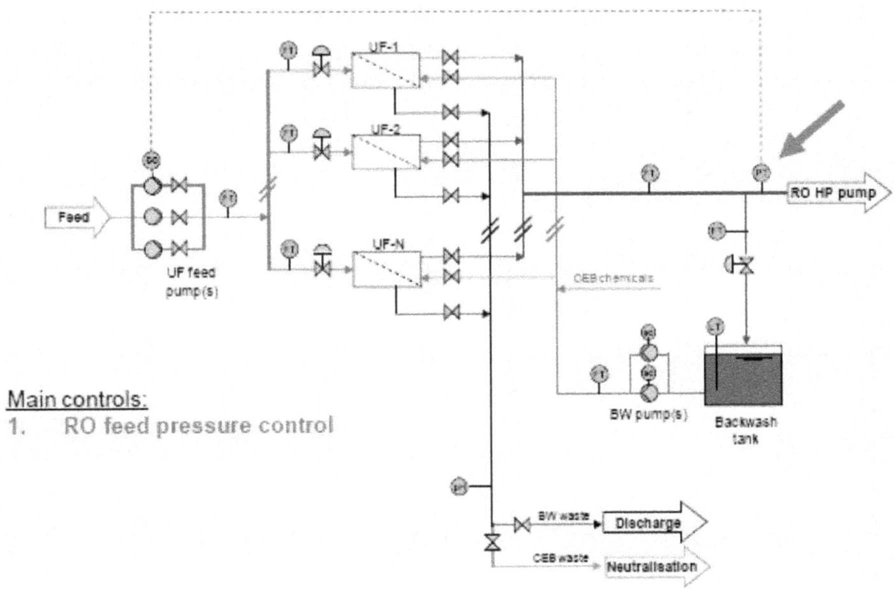

Figure 7.6. P&ID RO feed pressure control.

Explanation:

- The RO system is operated at a pre-defined flow set-point. This flow set point can be either fixed or variable Typically this flow set point is set by the feed flow of each individual RO unit multiplied with the number of RO units on line.
- For stable process control, the RO feed pressure is controlled at a pre-defined pressure set-point with the UF feed pumps. This pressure set point can be fixed or variable (e.g. following a temperature curve).
- Limited pressure fluctuations will be controlled with the frequency controlled UF feed pump. In case the pressure has a large deviation from the pressure set-point and the required feed flow change is beyond the control band of the frequency controlled UF

feed pump, one or more fixed speed pumps are started/stopped. To minimize the risk of potential pressure shocks to the membranes, slowly opening/closing motor operated valves in the feed pump discharge lines could be considered.

- The pressure control response by the UF feed pumps has to be faster than the flow changes of the RO system. This will ensure a stable RO feed pressure under varying RO flow conditions.

Note:

Alternative control could be considered, as long as above functionality is met, the full capacity range of the UF plant is covered and safe operation of the membranes can be guaranteed (no pressure shocks/surges).

- Frequency controlled feed pumps are important because of the following reasons:
- To adequately control RO feed pressure changes, as explained above.
- To enable flexible control over the full flow range of the UF plant (~10 to 100% of design flow).
- To compensate fluctuating hydraulic resistance of the UF membranes depending on membrane fouling status or fluctuating water temperatures. E.g. the Trans Membrane Pressure (TMP) of the membranes at 5°C is more than 100% higher than at 30°C.
- To enable adequate start-up of the UF plant at reduced plant capacities.
- To ensure minimum power consumption.

Flow control backwash tank filling.

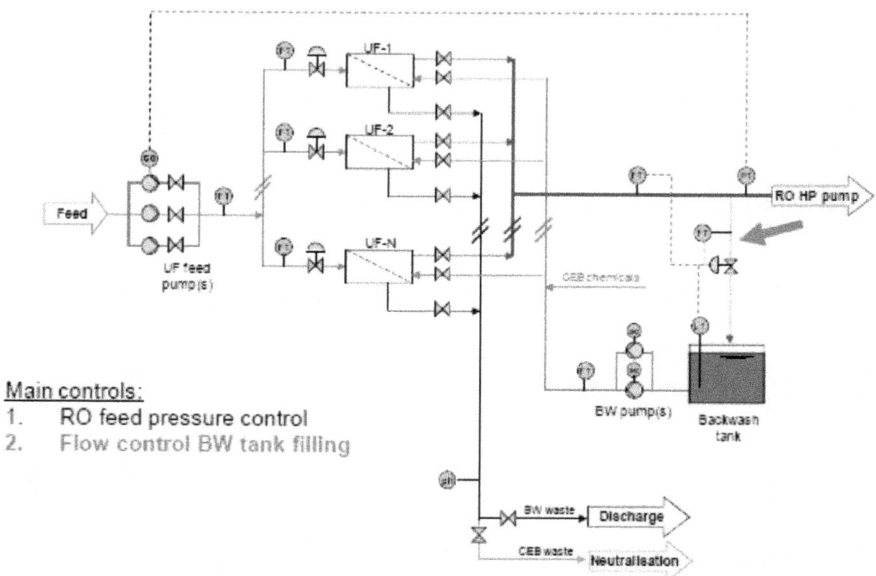

Figure 7.7. P&ID flow control back wash tank filling.

Explanation:

- Permeate flow to the backwash tank is controlled on flow ratio with the total UF permeate flow.

- Initial set-point for the flow ratio will be based on the expected UF plant recovery. To avoid overflowing or draining of the tank, this set-point is automatically adjusted based on the actual level in the tank.
- Because the total UF permeate flow is constant. Also the permeate flow to the backwash tank will be constant.

Flow distribution UF units.

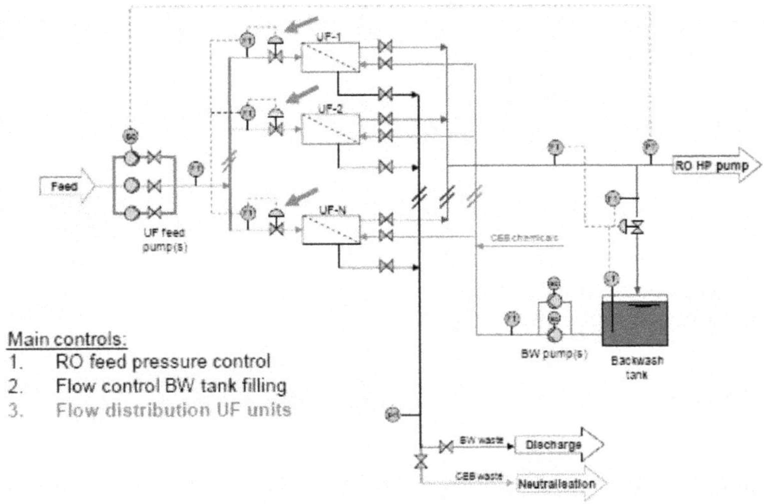

Figure 7.8. P&ID. Control of uniform distribution.

Explanation:

- The total feed flow from the UF feed pumps to the UF units is resulting from the RO feed pressure control The UF feed pumps will dictate the total feed flow to be treated by the UF units.
- The total feed flow is distributed evenly over all UF units in filtration by using the flow meters and flow regulating valves of the individual UF units.
- Since the UF units are switching between filtration, backwash and CEB, membrane fouling status and consequently TMP of the UF units will be fluctuating. Main purpose of the feed flow regulating valves is to compensate the TMP differences between the UF units in order to have the same total headloss for all UF units. In this way all UF units will see the same feed flow, i.e. filtration flux. Since filtration TMP of the X-Flow UF membranes is relatively low, typically 0.2 – 0.8 bar, pressure difference for the feed flow regulating valves is typically 0–0.6 bar. For this reason simple regulating butterfly valves1) are fit for purpose.
- Another purpose of the regulating butterfly valves is to pressurize the off line (atmospheric) UF units prior to start inline operation. Moreover to provide a minimum feed flow during execution of a Filling program. This program is executed at initial start-up and whenever the UF unit has been integrity tested with air or opened for maintenance. During the filling program, the applicable UF unit is filled with water at minimum feed flow to expel the air from the membranes in a controlled way.

Back wash flow control

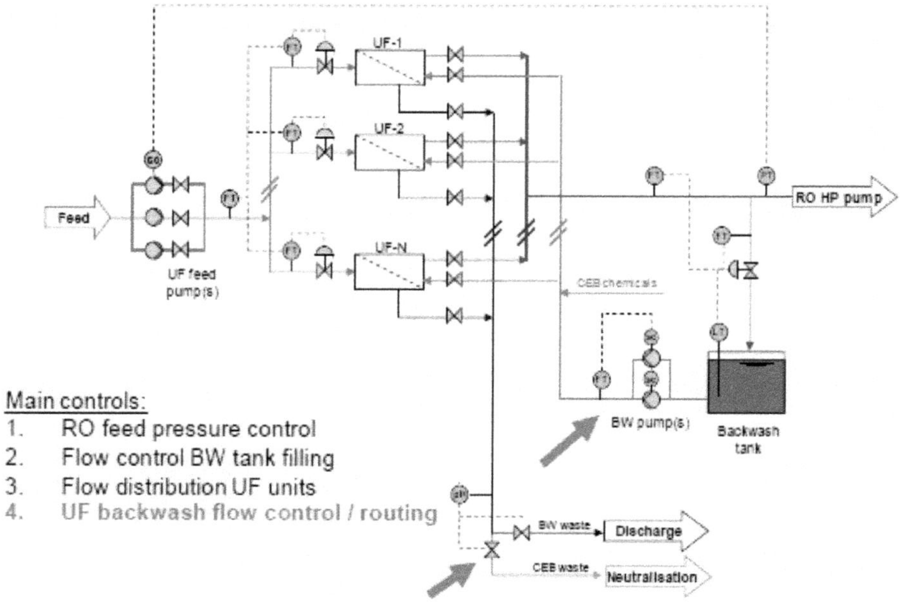

Figure 7.9. Brine back wash flow control.

Explanation:

- During backwash and CEB the backwash flow is controlled with frequency controlled backwash pumps.
- Since the backwash flow (flux) is relatively high (3 to 4 times higher than the feed flow (flux)), and duration of the backwash is short (less than 30 seconds for full backwash flow), frequency controlled backwash pumps are important because of the following reasons:
- Hydraulic resistance (TMP) of the membranes drops quickly and adequate, quick acting control is required to maintain the backwash flow (flux) at the required set-point.
- To ensure safe operation of the membranes (avoiding pressure shocks).

For big UF plants, the backwash is executed as a Train backwash (TBW), which means that all UF units, being served by the same backwash pump, are backwashed consecutively. This will save time and reduces the number of starts/stops of the backwash pump. Depending on the size of the system, the total UF plant is divided in one or more UF trains, each with a dedicated backwash pump. Multiple trains can be backwashed simultaneously.

7.3.8 Cost of UF versus conventional pretreatment.

Pretreatment is approximately 20% of the overall cost of a SWRO plant. CAPEX of UF/MF exceeds the cost of conventional pretreatment.

However, in terms of running cost, the pretreatment power and chemical costs are much lower than those of RO. UF/MF has slightly higher power and somewhat lower chemical cost than conventional pretreatment). UF/MF will save some of the RO chemicals cost, used in the process. Since this is significant, a modest saving will justify the additional UF CAPEX in a whole life evaluation. This would need to be confirmed in a real case study, but the payback of the additional capital cost will be in the region of 5-10 years.

Another factor which will have a major influence on the complete system is the influence of UF/MF pretreatment on RO flux. A modest increase, if allowed by the feed salinity, would easily pay for the additional CAPEX/OPEX.

7.4 Performance characteristics.

7.4.1 Flux.

Flux in a MF/UF membrane system is not a characteristic of a specific manufacturer so much as it is limited by the following parameters:

- Raw water quality (temperature, solids content, NOM content, etc.).
- Efficiency or existence of the pretreatment process.
- Maximum transmembrane pressure as dictated by the physical properties of the membrane polymer and limitations of manufacturer's equipment.
- Acceptability of the resulting cleaning frequency.

All of these factors are interrelated and will therefore dictate the flux rate achievable on a case by case basis. Flux rates though can be as low as 28 $l/h/m^2$ for challenging waters, and flux rates have been documented over 150 $l/h/m^2$ for very clean feed waters. Average flux rates range from 45 –70 $l/h/m^2$

7.4.2 Recovery.

Similar to flux, recovery rates in a MF/UF membrane system are not a characteristic of a specific manufacturer so much as recovery is limited by the following parameters:

- Raw water quality (temperature, solids content, NOM content, etc.).
- Efficiency or existence of the pretreatment process.
- Maximum transmembrane pressure as dictated by the physical properties of the membrane polymer and limitations of manufacturer's equipment.
- Acceptability of the resulting cleaning frequency High recovery rates generally require high trans membrane pressures to operate higher cleaning frequencies because the membranes are forced to run in a higher solids environment. All of the above factors are interrelated and will therefore dictate the recovery rate achievable on a case by case basis.

Recovery rates can be as low as 50% for challenging waters, while recovery rates have been documented at up to 99% for very clean feed waters. Average recovery rates range from 92% to 95%.

7.4.3 Particle rejection.

The main difference between membrane filtration and conventional filtration is that membranes reject particles based on size exclusion as opposed to media depth filtration. For all practical purposes, MF/UF membranes will produce a constant permeate water quality regardless of feedwater conditions because they act as an absolute barrier. This can be expected since most suspended solids present in raw water streams are larger than the pore size of most MF/UF elements. Permeate turbidity can be expected to be less than 0.1 NTU or lower and SDI less than 0.4. However, breaches in integrity such as a broken fiber or ruptured seal, or imperfections in the membrane due to the manufacturing process can allow particles to enter the permeate stream.

Most membranes are capable of up to 6- log pathogen (protozoan cysts and bacteria) removal, though this will vary depending on the membrane manufacturer. Virus rejection varies significantly between manufacturers with some systems removing virtually no appreciable amount of viruses while others report greater than 6-log removal.

7.4.4 Factors affecting fouling and cleaning frequency.

Fouling is the most serious disadvantage of pressure-driven membrane separation processes. Membrane fouling results in a decrease in flux and an increase in energy consumption and feed pressure. Fouling will occur in any MF/UF system, regardless of the membrane polymer, system manufacturer, and mode of operation. Fortunately, fouling can be effectively controlled through the proper use of pretreatment processes, chemical additions, and proper system design and operation. The following factors are the causes of fouling and can be controlled by proper design and operation.

Equipment Malfunctions. Malfunctioning equipment can increase cleaning frequencies.

Pretreatment Chemicals. Pretreatment chemicals must be evaluated for their compatibility with the membrane materials. For example, oxidants such as chlorine or potassium permanganate used for iron and manganese removal may cause permanent damage to some membrane materials, such as PP.

Polymers that are used in conventional treatment processes to enhance coagulation may create a slime layer on membrane fibers resulting in accelerated fouling. Furthermore, feeding excess amounts of coagulant in a pretreatment process results in no significant fouling impacts as long as the coagulant is compatible with the membrane material. However, the membranes can rapidly foul during coagulant underfeeds. The coagulation process produces a pin floc that is substantially larger than the membrane pores. When the size of the suspended solids is approximately the same as the membrane pores, membrane fouling increases because it becomes more difficult to dislodge the solids from the membrane surface.

Temperature. The effect of temperature on permeate flux is well documented in the membrane industry. Decreases in temperature decrease flux, increase TMP, and increase fouling rates.

Flux. Increasing flux increases cleaning frequency.

Recovery. Increasing recovery increases cleaning frequency due to higher solids concentrations in the feed water. For example, the solids concentration factor is 12.5 at 92% recovery and is 100 at 99% recovery.

Cake Formation. The cake layer can enhance the particle rejection characteristics of the membrane by depth filtration and adsorption. However, the cake layer increases the resistance to flux and leads to higher transmembrane pressures.

Adsorption. Organic matter present in the feed water can lead to membrane fouling either be adsorbing into the cake formation or by adsorbing into the membrane itself. Some membrane polymers exhibit higher adsorptive rates than others, depending on the hydrophobicity of the polymer. Pretreatment can be optimized to remove as much organic matter from the feed to extend membrane life, and chemical cleaning can be optimized to remove adsorbed foulants.

Metals Precipitation. Though metals precipitation in MF/UF membranes is not as prevalent as in NF/RO processes, fouling due to metals precipitation has been documented. Calcium carbonate scale can form on the permeate side of membranes in scaling waters where carbon dioxide is driven off as part of the process. In operations using membrane filtration for iron and manganese removal, precipitation of these metals on the concentrate and permeate side can occur. Postprecipitation of iron and aluminum associated with metallic-based coagulants has been observed and can lead to fouling on the permeate.

7.5 Proprietary systems.

The table below lists a summary of several of the major UF/MF manufacturers.

7.5.1 Memcor/US filter.

Memcor offers two different alternatives. A submerged configuration (CS/CMF-S) and a pressurized option (CP). They are outside-in systems which can work in dead-end mode with a pore size of 0.1 microns.

Figure 7.10. Photography of CP-Memcor system Courtesy of Memcor.

Typical limitations on feed quality for submerged CS system include a feed temperature of greater than 0 degrees Celsius (non-freezing) and less than 40° Celsius, turbidity of less than 150 NTU (although higher turbidities may be handled for short periods) and pH between 5.5 pH and 8.5 pH.

Memcor uses a patented backwash process. This process uses low pressure aeration to scour and agitate the Hollow Fiber Membranes and, when combined with a short reverse flow of filtrate, removes the retained solids from the membrane fiber bundle. Liquid backwash waste is then drained by gravity from the Cell to the Backwash Waste Disposal System. The Cell is then refilled with feed and the Unit returns to service. The control system typically initiates an automatic Backwash every 20 to 60 minutes of filtration (depending on feed quality). The Backwash cycle takes about two and a half minutes to complete.

Table 7.1 Comparison of different proprietary UF systems.

System manufacturer and model	System type	Pore size	Mode of operation	Cleaning method	Flux l/h/m2	Recov.
Legend	S---Submerged E-encased MF-Microfiltration UF-Ultrafiltration	Microns or daltons	I/O-Inside out O/I-Outside in CF—Crossflow DE-Deadend			
Hydranautics	E 8"X40" or 8" x 60" UF	150,000 daltons (0.9 μm)	I/O DE and CF	Chemical soak	59-120	95-98%
Hydranautics	S 17550 SF per 10 bundle module MF	N/A	O/I	Chemical soak and air scrub	N/A	N/A
Inge	E 485 m^2 per el 8"x40" or 8"x60" UF	100,000 daltons	I/O annual permeate collection DE and CF	Chemical soak and flush LP air scour	68-145	85-98%
Koch	E 8"x48" (32 m^2) Or 8"x72" (50 m^2) UF	100,000 daltons	I/O DE and CF	Chemical soak	N/A	N/A
Pall	E UF-5"x80" MF-6"x80"	UF-13000 or 80000 daltons MF-0.1 μm	Double skin – I/0 and O/I	Chemical soak and air scrub	UF-96-180 MF- 440	>95%

System manufacturer and model	System type	Pore size	Mode of operation	Cleaning method	Flux l/h/m2	Recov.
US-filter	S.E 5x40” 372 sf	MF-0.1 μm	O/I DE	Chemical caustic and acid air scour	25-70	90%-98.5%
x-flow	E 8”x60” Other diameters available	UF-50,000 daltons (0.2 μm) MF-150,000 Daltons (0.9 μm)	I/O CF and DE	Backwash and soak only–no air	71,4-172	90-96%
Zenon-zeeweed 500	S	(0.04 μm)	O/I CF	Continuous air scour, air and water backwash	17-68	85-99%
Zenon Zeeweed 1000	S	(0.04 μm)	O/I DE	Continuous air scour and water backwash	17-68	85-99%
Toray-TORAYFIL	S E	UF-150,000 Daltons	O/I			
Hyflux–Kristal 2000 E Kristal 600E	E	UF	O/I CF and DE	Continuous air scour and water backwash	55-75	85-99%

7.5.2 Zenon

7.5.2.1 ZeeWeed 500 series

The ZeeWeed_500 series are built with a reinforced, large diameter hollow fiber. The 1.9 mm outside diameter hollow fibers are flexible and have a very high tensile strength, two properties that allow vigorous air scouring in difficult applications.

Modules are rectangular frames containing thin bundles of hollow fibers. The hollow fibers are mounted vertically between headers with some slack to allow movement, air penetration and water renewal within the bundle.

Modules are assembled side by side into cassettes, leaving space for water circulation and air scouring. Cassettes have integrated headers for permeate collection and air distribution.

Cassettes are the building blocks that are immersed into the filtration tank and connected to permeate and air headers.

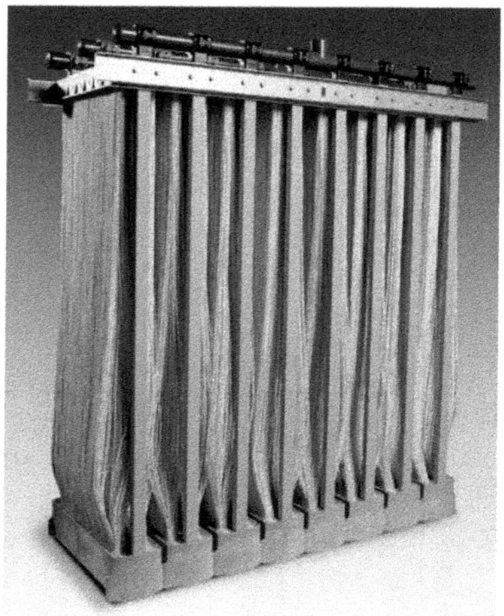

Figure 7.11. A ZeeWeed_ 500 cassette. Courtesy of Zenon.

7.5.2.2 ZeeWeed 1000 series.

The building block of a ZeeWeed 1000 filtration system is a parallel piped membrane element with hollow fibers mounted between two vertical headers. Fine hollow fibers provide high module packing density. Shrouds enclose the fibers, leaving only the bottom and top open to create a vertical flow channel through the fiber bundles.

Cassettes are built by assembling elements in the vertical and horizontal dimensions. In the vertical dimension, a standard stack has three elements, but stacks of one and two elements can be easily assembled to fit in shallow tanks. The cassette illustrated in Figure 7.11 contains 48 elements in three rows. An element can be inserted into, or removed from the cassette by sliding it like a book into a bookcase. Each stack of elements is connected to a permeate manifold that runs horizontally above the cassette.

When compared to the ZeeWeed 500 series, the ZeeWeed 1000 represents a significant reduction in footprint and energy consumption and is targeted at low solids applications including seawater pretreatment.

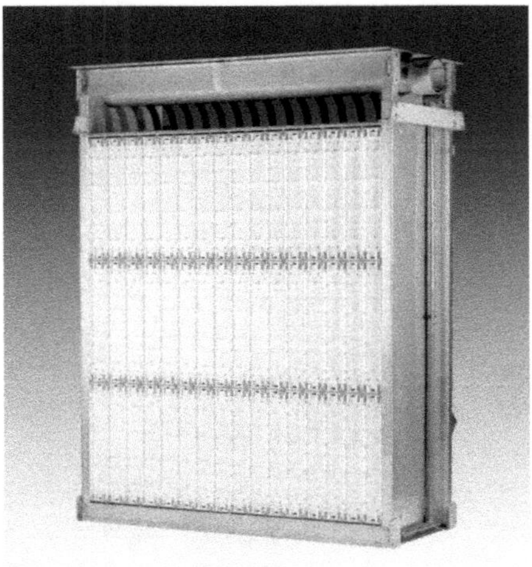

Figure 7.12. A ZeeWeed 1000 cassette. Courtesy of Zenon.

The flux rate for seawater pretreatment ranges from 42 to 70 l/m^2/h. The recovery rate ranges from 85% to 92%.

7.5.3 Norit/ X-Flow.

7.5.3.1 Horizontal dead end UF. Sea Guard 55.

Usually employed for large scale desalination plants with turbidity less than 30 NTU.

1. The membrane elements.

Fibers are operated inside-out and the feed suspension is pumped into the lumen (tube side) of the fibers.

The elements are 220 mm in diameter and 1.5 meter long. The Seaguard 55 element contains 55 m^2 of membrane surface area. The element has a central permeate connection in the middle and low trans membrane pressure of 0.2 – 0.8 bar that these systems are typically operated on.

Both membrane interior as well as the outer casing of the element are made of thermoplastic polymers ensuring identical expansion and contraction as a result of seasonal temperature changes of the water to be treated. The membranes are potted in a resin at both ends, to provide an absolute barrier between the feed side and filtrate side of the element.

The membranes are operated at relatively high fluxes. Depending on the application, the flux can range from 60 to 130 l/h/m^2. For turbidities less than 0.2 NTU, the flux can be more than 100 lmh with a global recovery of 98 %. For a turbidity of 30 NTU, the flux will be from 50 to 60 l/h/m^2 with a global recovery of 83%.The pressure needed t operate the system is extremely low, typically 0.3 – 0.8 bar due to the high porosity of the X-Flow membranes. This results in extremely low energy requirements, typically 0.1 – 0.2 kWh/m^3.

Moreover, the system can be controlled by either frequency controlled pumps or, if pressure is available, by means of flow control valves. In this case, there is no need for re-pumping and the feed pump section can be completely omitted.

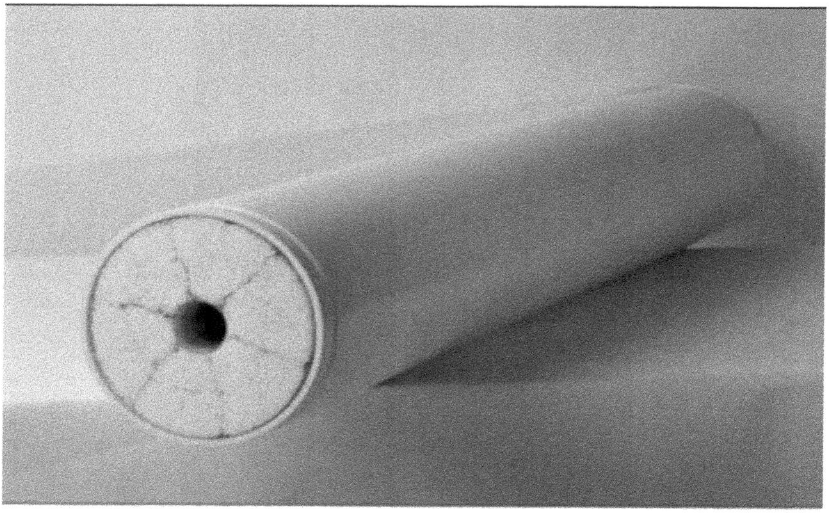

Figure 7.13. Membrane element Seaguard 55. Courtesy of Pentair.

2. Membrane housings.

Up to four (4) elements can be mounted in a 9" membrane housing, which guarantees a very compact way of construction. The membrane housings are similar to the ones that are used for Reverse Osmosis, but are optimized for low pressure ultrafiltration performance:

The operating pressure of a Seaguard system is typically very low (below 3 bar), the pressure rating of the membrane housings consequently is low as well (typically 5.1 bars). Since the operating pressure requires only a thin membrane housing, the cost for these types of housing is significantly lower than a corresponding RO housing. The flows during filtration and during backwash are higher than the flows observed typically in RO systems. Unlike reverse osmosis, the permeate flow is equal to the feed flow (dead-end operation).

Because of low pressure dead-end operation and because of discontinuous operation (filtration/backwash cycles), system venting is more critical in ultrafiltration than in reverse osmosis. Therefore, all membrane housings are equipped with vertically aligned integral venting connections.

The Seaguard 55 ultrafiltration elements are operated hydraulically in parallel. This is fundamentally different from standard spiral wound reverse osmosis, where each membrane element is fed by concentrate from the upstream membrane element (serial operation). Brine seals are eliminated and flow distribution inside the membrane housings is optimized by use of X-rings (Figure 7.13) that create an equal flow distribution of feed water inside each membrane housing.

The housings are aligned horizontally, comparable to reverse osmosis systems. The fact that up to four elements can be mounted into one housing provides a major advantage to vertical systems, where every single element needs to be connected to the plant piping.

Membrane elements are connected to each other by using interconnectors This connection and the use of the X-rings is shown in figure below.

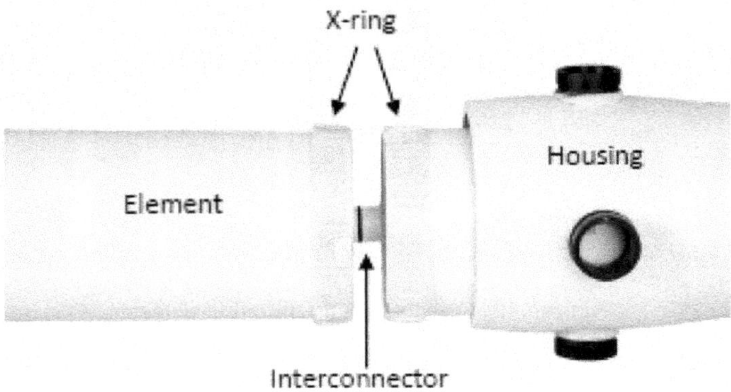

Figure 7.14. Detail of X-Rings of Seaguard 55 housing. Courtesy of Pentair.

The housing is fed through 2.5" side ports. Feed is pumped into the feed chamber between the end cap and the membrane element. It is then forced into both the tube side of the membrane fibers as well as in the circular gap between element and membrane housing (bypass arrangement).

3. Membrane units.

The result is a range of standardized skids which offer a reliable plant performance and the highest membrane area per m^2. The footprint is 60% less than vertical systems.

The range of standard skids and their characteristics are summarized below.

Membrane skid		**412-55**	**409-55**	**406-55**	**212-55**	**209-55**	**206-55**
Membrane housing	–	48	36	24	24	18	12
Elements/unit(max)	–	192	144	96	96	72	48
Membrane are per element	m^2	55	55	55	55	55	55
Mem. are per skid	m^3	10560	7920	5280	5280	3960	2640
Size vertical header	ND	300	250	200	300	250	200
Max.feed flow	m^3/h	1056	792	528	528	396	264

Figure 7.15. Seaguard 55 ultrafiltration elements.

7.5.3.2 Vertical cross-flow UF. Aquaflex 55.

Usually employed for large scale desalination plants with a turbidity more than 30 NTU and less than 100 NTU. If the turbidity is higher than 100 NTU an additional pretreatment would be required This system works in cross-flow mode and with more flexible cleaning procedures

The membranes are operated at medium fluxes. Depending on the application, the flux can range from 50 to 60 $l/h/m^2$. For turbidities of 30 NTU, the flux can be approx. 60 with a global recovery of 91 %. For a turbidity of 100NTU, the flux will be from 50 $l/h/m^2$with a global recovery of 86%.

The vertical system allows to work under pressure but changing the vessels with the associated extra cost. Consequently, for pressure of 2-4 bars at the UF outlet, XIGA horizontal system is the preferred alternative.

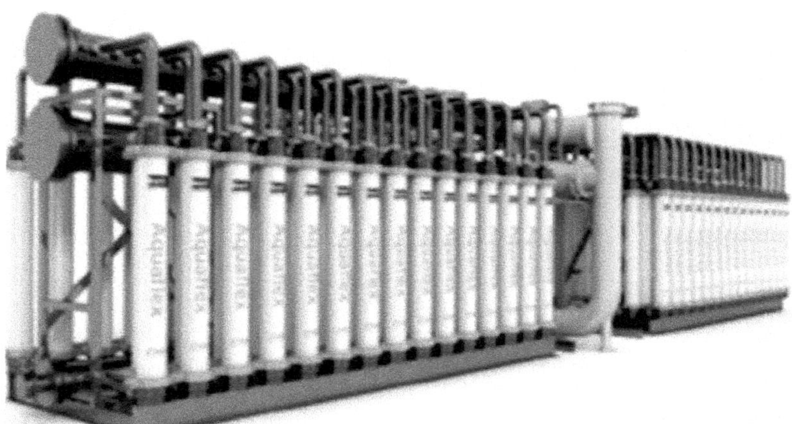

Figure 7.16. Vertical UF System-Norit. Courtesy of PENTAIR.

7.5.4 Hydranautics.

Hydranautics offers microfiltration modules (HYDRACAP MAX ™) made from hollow fiber membranes that offer the following features:

- 0.1 μm nominal pore diameter .
- PVDF polymeric hollow fibers for high strength and chemical resistance.
- Outer diameter: 1.2 mm / Inside diameter: 0.6 mm.
- Outside-In flow configuration for high tolerance to feed solids and the use of air scour cleaning.
- U-PVC housings.
- They can work in cross-flow or dead end mode of operation.

The process conditions of Hydranautics ultafiltration membranes are:

- Feed pressure.(Max at inlet 20°C): 5 bars
- Operating TMP: 1.52 bar (max.)
- Backwash pressure: 2.42 bar
- Filtrate flux: 59-150 l/m²/h
- Backwash flux: 298-340 l/m²/h
- Back wash frequency: Once every 15-60 min
- Back wash duration: 30-60 seconds
- Operating pH range: 2-11

7.5.5 Dow.

The DOW Ultrafiltration modules are made from high strength, hollow fiber membranes that offer the following features:

- 0.03 μm nominal pore diameter for removal of bacteria, viruses, and particulates including colloids.
- PVDF polymeric hollow fibers for high strength and chemical resistance.
- Hydrophilic PVDF fibers for easy cleaning and wettability that help maintain long term performance.
- Outside-In flow configuration for high tolerance to feed solids and the use of air scour cleaning.
- U-PVC housings eliminate the need for pressure vessels and are resistant to light.

The hollow fiber membranes are 1.3 mm outside diameter and 0.7 mm inside diameter and are made from PVDF polymer.

The operating conditions of DOW ultafiltration membranes are:

- Feed pressure.(max at inlet 20°C): 6.25 bars.
- Operating TMP: 2.1 bars.
- Filtrate flux: 40-120 l/m²/h.
- Operating pH range: 2-11.

An operating cycle ranges from 20 to 60 minutes. While operating 100% of the feed water is converted to filtrate. As contaminants are removed during the operating step the transmembrane pressure will rise. At the end of the preset operating cycle time, a backwash sequence is triggered. The back wash includes an air scour with a of 40 to 120 seconds and a flow per module of 5 to12 Nm³/h and water back wash in two steps at 100 to 150 l/m²/h.

Figure 7.17. UF skid-Dow. Courtesy of Dow.

7.5.6 Koch.

The KOCH pressurized ultrafiltration modules are made of, hollow fiber membranes that offer the following features:

- 100,000 daltons of nominal molecular weight cut-off.
- Polyethersulfone hydrophilic membranes.
- U-PVC housings eliminate the need for pressure vessels and are resistant to light.
- The hollow fiber membranes are 0.97 mm inside diameter.

The operating conditions of membranes are:

- Operating TMP: 2 bars.
- Operation mode: Inside-out.
- Filtrate flux: 100 l/h/m^2 for beach well water, 68 for open intake (NTU less than 5) and 48 for open intake with a turbidity higher than 5 NTU.
- Operating pH range: 5-8
- Typical overall water recovery: Beach well (88-93%); Open intake (80-90%)
- Prescreening requirement: Minimum 200 microns. For open intake, it should be followed by a 50 micron disk filter.
- Treated water turbidity: Lower than 1NTU.

7.5.7 Inge

The Hydranaitics offers UF modules (Dizzer XL ™) made with multibore membranes. Mulibore membranes offer the following features:

- Modified PES.
- 7 single capillaries combined into one fiber.
- Moderately hydrophilic, easily wettable.
- Pore size; UF 10 - 25 nm.
- Fiber ID, 0.9 mm; od 4.3 mm.

- Cl_2 resistance moderately high.
- pH tolerance 1.5 – 13.

Each Dizzer XL ™ module has a membrane area of 70 m².

Inge provides for large scale SWRO skids consisting of 136 modules with a total filtration surface of 9520 m^2. , Assuming a flux rate of 80 $l/h/m^2$, it would correspond to a water treatment capacity of 760 m^3 /h. The system would be expandable adding skids to match with requirements.

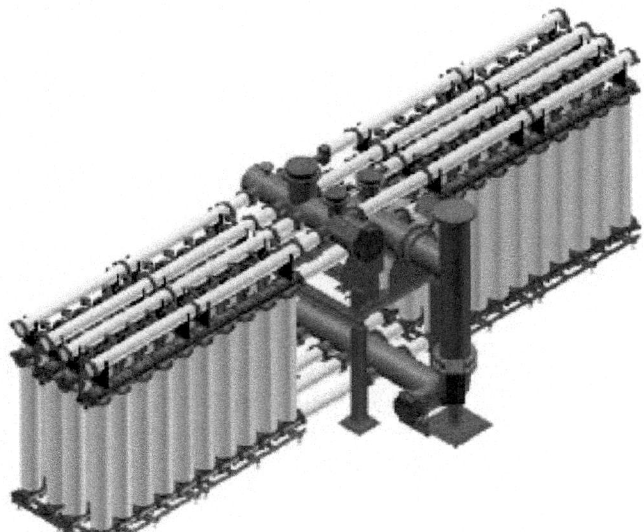

Figure 7.18. Figure of rack consisting of 136 modules.

7.6 Operational SWRO plants with (UF) or (MF) pretreatment.

Table 7.2 shows some examples of SWRO plants wit UF/MF pretreatment

Table 7.2 Examples of MF/UF plants.

Outside feed in housing and pressurized, mostly PVDF (Pall/Asahi, Dow/Omexell, Memcor), some PS and PAN			
Plant characteristics	**Coagulant**	**CEC/CIP**	**Remarks**
Yueqing, Zhejiang, China Yueqing power station 21,600 m^3/day RO	0.3-1.5 mg/L $FeCl_3$, flocculation, sedimentation	not specified	driver: reliability, difficult feed water MF: Asahi
Penghu, Taiwan 13,000 m^3/day RO	no coagulant	not specified	open intake, driver: difficult feed water UF: Dow

Outside feed in housing and pressurized, mostly PVDF (Pall/Asahi, Dow/Omexell, Memcor), some PS and PAN			
Wang Tan Power Station, China 9600 m³/day UF 3120 m³/day RO	no coagulant	15 mg/L NaOCl every backwash (1x/hour) (equivalent to 0.5 mg/L cont.) CIP: 0.05% NaOH, 0.2% NaOCl, followed by 0.36% HCl, 1x/year	UF operated at 50% flux; disc filter prior to UF; UF selected after pilot tests with UF and CP due to a consistent filtrate quality and lower operating and land use costs despite a wide ranging raw water quality, open intake industrial area, UF: Dow
Magong, Taiwan 23,000 m³/day UF 5,500 m³/day RO	no coagulant	BW: 1x/hour (with air scrub) no CEB, only CIP 1x/month, 2%w=w $H_2C_2O_4$ equal to 20,000 mg/L followed by 0.2%w/w NaOCl (2000 mg/L) (equal to a continuous dose of 0.4 and 4.0 mg/L respectively)	CP since 2002, replaced by UF in 2008, open intake and self-clean filters (130 μm), SDI_{15} and MFI < 2:1 achieved despite poor feed quality (SD_{15} = 20), UF: Dow
Outside feed uncontained, submerged and vacuum driven, mostly PVDF (Zenon/Zeeweed, Siemens/Memcor), some PAN			
Plant characteristics	**Coagulant**	**CEC/CIP**	**Remarks**
Zhejiang, China 76,800 m³/day UF 34,560 m³/day RO	-10 mg/L Fe, lamella sedimentation	not specified	open intake, coagulation, flocculation, clarification prior to UF to remove organics/floatable substances, UF: Zenon

Inside feed ⊠ always in housing, always pressurized, mostly PES (Norit, Hydranautics, Inge membranes), some PS (Koch) and CA (Aquasource)			
Plant characteristics	**Coagulant**	**CEC/CIP**	**Remarks**
Palm Jumeirah, UAE 64,000 m³/d RO (2 x 32,000 m³/day RO)(2 x 92,000 m³/day UF)	0.3 mg/L as Fe (FeCl3 used)	CEB: 200 mg/L Cl_2 1x/day CIP: 0.5% oxalic acid + 0.25% ascorbic acid once every 9 months	chemical wastes neutralized by acid/alkali, CIP and BW mixed with RO reject and tertiary sewage before discharge, SWRO CIP anticipated once every 18 months, UF: Norit
Qingdao power plant, China 13,000 m³/day RO	coagulation, flocculation,	not specified	UF selected because of reliability, difficult feed water, UF: Hydranautics/Hyflux
Qingdao power plant, China 5,600 m³/day RO	coagulation, flocculation, lamella sedimentation	not specified	open intake, UF selected because of reliability, UF: Norit
Jeddah, Saudi Arabia 56,500 m³/day UF 25,500 m³/day RO	design dose of 1 mg/L as Fe is not necessary, no coagulant aid used either	20 mg/L Cl_2 every 8-20 hours, CEB with H_2SO_4	2 stage media filters prior to UF, stable operation without $FeCl_3$, combination of BW and CEB sufficient, one CIP (of UF and RO) in 6 months, UF: Hydranautics

8

MATERIALS.

Four general areas of service will be discussed for analysis of materials in a SWRO : plant, seawater service, concentrate service, and permeate service. The difference in water chemistry and pressure in each of these three areas necessitates category specific materials selection. They are characterized as follows:

- **Seawater:** High chloride level (~19,000 mg/L), high TDS (~36,000 mg/L), microbiologic. contaminants, neutral to alkaline pH (8). Pressures up to 1,000 psi.
- **Concentrate:** Extremely high chloride levels (34,000 - 45,000 mg/L), high total dissolved solids (>63,000 mg/L), slightly alkaline pH (8.5) Pressures up to 1,000 psi.
- **Permeate:** Low chloride levels (lower than 150 mg/L), low total dissolved solids (less than 300 mg/L), slightly acidic to near neutral pH (6.3-7.3). Finished water will be slightly less aggressive after recalcification and the addition of corrosion inhibitors.
- **Chemical Additives:** Exposure conditions will vary per chemical, but range from acidic to basic.

Standard materials used in reverse osmosis plants are:

1. **Plastics**: GRP (Glass fiber reinforced plastic), PE (polyethylene), PP (polypropylene) and PVC.
2. **Metallic**: Stainless steel. And carbon steel+ebonite.

8.1 Plastic.

- **PVC.**

UPVC has excellent chemical resistance across its operating temperature range, with a broad band of operating pressures. Due to its long-term strength characteristics, high stiffness and cost effectiveness, UPVC systems account for a large proportion of plastic piping installations CPVC is resistant to many acids, bases, salts, paraffinic hydrocarbons, halogens and alcohols. It is not resistant to solvents, aromatics and some chlorinated hydrocarbons.

It can be used in piping of chemical dosing systems installed indoors. When piping is installed outdoors, it will be painted or PP will be used instead of PVC.

It can be used for all the drainage piping and low pressure piping (lower than 3.5 bars) and piping of diameter less than 100 mm.

- **PP.**

It can be used for reactives installed outdoors, diluted acid and permeate of racks.

- **PE.**

There are different densities (low or high). It can be used for sea outfalls, buried pressurized pipes and water pipes installed indoors. The disadvantage of PE is the lack of accessories.

Chemicals storage tanks can also be manufactured in PE. It is a cheaper alternative than GRP.

- **GRP.**

It is used for high pressure piping. and reactives dosing. Some EPC have started to use this material in pressure sand filters

The GRP material is a thermosetting resin, once cured, it cannot be softened and reshaped by heating. The basic raw materials comply with the requirements of ISO standards and are tested prior to use:

Glass Fiber:

Direct Roving Fibers; Chopable Fibers and Reinforcing Fiber Filaments. Fibers vary in thickness, weight and width and when combined with resin provide significant hoop strength and axial reinforcement.

Resins:

The pipe's chemical stability depends on the type of unsaturated polyester resin used in the pipe structure. Isophthalic polyester and Vinyl Ester resins are used. The first one gives mechanical resistance and the second one gives chemical resistance.

Sand:

Silica Sand is used to provide rigidity in the pipe and is applied to the GRP is manufactured with glass fiber resins (typically polyester but sometimes polyester).

Deposits which will contain chemical products will be manufactured in Vinyl ester resins.

It must be taken into account that there are two type of GRP pipes. The buried pipe with rubber seal lock joint with glass fiber (due to the high content of sand chemical connections are not possible), sand and resins as raw material and the above ground installed & industrial pipes with only glass fiber and resins.

Buried piping will be manufactured for pressures NP 4, NP6, NP10, NP16 and NP25. The Maximum pressure used has been NP 10 and sometimes NP16. *Specific Circumferential Stiffness will be between* 1250 N/m^2, 2500 N/m^2, 5000 N/m^2 and 10,000 N/m^2

The raw materials applied during the process vary to produce a series of layers that give protection, pressure, resistance, structural integrity and stiffness.

- **Internal & External Protective Layer:**

The inner and outer layers contain a high concentration of polyester resin that provides a protective layer that has resistance to chemicals, and abrasion.

- **Barrier Layer:**

The Barrier Layer is reinforced polyester resin layer that prevents penetration of materials into the structural layer of the pipe.

- **Structural Layers:**

The Structural Layers provide hoop strength, axial reinforcement and structural integrity. This consists of Glass Fibers of different types and sizes together with thermosetting resins.

- **Core Layer:**

The centre core layer provides strength, reinforcement and stiffness to the product. This layer contains chopped glass fibers, sand and resins.

GRP piping needs many supports and it is very sensible to temperature changes. A medium sized desalination plant may need 20,000 kgs of supports only for GRP.

Areas where GRP is being installed are:

- Seawater collection pits and pumping to desalination plant.
- Conductions and connections corresponding to pretreatment .
  - o Flotation.
  - o Front sand filters.
  - o Cartridge filtration.
- High pressure pumping suction.
- Flushing and backwashing .
- Post treatment.

See below schematic of seawater desalination plant indicating the areas where GRP piping is installed.

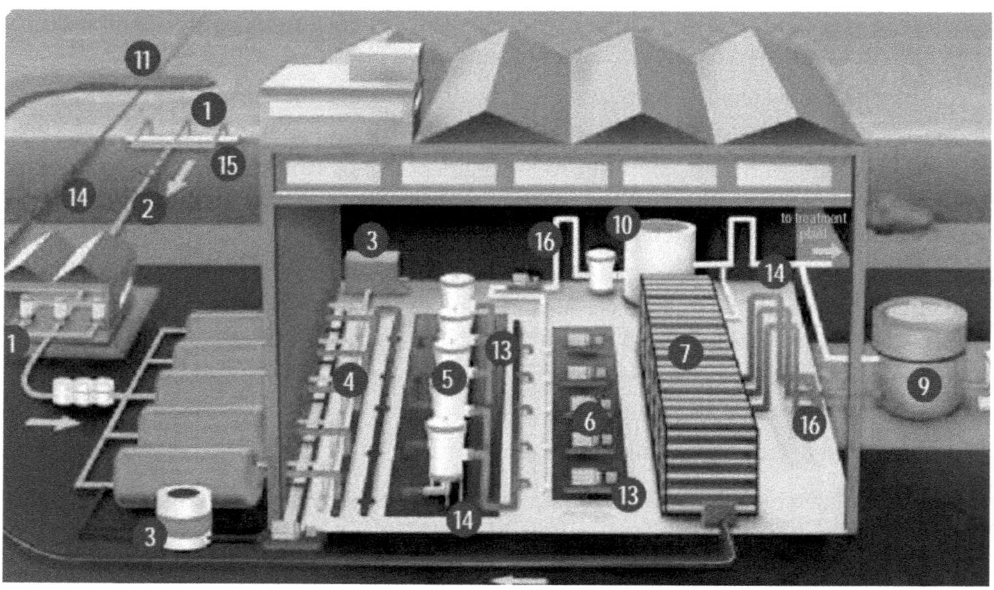

Figure 8.1. Location of GRP in an SWRO.

8.2 Metallics.

- <u>*Carbon steel+ebonite.*</u>

It is used only for pressure filters (sand, cartridge or precoat).

They are manufactured of S275JR or 516 GR70. Hard rubber lining (ebonite) will be 3 mm thickness, FDA approved.

- <u>*Stainless Steel piping.*</u>

Stainless steel is used mainly for high-pressure piping. Pressure > 16 bars, and it covers membranes, feed piping and recovery systems. However, stainless steel used for high-pressure piping is exposed to two very similar environments; i-e seawater and brine at ambient temperature and consequently corrosion aspects must be taken into consideration.

The corrosion types which could be potentially involved are:

1. **Crevice** refers to corrosion occurring in confined spaces to which the access of the working fluid from the environment is limited. These spaces are generally called crevices. Examples of crevices are gaps and contact areas between parts, under gaskets or seals, inside cracks and seams, spaces filled with deposits. Crevices are found in RO plants at victaulic-type couplings and flanged joints.

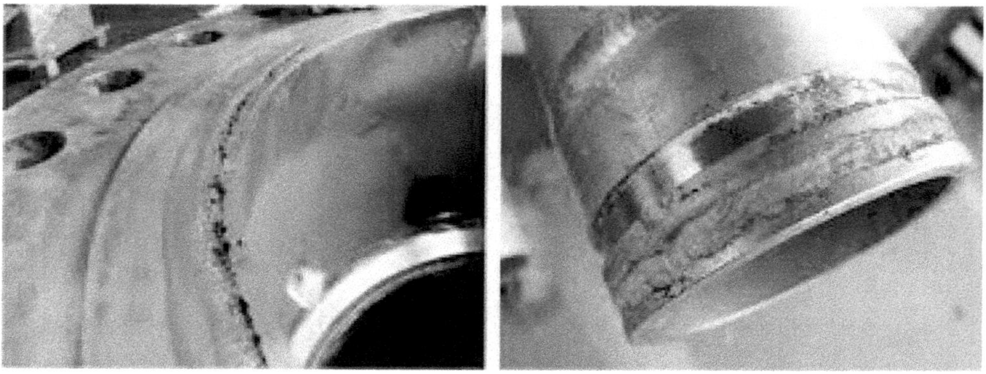

Figure 8.2. Photography of crevice corrosion.

2. **Pitting.** Pitting corrosion is a localized form of corrosion by which cavities or "holes" are produced in the material. It is caused by chlorides and temperature. It has not been detected in stainless steels.
3. **Erosion.** It is produced by the high velocity in piping especially from changes of directions.
4. **Galvanic** It is an electrochemical process in which one metal corrodes preferentially to another when both metals are in electrical contact and immersed in an electrolyte (sea water). It is not present between stainless steels, but is a potential type of corrosion in copper-nickel & stainless steel joints. It is very unusual. .
5. **Intergranular.** The causes of intergranular corrosion are welding, stress annealing, improper heat treating or overheating in service. However, pitting and crevice corrosion are the only corrosion types which could be involved in a SWRO, and hardly others.

To classify the resistance to corrosion of the different stainless steels, different coefficients have been developed:

- PREn (Pitting resistance equivalent number).
- CPT (Critical pitting temperature).
- CCT (Critical crevice temperature).

The values of the different coefficients can be obtained with the below indicated formulas with concentrations of chromium, molybdenum and nitrogen:

$$PREn = C_r + 3.3 \times Mo + 16 \times N$$

PREn can be expressed with alternative formulas, but this is usually employed by stainless steel manufacturers. Other alternatives is:

$$PREn = C_r + 3.3 \times Mo + 30 \times N$$

Tungsten (W) is known to have an effect on the pitting resistance and for some grades, so a modified formula is used:

$$PREn = C_r + 3.3 \times Mo + 2 \times Cu + 2 \times W + 16 \times N$$

The PRE-number can be used only for a rough ranking of alloys if no reliable test data are available, but a safer ranking is achieved by pitting and crevice corrosion tests:

$$CPT = 3.1835 \times (PREn - 23.2)$$

$$CCT = 1.92 \times (PREn - 25.6)$$

There is generally a good correlation between, PRE, CPT and CCT values, with the exception of a slightly strange and reverse positioning of 2205 and 904 L with respect to pitting and crevice corrosion.

PREn higher than 40 are required in seawater reverse osmosis plants to guarantee corrosion resistance, but it must also be taken into consideration temperature and salinity. The experience reported in the red sea and Arabian Gulf coast indicates that PREn of 44 or even more could be required in this area.

With brackish water, it is necessary to be careful with chlorides. Stainless steel AISI 316 is insufficient for chloride levels above 4000 ppm. In reject lines, these values are reached very easily, so duplex is recommended.

The most common stainless steels used according to their composition and use are as follows:

- **Piping and accessories.**
  - Austenitics. Higher content of Ni.
    - 1.4404-AISI-316L
    - 1.4434-AISI-317LN
    - 1.4539-AISI-904L.
  - Ferrítics-Duplex
    - 1.4462-S-31803
  - Ferrítics-Super-duplex. PREn > of 40.
    - 1.4410-S-32750-SAF-2507
    - 1.4501-S-32760- Zeron 100
  - Super-Austenitics
    - 1.4547-S-31254-254-SMO
    - N-08367-AL-6X
- **Stainless Steel casting**
  - Austenitics
    - 1.4404-AISI-316L
    - 1.4434-AISI-317LN
    - 1.4539-AISI-904L
  - Ferritic-Duplex
    - A-890-3A

        - 1.4468-S-31803
        - 1.4517
        - 1.4593
        - A-351-CD4Mcu
    - Ferritic-Super-duplex
        - A-890-5A
        - IR-885
        - Ferralium 255-3SC PREn>40
        - 1.4501-S-32760- Zeron 100
        - 1.4469
    - Super-Austenitics
        - 1.4547-S-31254-254-SMO

Seamless tubes are mainly required for smaller diameters (OD) though, less than 4" with wall thickness of Sch 40 (6 mm). For diameters more than 6", welded stainless steel will be ordered. A welded tube is in most cases less costly that its counterpart, it also offers good tolerances, e.g. the wall thickness of a welded tube is actually more even around the tube, whereas a seamless tube is slightly eccentric due to the manufacturing process. Design norm will be ASME/ANSI B-36.10.

Piping accessories are the below indicated:

- Forged flanges. Design following norm ASME/AMSI B16.6 and material according norm A-182, class 600.
- Elbows (welded for diameters higher than 3", norm A-403). It will be manufactured in two pieces and welded.
- Tees (welded for diameter bigger that 3", norm A-403). where standard tee is not available due to size restrictions (For diameters below 1 ½").
- Bolts in AISI 316L.
- Nozzle ASME/ANSI B-16.9.

8.2.1 Most used stainless steels.

A highly alloyed stainless Steel used in SWRO plants for piping and accessories has been 904 L. However crevice corrosion has been reported in threaded connections and manifolds so it has been progressively substituted by superduplex. (PREN > 40)

As commented the use of superduplex material has consolidated in desalination. 1.4547-S-31254-254-SMO is very common for high pressure piping. However, ZERON 100 is actually the most common material because of their benefits and price.

ZERON 100 is a superduplex stainless steel i.e. it is 50% austenite and 50% ferrite. This gives the alloy the benefits of the austenitic microstructure, such as ductility, and also those of the ferritic microstructure, such as strength. ZERON 100 is a high strength alloy with good ductility and formability.

The use of high strength stainless steel like ZERON 100, would allow the customer to reduce the wall thickness; 20 instead of Sch 40 (wall thickness for 904L), except in Victaulic connections where it must be used Sch 40.

Anyway, for pressures above 80-85 kg/cm^2, schedule must be analyzed in detail.

The velocity in high pressure piping Sch must be less than 3 m/s.

In **high pressure pumps** it has always been used **superduplex.** SULZER offers Ferralium 255, and a PREn more than 40 should be required by the designer. If not, a A-890 Gr 5A would be acceptable.

In **high pressure pumps, super duplex** should be used.

In **Intermediate pumping, super duplex** should be used. If duplex (1.4517) is used, cathodic protection should be installed.

In **seawater pumps, super duplex** should be used. If duplex (1.4517) is used, cathodic protection should be installed.

In **ultrafiltration backwashing pumps, superduplex** should be used as backwashing is carried out with brine. If not, cathodic protection should be installed.

In **filters back washing pumps, superduplex** should be used as backwashing is carried out with brine. If not, cathodic protection should be installed.

In **reject pumps**: Use cast iron pumps+painting+sacrifice anode.

Tables 1,2,3 and 4 that follow present the recommended materials by general system component.

Seawater equipment

Equipment	*Recommended materials*
High Pressure:	
Piping	AL-6XN, SAF 2507, 254 SMO, Zeron 100
Valves	AL-6XN, SAF 2507, 254 SMO, Zeron 100
Fittings	AL-6XN, SAF 2507, 254 SMO, Zeron 100
Pump Bodies-RO High Pressure Booster Pumps, ERD Pressure Booster Pumps	AL-6XN, SAF 2507, 254 SMO, Zeron 100
RO Membrane Modules	FRP
Energy Recovery Devices	FRP3, AL-6XN, SAF 2507,
(Seawater and Concentrate)	254 SMO, Zeron 100
Low Pressure:	
Plant Piping-Raw, Coagulated, Clarified, MF/UF Filtered, Used Washwater, Residuals	Schedule 80 PVC, FRP
Valves	PVC, double fusion epoxy coated steel
Fittings	Schedule 80 PVC, FRP_3

Equipment	*Recommended materials*
Pump Bodies-Flash Mix, MF/UF Feed, RO Feed Water Transfer Pumps, Reclaimed Water, Solids Transfer, Sludge Disposal	AL-6XN, Zeron 100, SAF 2507, 254 SMO
DAF Piping and Equipment	AL-6XN Zeron 100, SAF 2507, 254 SMO
Submerged Mixers in Equalization Basins	AL-6XN Zeron 100, SAF 2507, 254 SMO
Strainers	FRP, AL-6XN Zeron 100, SAF 2507, 254 SMO,
MF/UF Membrane Modules	FRP
Cartridge Filters Housing	FRP, RTR, PP, GRP
Solids Clarifier/Thickener	AL-6XN Zeron 100, SAF 2507, 254 SMO
Submerged Equipment	AL-6XN Zeron 100, SAF 2507, 254 SMO
Solids Transfer and Disposal Pump Bodies	AL-6XN Zeron 100, SAF 2507, 254 SMO
Equalization Basins Raw/Coagulated/Clarified Water; Used Washwater Solids	Concrete
Clarifier/Thickeners Basins	Concrete
DAF Basins	Concrete

Permeate equipment

Equipment	*Recommended materials*
Piping-RO Permeate, Plant Service water	317L, 316L, Schedule 80 PVC, HDPE, FRP
Valves	PVC, FRP, 317L, 316L
Fittings	317L, 316L
Pump Bodies-High Service Pumps	317L, 316L, Suitable non-metallic
Basins-Chlorine Contactor/Clearwell, High Service Pump Station Wet Well	Concrete

Concentrate equipment

Equipment	Recommended materials
High pressure (energy recovery)	
Piping	AL-6XN, SAF 2507, 254 SMO, Zeron 100
Valves	AL-6XN, SAF 2507, 254 SMO, Zeron 100
Fittings	AL-6XN, SAF 2507, 254 SMO, Zeron 100
Pump bodies-High pressure RO feed	AL-6XN, SAF 2507, 254 SMO, Zeron 100
Energy recovery devices (Seawater and concentrate)	AL-6XN, SAF 2507, 254 SMO, Zeron 100
Low pressure (Disposal):	
Piping	Schedule 80 PVC, FRP2, AL-6XN, SAF 2507, 254 SMO, Zeron 100, HDPE
Valves	Schedule 80 PVC, double fusion epoxy coated steel
Fittings	Schedule 80 PVC, FRP2, AL-6XN, SAF 2507, 254 SMO, Zeron 100
Pump Bodies-Transfer Pumps	AL-6XN, SAF 2507, 254 SMO, Zeron 100, or suitable non-metallic
Equalization basins	Concrete

Finished water equipment

Equipment	*Recommended material*
Piping-Plant Service Water, Finished (Potable) Water	Schedule 80 PVC, HDPE, FRP, Ductile Iron, Welded Steel
Valves	Ductile Iron, Cast iron, suitable non
Fittings	Ductile Iron, Cast iron, brass
Pump Bodies-High Service Pumps	Cast iron, Suitable non-metallic
Basins-Chlorine Contactor/Clearwell, High Service Pump Station Wet Well	Concrete1

9

REVERSE OSMOSIS MEMBRANES.

9.1 Type of membranes used in reverse osmosis.

The most common RO membrane used in desalination are spiral wound Thin Film Composite. They consist in a flat sheet sealed like an envelope and wound in a spiral. The envelopes are rolled like a paper towel roll and a fiber glass is wrapped around the envelope bundle. The result is a spiral wound membrane.

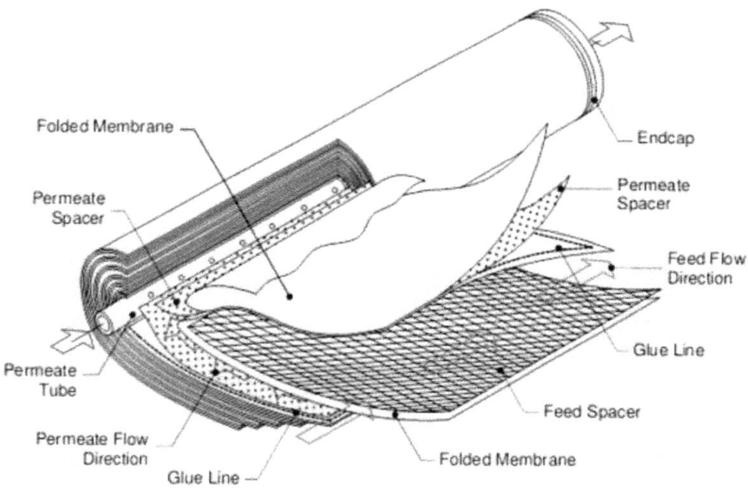

Figure 9.1. Detail of a spiral wound RO membrane.

Feed water enters the flow path provided by the spacers, at the upstream end. Due to the pressure supplied by a high pressure pump, a portion of the feed water is forced through the membrane. The water that does not pass through the membrane, plus the dissolved and suspended solids, exit the element at the downstream end of the flow path provided by the sheets of spacer material.

The water which passes through the membrane travels in the water carrier material inside each envelope and is collected in a central perforated tube. Figure 9.2 illustrates two envelopes, with two sheets of permeate water carrier attached to a perforated permeate tube.

Thin film membranes differ from asymmetric membrane. The salt rejecting layer and the porous support layers are made of different materials. The support material is a uniformly porous material (polysulfone) and the backing material (polyester). The salt rejection material is manufactured of polyamide.

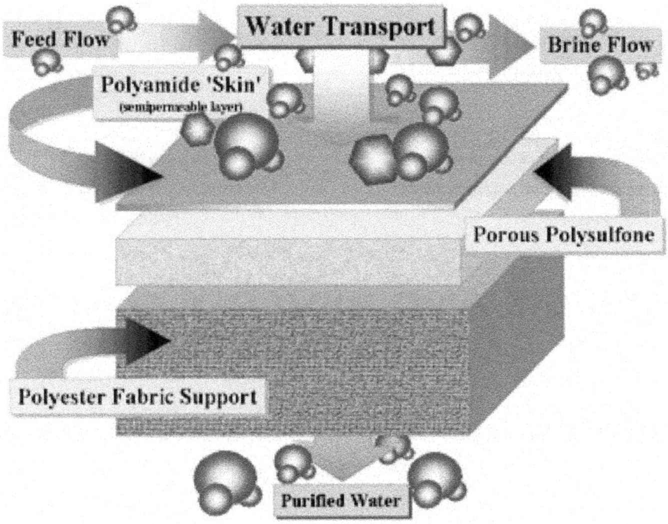

Figure 9.2. Polyamide membrane section.

The outer wrap of elements is of fiberglass and each element has a brine seal. Eight-inch elements that are 40 inches in length are used in large scale desalination plants.

9.2 Reverse osmosis basic concepts.

9.2.1 Osmotic pressure.

The osmotic pressure, P_{osm}, of a solution can be determined experimentally by measuring the concentration of dissolved salts in solution:

$$P_{osm} = 1.19 \times (T + 273) \times \Sigma mi$$

where P_{osm} = osmotic pressure (in psi), T is the temperature (in °C), and Σ(mi) is the sum of molal concentration of all constituents in a solution. An approximation for P_{osm} may be made by assuming that 1000 ppm of Total Dissolved Solids (TDS) equals about 11 psi (0.76 bar) of osmotic pressure.

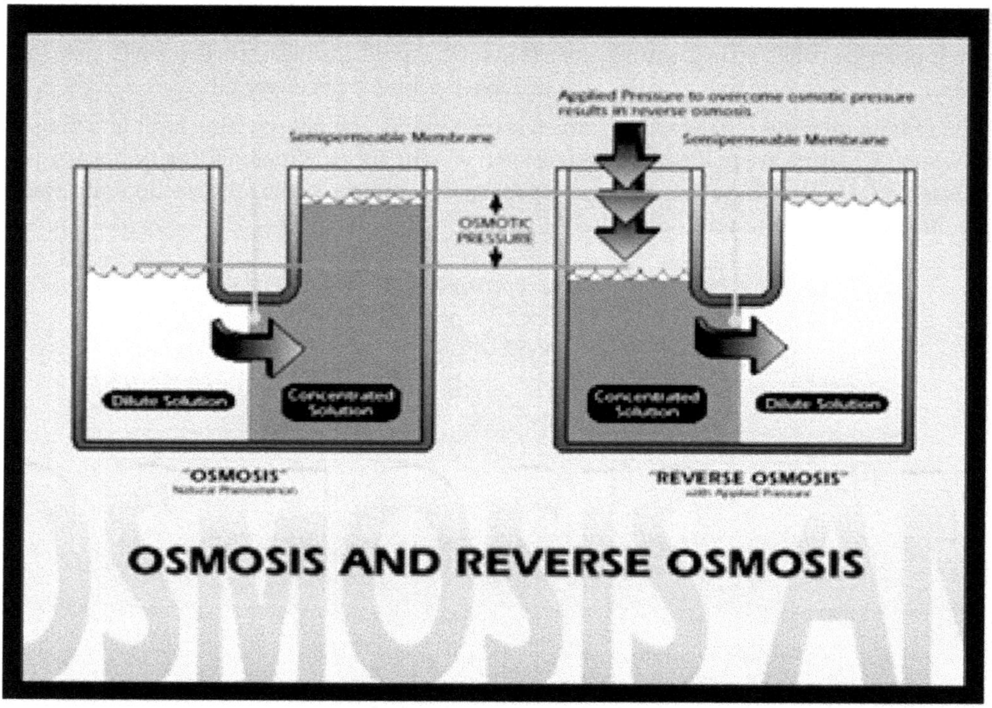

Figure 9.3. Osmotic pressure.

The mechanism of water and salt separation by reverse osmosis is not fully understood. Current scientific thinking suggests two transport models: porosity and diffusion. That is, transport of water through the membrane may be through physical pores present in the membrane (porosity), or by diffusion from one bonding site to another within the membrane. The theory suggests that the chemical nature of the membrane is such that it will absorb and pass water preferentially to dissolved salts at the solid/liquid interface. This may occur by weak chemical bonding of the water to the membrane surface or by dissolution of the water within the membrane structure. Either way, a salt concentration gradient is formed across the solid/liquid interface. The chemical and physical nature of the membrane determines its ability to allow for preferential transport of solvent (water) over solute (salt ions).

9.2.2 Net driving pressure (NDP).

Working net pressure is the applied pressure to the membrane minus the osmotic pressure, minus the counter pressure and minus half of the head loss through the membrane length.

$$P_{EFF} = P_I \, {}^{1}/_{2} \Delta P - P_B - P_\pi$$

where P_i is the feed pressure before the membranes, P_B is the counter pressure and $P\pi$ is the osmotic pressure.

9.2.3 Water transport

The rate of water passage through a semipermeable membrane is defined in Equation.

$$Q_w = (\Delta P - \Delta P_{osm}) \times K_w \times {}^{S}/{}_{d}$$

where Q_w is the rate of water flow through the membrane, ΔP is the hydraulic pressure differential across the membrane, ΔP_{osm} is the osmotic pressure differential across the membrane, K_w is the membrane permeability coefficient for water, S is the membrane area, and d is the membrane thickness. This equation is often simplified to:

$$Q_W = A \times (NDP)$$

Where A represents a unique constant for each membrane material type, and NDP is the net driving pressure

9.2.4 Salt transport.

The rate of salt flow through the membrane is defined by Equation:

$$Q_s = \Delta C \times K_s \times \frac{S}{d}$$

where Q_s is the flow rate of salt through the membrane, K_s is the membrane permeability coefficient for salt, ΔC is the salt concentration differential across the membrane, S is the membrane area, and d is the membrane thickness. This equation is often simplified to:

$$Q_s = \Delta C \times B$$

Where B represents a unique constant for each membrane type, and ΔC is the driving force for the mass transfer of salts. Equations above shown show that for a given membrane:

a) Rate of water flow through a membrane is proportional to net driving pressure differential (NDP) across the membrane.

b) Rate of salt flow is proportional to the concentration differential across the membrane and is independent of applied pressure

Salinity of the permeate, C_p, depends on the relative rates of water and salt transport through reverse osmosis membrane:

$$C_p = \frac{Q_s}{Q_w}$$

The fact that water and salt have different mass transfer rates through a given membrane creates the phenomena of salt rejection. No membrane is ideal in the sense that it absolutely rejects salts; rather the different transport rates create an apparent rejection. The equations explain important design considerations in RO systems. For example, an increase in operating pressure will increase water flow without changing salt flow, thus resulting in lower permeate salinity.

9.2.5 Salt passage (SP).

Salt passage is defined as the ratio of concentration of salt on the permeate side of the membrane relative to the average feed concentration. Mathematically, it is expressed as:

$$SP = 100\% \times \left({}^{C_p}\!/_{C_{fm}} \right)$$

where SP is the salt passage (in %), C_p is the salt concentration in the permeate, and Cfm is the mean salt concentration in feed stream.

Applying the fundamental equations of water flow and salt flow illustrates some of the basic principles of RO membranes. For example, salt passage is an inverse function of pressure; that is, the salt passage increases as applied pressure decreases. This is because reduced pressure decreases permeate flow rate, and hence, dilution of salt.

9.2.6 Permeate recovery rate & Conversion.

Permeate recovery is another important parameter in the design and operation of RO systems. Recovery or conversion rate of feed water to product (permeate) is defined by Equation

$$R = 100\% \times \left({}^{Q_p}\!/_{Q_f} \right)$$

where R is recovery rate (in %), Q_p is the product water flow rate, and Q_f is the feed water flow rate. The recovery rate affects salt passage and product flow. As the recovery rate increases, the salt concentration on the feed-brine side of the membrane increases, which causes an increase in salt flow rate across the membrane. Also, a higher salt concentration in the feed-brine solution increases the osmotic pressure, reducing the NDP and consequently reducing the product water flow rate

9.2.7 Salt rejection (SR)

Salt rejection is the opposite of salt passage, and is defined by Equation

$$SR = 100\% - SP$$

where SR is the salt rejection (in %), and SP is the salt passage as defined before.

9.2.8 Concentration factor (C_f).

Increase of salts in reject.

$$C_f = {}^{100}\!/_{(100 - R)} \text{ expressed in number}$$

9.2.9 Concentration salinity

It is the calculation of reject salinity based on concentration factor.

$$S_c = S_a \times C_f$$

9.2.10 Concentration polarization (CF).

As water flows through the membrane and salts are rejected by the membrane, a boundary layer is formed near the membrane surface in which the salt concentration exceeds the salt concentration in the bulk solution. This increase of salt concentration is called concentration polarization. The effect of concentration polarization is to reduce actual product water flow rate and salt rejection versus theoretical estimates. The effects of concentration polarization are as follows:

1. Greater osmotic pressure at the membrane surface than in the bulk feed solution, ΔPosm, and reduced Net Driving Pressure differential across the membrane (ΔP -ΔPosm).
2. Reduced water flow across membrane (Q_w).
3. Increased salt flow across membrane (Q_s).
4. Increased probability of exceeding solubility of sparingly soluble salts at the membrane surface, and the distinct possibility of precipitation causing membrane scaling.

The Concentration Polarization Factor (CPF) can be defined as a ratio of salt concentration at the membrane surface (C_s) to bulk concentration (C_b).

$$CPF = {}^{C_s}\!/_{C_b}$$

An increase in permeate flux will increase the delivery rate of ions to the membrane surface and increase C_s. An increase of feed flow increases turbulence and reduces the thickness of the high concentration layer near the membrane surface. Therefore, the CPF is directly proportional to permeate flow (Q_p), and inversely proportional to average feed flow (Q_{favg}).

$$CPF = K_p \times exp\left({}^{Q_p}\!/_{Q_{favg}}\right)$$

Where K_p is a proportionality constant depending on system geometry.
Using the arithmetic average of feed and concentrate flow as average feed flow, the CPF can be expressed as a function of the permeate recovery rate a of membrane element (R_i).

$$CPF = K_p \times exp\left({}^{2 \times R_i}\!/_{(2 - R_i)}\right)$$

The value of the Concentration Polarization Factor of 1.20, which is the recommended limit, corresponds to 18% permeate recovery for a 40" long membrane element.

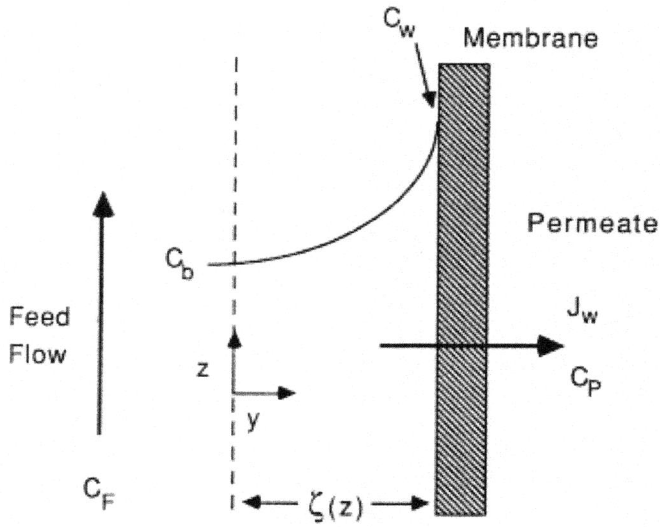

C_F = initial feed concentration

C_b = bulk feed concentration

C_w = wall concentration

Figure 9.4. Detail of concentration polarization factor.

9.2.11 Permeability coefficient.

The flux a membrane can provide per unit of surface and unit of pressure. It is measured in $l/hm^2/bar$.

9.2.12 Permeate flux rate.

Permeate flux represents the water flow through a specified area of membrane surface. That is, part of the feed water passes through the membrane leaving on the feed side rejected dissolved constituents and water born particles. The value of the average flux rate (the flux rates averaged over the entire system) is a very important consideration in designing a system. The water flow through the membrane for a certain pressure is measured in $l/m^2/h$ or gfd gallon/ft/day.

9.2.13 Fouling factor.

It is a coefficient which ranges from 1 to 0.5 and tends to represent the loss or properties of membrane over the years.

- **Temperature factor.**

Temperature affects salt passage and water flow through membrane approx. 3% per °C.

- **Membrane surface.**

Filtration surface of a membrane in m^2 (ft^2)

- **Product flow.**

Production flow of a membrane in gpd or m^3/day.

- **Membrane age.**

It is the number of years a membrane has been installed before replacement. For projection of long term performance an estimation of average membrane life is necessary.

- **Replacement rate.**

% of membranes which needs to be replaced annually.

- **Guarantee.**

Guarantee provided by the membranes manufacturer. It is usually five (5) years with a certain replacement rate (5) expected for the 1st year.

9.3 Type of membranes manufactured.

1. Low salinity and low pressure membranes (lower than 3000 ppm).

Surface > 400 ft^2, low pressure of operation and high permeate flow 30-50 m^3/day. Salts rejection flux is low. From 99.2% to 99.6%.

2. Medium salinity membranes (between 3000 and 15,000 ppm).

Surface between 365 and 400 ft^2, medium operation pressure 15-25 bars, high permeate flow 20 to 40 m^3/day and high salts rejection percentage 99.5% to 99.7%.

3. High salinity membranes (higher than 15,000 ppm).

360 to 440 ft^2 surface membranes, high operation pressure 50-70 bar, medium permeate flow 15 to 30 m^3/day and high salts rejection percentage of 99.7 to 99.8%.

4. High pressure and high salinity membranes

Surface of 360-440 ft^2, high operation pressure 70-80 bar and medium permeate flow 15 to 30 m^3/day and high salts rejection of 99 to 99.8%.

They are similar to the high salinity membranes but can operate at higher pressure and are used in the second stage of systems.

5. Nanofiltration membranes.

High surface (400–440 ft^2), low operation pressure(7 bars) medium permeate flow 25 to 30 m^3/day and low salts rejection (80% for Ca and Mg)

6. Other type of membranes.

There are specific membranes with higher or lower boron rejection or low fouling membranes.

The seawater reverse osmosis commercially available can be classified in the following key groups:

- Standard rejection membranes.
- High rejection membranes.
- High productivity or low energy membranes.
- High pressure pumps.

Standard membranes are designed to remove up to 99.6% of salts.

High rejection membranes are designed with tighter membrane structure allowing the increased mass of rejected ions and to reject smaller size ions such as boron (see boron removal chapter for specific information these membranes). The higher rejection membrane capabilities of 99.75%- 99.85% come at a price 10-20% higher operation pressure.

The high productivity(low energy consumption) membranes are designed with features to yield more produced water per membrane element. The features are: (1) higher surface area and (2) denser membrane packing. Increasing active membrane leaf surface area allows you to gain productivity using same size of membrane element.

The main purpose of high pressure SWRO is to provide fresh water from concentrated seawater with salinities above 50,000-60,000 mg/L. These membranes are designed to operate at a 20-40% higher pressure than that of the standard type of membranes and to treat high salinity concentrated produced by the first stage of a two-stage SWRO system.

9.4 Membranes characteristics.

9.4.1 Membrane specifications.

Main characteristics of membranes are:

- Operating pressure.
- Feed salinity.
- Membrane diameter.
- Surface.
- Rejection percentage.
- Production flow.

All membranes are tested in the factory, and the values obtained are normalized to the following conditions: 32,000 ppm NaCl, 5 ppm boron, 800 psi (5.5 MPa), 77°F (25°C), pH 8 and 8% recovery.

High area sea water membranes (440 ft^2) are available in the market, and the tendency is to install higher surface membranes in order to reduce the number of membranes and associated equipment.

- **Design guidelines of membranes.**

Operating limits of membranes depend on the feed water quality. The parameters to be controlled are:

- **Average flux:** Average flux of all the membranes of vessel.
- **Maximum flux:** Usually in the first membrane.
- **Maximum recovery:** Usually in the first membrane.

- **Minimum flow:** Usually in the last membranes.
- **Maximum flow:** Feed flow to the membranes.
- **Relationship between concentrate flow and product flow in last membrane:** It must be higher than a value which will range from 3 to 10.

The below shown table show the recommended guidelines for designing RO systems with 8-inch elements, arranged according to feed water type:

Table 9.1. Recommended guidelines for designing RO systems.

Feed source	Well water/ MF	Open intake	Concentrate (2nd stage)	RO permeate (2nd pass)
Feed silt density index	<3	<5	SDI< 1	SDI<1
Average flux (l/h/m^2)	15-20	12-15	7-10	35-40
Lead element permeate flux (l/h /m^2)	<35/32	<28	<26	<48
Max feed flow rate 400 ft^2 (m^3/h)	19	19	19	19
Max element recovery%	15	13	12	30
Minimum concentrate flow rate (m^3/h)	3,4	3,6	3,6	3

The flux decline rate will be higher than ten (10) for open intakes. For RO permeate will be more than three (3) and seven(7) for beach wells.

Saturation Limits at 100% for Soluble Salts in the Concentrate are limited to below indicated values.

Salt Saturation %

$CaSO_4$ 230

$SrSO_4$ 800 (using antiscalant)

$BaSO_4$ 6000 (using antiscalant)

SiO_2 100

Limits of Saturation Indices

Condition* LSI Value

LSI and SDSI without scale inhibitor < -0.2

LSI & SDSI with SHMP < 0.5

LSI & SDSI with organic scale inhibitor < 1.8

The maximum Fe reported in membranes is 5 ppm and the maximum value for aluminum is 4 ppm.

9.5 Steps to design a membrane system.

1. **Select the flow configuration and number of passes:** The standard flow configuration for water desalination is a plug flow, where the feed volume is passed once through the system.
2. **Select membrane element type:** Elements are selected according to feed water salinity, feed water fouling tendency, required rejection and energy requirements. The standard element size for systems greater than 10 gpm (2.3 m^3/hr) is 8-inch in diameter and 40-inch long.
3. **Select average membrane flux**: Select the design flux, f, (gfd or l/m^2/h) based on pilot data, customer experience or the typical design fluxes according to the feed source found in above shown table 9.1. In some cases, the design flux value is determined by pilot experiment data or customers experience.
4. **Calculate the number of elements needed:** Divide the design permeate flow rate Q_p by the design flux and by the membrane surface area of the selected element SE (ft^2 or m^2) to obtain the number of elements.

$$NE = {}^{Q_p}\!/\!_{(JV_{ave} \times (MA)E)}$$

In which:

NE = total element numbers.

Q_p = product flow rate.

$JV,_{ave}$ = average permeate flux.

(MA)E = membrane area of element.

The calculated number of RO elements may be a slightly changed based on the decision of element arrangement, that is, the number of pressure vessels and RO elements per pressure vessel.

5. **Calculate the number of pressure vessels needed:** Divide the number of elements by the number of elements per pressure vessel, to obtain the number of pressure vessels, NV – round up to the nearest integer.
6. **Select number of stages:** The number of stages is a function of the planned system recovery, the number of elements per vessel, and the feed water quality. The higher the system recovery and the lower the feed water quality, the longer the system will be with more elements in series. Typically, the number of serial element positions is linked with the system recovery and the number of stages.

For a global recovery of 45% and 7 elements per vessel, one stage will be enough. For a global recovery of 50% or higher and 7 elements per vessel, two stages will be required unless excellent quality of feed water and pretreatment. With 8 membranes per vessel it would be possible to get a recovery of 50% with only one stage.

Table 9.2. Number of stages of seawater system.

System recovery (%)	No of serial element position	No of stages 6-element vessel	No of stages 7-element vessel	No of stages 8-element vessel
35-40	6	1	1	-
45	7-12	2	1	1
50	8-12	2	2	1
55-60	12-14	2	2	-

7. **Select the staging ratio:**
8. **Balance the permeate flow rate:** The permeate flow rate of the tail elements of a system (the elements located at the concentrate end) is normally lower than the flow rate of the lead elements. This is a result of the pressure drop in the feed/brine channel and the increase of the osmotic pressure from the feed to the concentrate. Under certain conditions, the ratio of the permeate flow rate of the lead element and the tail element can become very high.

- High system recovery.
- High feed salinity.
- Low pressure membranes.
- High water temperature.
- New membranes.

The goal of a good design is to balance the flow rate of elements in the different positions. This can be achieved by the following means:

- Boosting the feed pressure between stages: preferred for efficient energy use.
- Apply a permeate backpressure only to the first stage of a two-stage system: low system cost alternative.
- Hybrid system: use membranes with lower water permeability in the first positions and membranes with higher water permeability in the last positions: For example, high rejection seawater membranes in the first and high productivity seawater membranes in the second stage of a seawater RO system.

9. **Analyze and optimize the membrane system:** The chosen system should then be analyzed and refined using the Reverse Osmosis System Analysis computer program.

9.5.1 Parameters which affect the efficiency of a membrane.

Parameters which affect the efficiency are:

- **Salinity:** Changing the feed water salinity will require increasing the required feed water pressure and permeate water salinity.
  The rate of increase in permeate salinity is higher than the rate of increase in feed pressure. If the different feed water also contains concentrations of sparingly soluble salts higher than in the design feed water, the recovery ratio may have to be reduced to avoid the possibility of precipitation of scale from the concentrate stream.

- **Temperature:** Change in feed water temperature results in a change in the rate of diffusion through the membrane. The change in the permeate flux rate with the temperature is described by the following equation:

$$TCF = e^{(k\times(1/(273+t)-1/298))}$$

Where TCF is temperature correction factor, K is a constant characteristic for a given membrane material, and t is feed water temperature in degrees Celsius. In this equation, a temperature of 25°C is used as reference point, with TCF equal to 1. The change in flux rate with temperature for polyamide membranes is shown in the following graph.

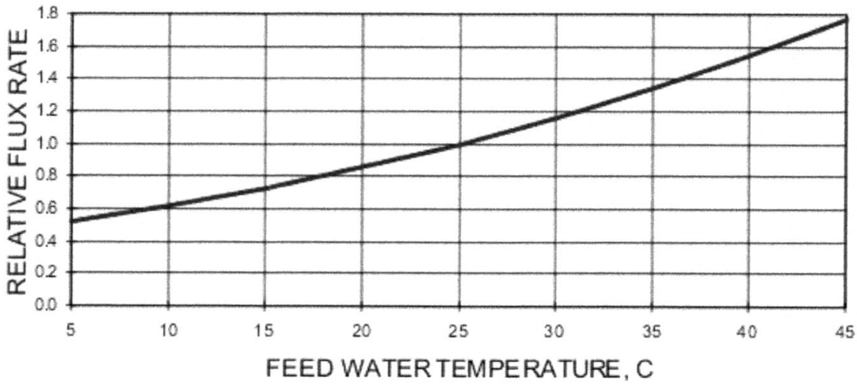

Figure 9.5. Graph of influence of flux rate with temperature.

The rate of change is about 3% per degree. Since RO systems are design to operate at constant output, feed pressure is adjusted to compensate for water flux changes when temperature changes. The salt diffusion through the membrane changes with temperature approximately at the same rate as water flux. Because permeate flux is maintained constant, permeate salinity changes accordingly to the temperature fluctuations.

- **Recovery:** Increases the operation pressure and decrease the quality of production.
- **Fouling factor or membrane age:** It increases operation pressure and permeate water salinity. There are two parameters which need to be considered:
- **Flux reduction:** It is measured in % and ranges from 2% to 10%. It is usually 7% per year for open intakes.
- **Salts passage (SP):** It is measured in % per year and ranges from 3-17% per year for polyamide membranes. It is usually 10% per year for open intakes. Assuming that salt passage increase is lineal.

$$S_{py} = S_{Po} \times (1+S_{PI} \times Y)$$

Where S_{py} is salt passage at the year Y, S_{Po} is the initial salt passage and S_{PI} is salt passage increase factor expressed as a decimal fraction.

For projections of long term performance an estimation of average membrane life is necessary. The estimation is based on assumed membrane replacement rate. An example of the calculation of average membrane age is given in table 9.3.

In the table the average age is calculated based on setting the membrane replacement event at the beginning of the given year of plant operation. The average age is used as a parameter for calculation of membrane performance, which eventually is the basis for the terms of system performance warranty.

Table 9.3. Average membrane age table for a replacement rate.

	3	4	5	6	7	8	9	10
5%	2,70	3,50	4,25	4,95	5,60	6,20	6,75	7,25
6%	2,64	3,40	4,10	4,74	5,32	5,84	6,30	6,70
7%	2,58	3,30	3,95	4,53	5,04	5,48	5,85	6,15
8%	2,52	3,20	3,80	4,32	4,76	5,12	5,40	5,60
9%	2,46	3,10	3,65	4,11	4,48	4,76	4,95	5,05
10%	2,40	3,00	3,50	3,90	4,20	4,40	4,50	4,50
11%	2,34	2,90	3,35	3,69	3,92	4,04	4,05	3,95
12%	2,28	2,80	3,20	3,48	3,64	3,68	3,60	3,40
13%	2,22	2,70	3,05	3,27	3,36	3,32	3,15	2,85
14%	2,16	2,60	2,90	3,06	3,08	2,96	2,70	2,30
15%	2,10	2,50	2,75	2,85	2,80	2,60	2,25	1,75

For example, after 5 years with a replacement rate of 10%, the number of membranes replaced will be of 40%, so 40% of membranes will have less that 5 years of life and the remaining 60% of membranes 5 years, so the average life will be 3.5 years.

9.5.2 Use of computer programs for projection of membrane performance.

Each manufacturer has software that provides a definition of system flow rates, recovery, operation pressure, and product quality.

Commercial softwares are:

FILMTEC	ROSA
HYDRANAUTICS	IMSDesing
TORAY	CAROL
GE WATER & PROCESS	WINFLOWS

The calculation procedure usually starts with the input of feed water analysis. After completing input to the water analysis screen the next screen includes entry of process and system design parameters. These include:

- Feed temperature
- Feed water pH
- Membrane age

- Permeate flow
- Recovery rate
- Membrane model
- Membrane array
- Salt passage increase factor
- Flux decline coefficient

The element type is selected from elements look up table. The array and number of elements are manually adjusted targeting permeate flux to be within the range suitable for the feed water type being processed.

Water calculations start with calculations of specific flux for the selected element. These values are corrected for temperature and element age (or fouling factor). Next, the program conducts calculations of permeate flow and salinity for a single element, in an element by element mode, starting with the lead element and progressing towards the concentrate end of the system.

The calculations are based on projected feed salinity to a given element, specific permeability and value of feed pressure selected from a predefined range. The resulting permeate flow is used to calculate the recovery rate. The calculated recovery rate is used in turn to correct the average feed salinity, which again is used to calculate new permeate flow and permeate salinity. Then, the subsequent results of two calculations of permeate flow are compared, and if the difference does not exceed preset value, the calculations are conducted for the next element.

Most computer programs are flexible enough to enable calculations for multistage/multipass systems configurations. Additional configurations options may include permeate blending and permeate throttling.

Some programs may have auxiliary routines that enable calculation of power consumption associated with projected operating parameters, calculation of operating cost or conduit simulation of chemical treatment of permeate.

The computer projection program produces a summary of input and calculated parameters in the form of a printout which serves as a basis for RO process design.

9.6 Types of membranes. Manufacturers.

9.6.1 Hydranautics.

Table 9.4. Hydranautics manufacturer. Membrane models.

Model	Permeate flow (GPD)	Salts Removal (average/minimum)	Boron removal
SWC4Bmax	7200	99.8%/99.7%	95%
SWC4max	7200	99.8%/99.7%	93·%
SWC4B	6500	99.8%/99.7%	95%
SWC5max	9900	99.8%/99.7%	92%

Model	Permeate flow (GPD)	Salts Removal (average/minimum)	Boron removal
SWC5-LD	9000	99.8%/99.7%	92%
SWC6	12,000	99.8%/99.7%	91%
SWC6Max	13,200	99.8%/99.7%	91%
2º pass			
ESPA2-MAX	12,000	99.6%	99.3%
ESPAB-MAX	9000	99.3%	96%

9.6.2 Dow-filmtec.

Table 9.5. Filmtec manufacturer.Membrane models.

Model	Permeate flow (GPD)	Salts Removal average/minimum	Boron removal
SW30HRLE-400	7500	99.75%/99.6%	92%
SW30XLE-400i	9000	99.7%	91.5%
SW30ULE-400 i	11,000	99.7%	89%
SW30HRLE-440i	8200	99.8%/99.65%	92%
SW30XLE-440	9900	99.75%/99.65%	91,5%
SW30XHR-440 i	6600	99.82%/99.65%	92%
SW30ULE-440	12,000	99.7%/99.6%	89%
2º pass			
LE-440i	12,650	99.3%	68%
BWHRLE-440i	12,650	99.5%	70%

If the primary driver is low energy consumption, SW30XLE-400/440 is the first option. Single-pass designs with these type of membranes can achieve permeate quality levels of chloride of 200 ppm or more and TDS values of 350 ppm or more.

If highest rejection is required, SW3XHR or SW30HRLE for one pass systems or SW30X-LE-400 I for two pass systems (combined with brackish water membranes in the second pass) are preferred. Single-pass designs with permeate quality requirements of boron less than 1.0 ppm, chloride of less than 25-200 ppm and TDS of less than 350 ppm can be obtained with high rejection membranes in one pass.

9.6.3 Toray.

Table 9.6. Toray manufacturer. Membrane models.

Model	Permeate flow (GPD)	Salts Removal (average/minimum)	Type
TM820-370	6000	99.75%/99.5%	High boron rejection
TM820C-400	6500	99.75%/99.5%	High flux/boron reject
TM820E-400	7500	99.7%	High flux
TM820F-400	9000	99.7%/99.4%	Low energy/high flux
TM820R-400	8500	99.8%	High flux
TM820R-440	9400	99.8%	High flux
TM820M-440	7700	99.8%	High flow and rejection. B rejection (95%)
TM820M-400	7000	99.8%	High flow and rejection. B rejection (95%)
TM820V-400	9000	99.8%	Low energy
TM820V-440	9900	99.8%	Low energy
TM820K-400	5800	99.86%	Highest rejection. Boron rejection(96%)
TM820K-440	6400	99.86%	Highest rejection Boron rejection(96%)
TM820L-400	12,200	99.8	Best energy saving
TM820L-440	1300	99.8	

NannoH20 is a new membranes manufacturer.

All of the manufacturers work with similar parameters. The maximum operating pressure is about 83 bars. Minimum and average salts rejection values are about 99.6% and 99.7% respectively.

All of the manufacturers are doing intensive research to get membranes which can operate at lower pressures while keeping high rejection values. Examples of these membranes which combine high flow and high rejection are the models FILMTEC SW30HR LE-400*i* which offers a flow of 31 m^3/d and salt rejection levels of 99.8%, TM820V-400 from Filmtec and SW6 from Hydranautics.

9.7 Membranes installation.

Membranes are housed in a system specially designed for this purpose that can withstand a range of pressure. This piece of equipment which houses RO elements is called pressure vessel.

Fiberglass Pressure vessels can be manufactured to house a single element or multiple elements. The only difference is in the length of the pressure vessel. The range is from 1 to 8 elements per pressure vessel, but the most common are 6-7 membranes per vessel. Lately, engineering firms have started to use eight (8) membrane vessels for sea water.

We will now look at the elements and their connections within a pressure vessel (PV). Lets first look at a two element PV. Elements are loaded into the feed water end of a PV because each element has a brine seal on one end (usually on the feed water end). The purpose of the brine seal is to provide a seal between the outer surface of the element and the inner surface of vessel. If this area is not sealed, some feed water will bypass the elements by flowing around them instead of through them. This will reduce the ROs units performance and cause issues.

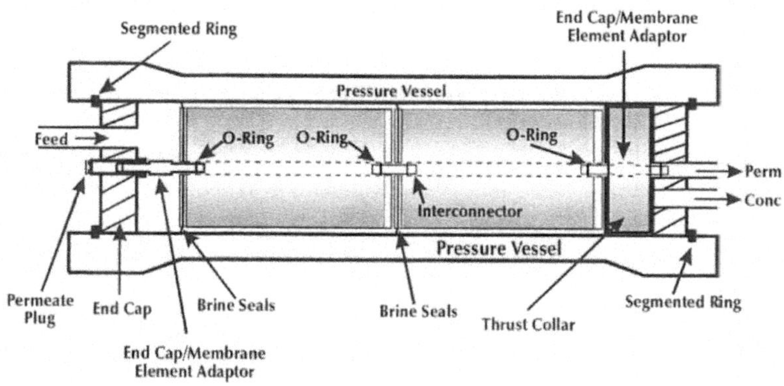

Figure 9.6. Section of pressure vessel.

The brine seal is placed into a groove in one end of the anti-telescopic device (atd). It is most common to place the brine seal in the feed water end. The rationale is that less suspended solids will be lodged between the elements and PV by placing the seal there.

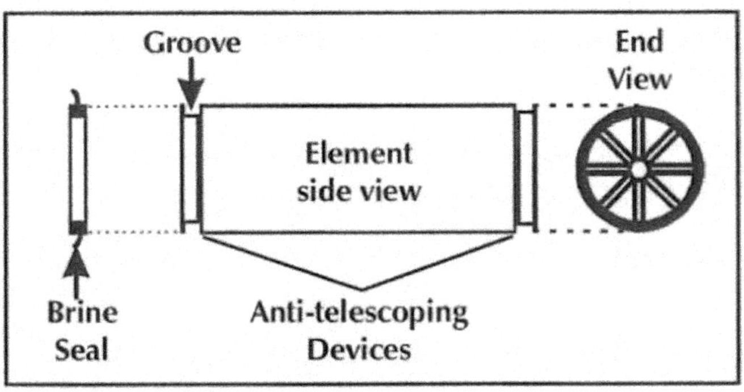

Figure 9.7. Detail of brine seals.

The brine seals acts as a "check valve". The "flap" of the brine seal will "open" with flow in one direction and "close" with flow in the opposite direction.

Because of the construction of the brine seals, the elements must be loaded and removed in one direction. If the elements are pushed in "backwards", the brine seals will open and make it impossible to load.

The elements must be connected to each other and to the permeate piping. The elements have an interconnector between them which connects and seals the internals permeate tube of each element. Interconnectors must have one or two o-rings on both ends.

FILMTEC has a patented design (*iLEC*™ technology) which consists of a single axially-compressed o-ring seal in combination with a rotational (vs. an axial or radial) mechanical connection Interlocking tabs around the periphery of each end-cap which provide a fixed mechanical connection between elements, compressing the o-ring upon installation and maintaining a tight seal.

With this system all the membranes work together and offer some advantages such as reduced permeate pressure drop, long term seal integrity and easy loading and unloading. However, this system is more expensive.

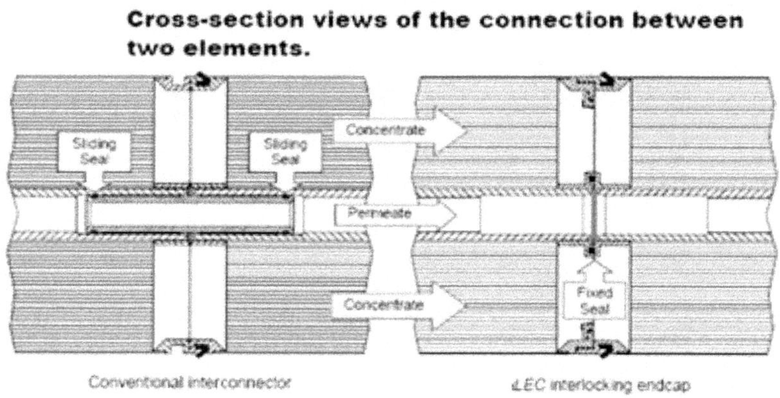

Figure 9.8. Detail of conventional interlock and iEC interlocking endcap.

The elements permeate tubes must also be connected to and sealed to the end caps. An end cap adaptor is the piece that connects the end cap with an element. It has o-rings to seal at the end cap and at the element.

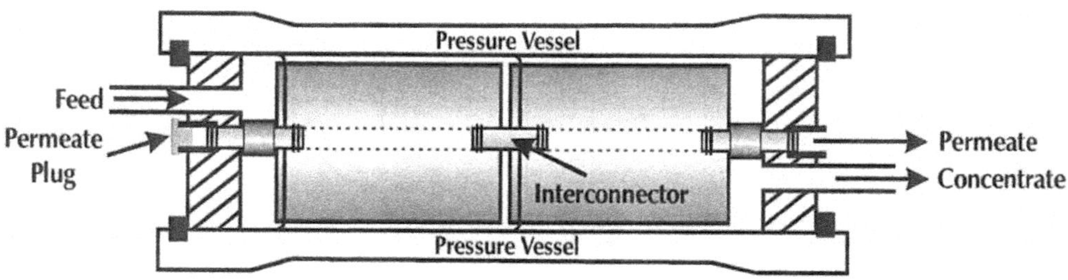

Figure 9.9. Section of pressure vessel.

At each end of a pressure vessel (PV) there is an end cap. The end cap seals the two open ends of the PV so it can be pressurized. Feed water enters a PV through a port in the end cap. Permeate and concentrate exit a PV through ports in the end cap.

When the high pressure pump starts there is surge of flow and pressure. This pushes the elements forward. The elements will move to the extent that the fittings and clearances allow. At the downstream (concentrate) end of each PV a thrust collar is installed in order to minimize the forward movement and also take the load off of the end cap adaptor.

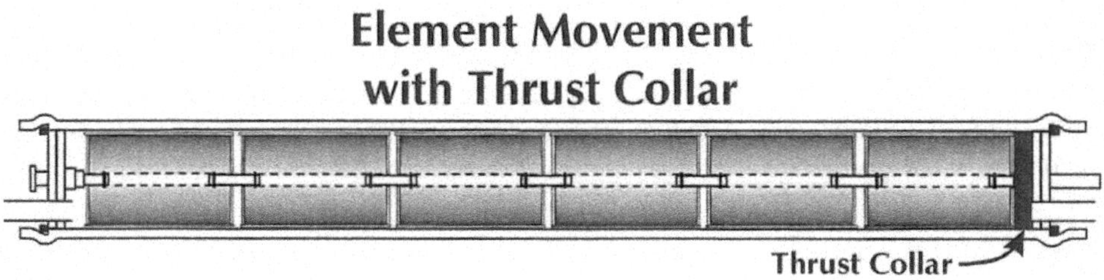

Figure 9.10. Detail of thrust Collar.

9.8 Flow configuration.

There are several flow configurations for RO that allow the user to get the most out of the system. This chapter will cover a few of the techniques used in RO design to optimize system performance.

9.8.1 Single-stage.

Majority of current seawater RO system designs are single stage with seven elements per vessel. In one single vessel there are two (2) alternatives; with one (1) or with two (2) product outlets. The last alternative will be used when there is a 2nd pass to improve product quality.

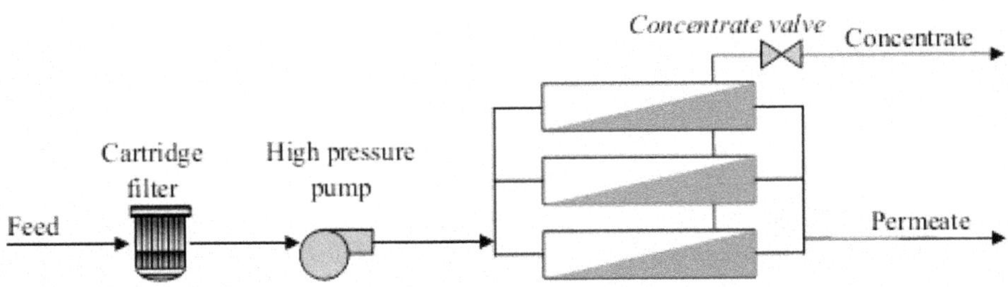

Figure 9.11. One stage configuration with recirculation.

Even if using the highest rejection RO membrane elements commercially available today (nominal minimum rejection of 99-75%), the single stage SWRO desalination systems typically cannot consistently yield permeate with TDS lower than 200 mg/L, chloride level of less than 100 mg/L and boron concentration lower than 0.5 mg/L.

9.8.2 Two stages in the same vessel.

It is used to reduce the flow of initial stages. One (1) or two(2) product connections can be installed.

9.8.3 Two stages with blind interconnector.

Alternative when (2) stages will operate at different inter-stage.

9.8.4 Two stage RO unit with different pressure vessels.

The concentrate from the first stage becomes the feed to the second stage; this is what is meant by the term "concentrate staging." The flows and pressures in the multistage unit are controlled with the feed and concentrate valves. The feed valve, after the high pressure pump, controls feed flow to the unit. The concentrate valve, at the outlet of RO block, controls the feed pressure.

In some cases it is necessary to equilibrate permeate flow between stages i.e. decrease permeate flow from the first stage and increase permeate flow from the last stage. This can be accomplished in one of two design configurations. One solution is to install a valve on the permeate line from the first stage, as shown.

By throttling this valve, permeate back pressure will increase, reducing net driving pressure and reducing permeate flux from the first stage. The differential permeate flux is produced from the second stage by operating the reverse osmosis unit at a higher feed pressure.

The other solution is to install a booster pump on the concentrate line between the first and the second stage, as diagrammed below. The booster pump will increase feed pressure to the second stage resulting in higher permeate flow.

The advantage of a permeate throttling design is the simplicity of the reverse osmosis unit and low capital cost. However, this design results in additional power losses due to permeate throttling and higher power consumption. The inter-stage pump design requires modification of the inter-stage manifold and an additional pumping unit. The investment cost is higher than in the first design, but the power consumption is lower.

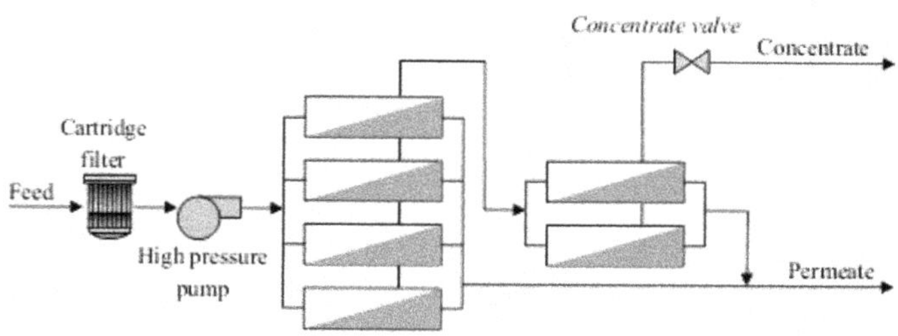

Figure 9.12. Configuration two stages.

The rationale behind two stage design is that it results in a high feed concentrate flow, which reduces concentration polarization. In seawater RO systems scaling is not a recovery limiting factor, but lower concentration polarization will result in lower ions concentration at the membrane surface and therefore somewhat lower permeate salinity. However, with higher feed flow, higher feed pressure is required due to increased pressure drop across the RO trains. Design efforts to reduce power consumption and system cost, resulted in transition of seawater plant design to a single stage configuration and increased number of elements per vessel. Two stage RO unit will usually only be used in the second pass of a two pass RO unit.

Global recovery may reach 60-65%. Booster pump pressure will be variable depending on selected recovery for the second stage and the type of membrane and can range from 10 bars for a global recovery of 55% up to 20-30 bars for both high global recovery and high salinity rejection membranes.

Designing the SWRO plant with a higher recovery will allow minimizing the size of the plant intake and pretreatment facilities and the capital expenditures for their construction and operation.

9.8.5 Two pass RO unit with different pressure vessels.

For some applications, the single pass RO system may not be capable of producing permeate water of a required salinity. Such conditions may be encountered in two types of SWRO applications:

- Seawater RO systems, which operate on a very high salinity feed water, at high recovery ratio and/or at high feed water temperature.
- Reduce level of Boron.

To achieve an additional reduction in permeate salinity, the permeate water produced in the first pass is desalted again in a second RO system. This configuration is called a two pass design, or "permeate staging." Depending on quality requirements, all or part of the first pass permeate volume is desalted again in the second pass system. The system configuration is known as a complete or partial two pass system depending on whether all of the permeate is fed to the second pass or not.

The first pass permeate is a very clean water. It contains very low concentrations of suspended particles and dissolved salts; therefore, it does not require any significant pretreatment. The second pass system can operate at a relatively high average permeate flux and high recovery rate. The common design parameters for the second pass RO unit are a recovery rate of 85% - 90%. In a two pass system, the permeate from the first pass flows through a storage tank or is fed directly to the suction of the second pass high pressure pump. There are number of possible configuration of the two pass RO units. One configuration, which is a partial two pass system, shown here, splits the permeate from the first pass into two streams.

One stream is processed by the second pass unit, and is then combined with the unprocessed part of the permeate from the first pass. Provided that the partial second pass system can produce the required permeate quality, this configuration results in smaller capital and operating costs, as well as higher combined permeate recovery rate (utilization of the feed water), compared to a complete two pass system.

It is a common procedure in a two pass systems to return concentrate from the second pass unit to the suction of the high pressure pump of the first pass unit. The dissolved salts concentration in the concentrate from the second pass is usually lower the concentration of the feed to the first pass unit. Therefore, blending feed water with the second pass concentrate reduces slightly the salinity of the feed, and increases the overall utilization of the feed water.

Membranes used in the second pass are brackish water membranes achieving a salinity of second pass permeate lower tan 20-30 ppm. First pass permeate salinity is about 400-500 ppm, so it can be obtained a salinity below 200 ppm.

As a conclusion, when project TDS limits are below 250-300 ppm, split-two pass system will be required. “Booster” pump between passes pressure will be between 8-12 kg/cm^2.

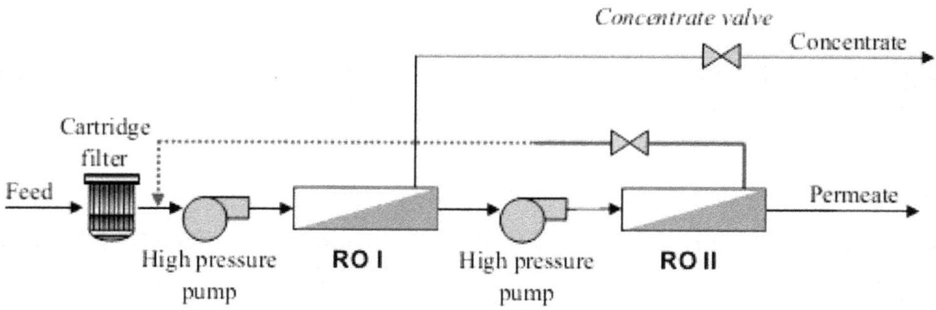

Figure 9.13. Configuration two pass system.

If 100% of first pass permeate is treated in second pass, considering a recovery of 45% for first pass and 90% for second pass, global recovery would be of approx. 40%. With a partial split of 20% (common value to achieve boron levels to be into limits) global recovery would be approx. 41-43%.

As an example of large scale two pass, two-stage SWRO plant is the 170,500 m^3/day Fujairah plant. The first pass of this plant consists of 17 duty plus one (1) stand by reverse osmosis trains using standard rejection SWRO membranes and producing permeate of TDS concentration of 400-500 mg/L. The overall recovery of the first stage is 43%. The second pass has eight (8) BWRO trains with a total recovery of 90%. The second pass BWRO trains have two stages and treat approximately 80% of the first pass permeate to TDS of 10-20 mg/L. The rest of the first pass permeate is blended with the second pass permeate to produce a finished water of TDS concentration less than 120 mg/L. The concentrated produced in the second pass has a salinity lower than that of the source water and is recycled to the feed of the first pass.

9.8.6 Split partial second pass RO design concept (SPSP)

The front RO elements are producing higher flow of low salinity permeate than elements at the back of the pressure vessel. As it can be seen in the chart below (Fig. 9.14) showing the test results conducted with a seawater membrane on a seawater feed, the first 4–5 elements in the pressure vessel are usually producing about 70–80% of the total permeate flow with a combined salinity of less than 300 mg/L. The rest of the elements are producing a remaining permeate flow with a combined salinity of above 1500 mg/L. This detailed knowledge of flow and salinity distribution within the pressure vessel is fully utilized in the SPSP design concept. In the SPSP design, the first pass permeate is collected from both sides of the pressure vessels. Front permeate with very good quality is directly sent to the product tank, while the rear permeate which has a high salinity is treated by a second pass RO.

The second pass RO permeate is then blended together with the front permeate from the first pass RO in the product tank. Basically, it means that the second pass treats a smaller, but more saline stream than a traditional partial two pass system. With such an arrangement, it is possible to maximize the size reduction of second pass RO when compared to a conventional partial second pass design. This also has the advantage of increasing the overall recovery and results in reducing the size of the first pass and the pretreatment.

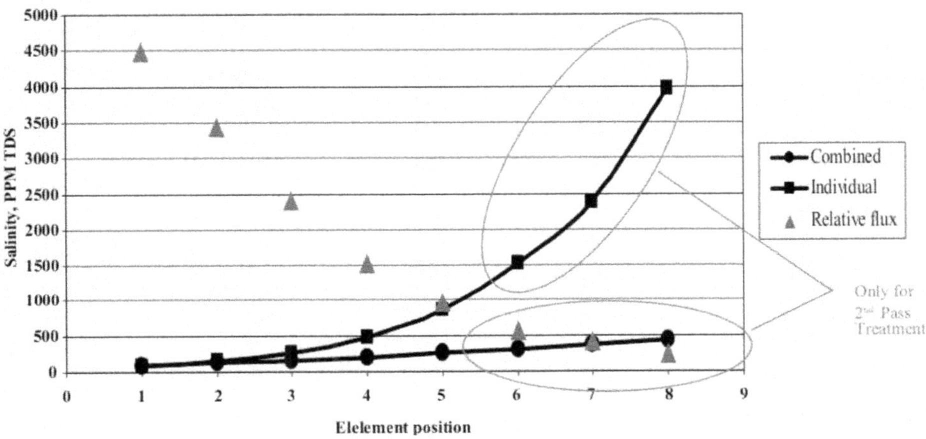

Figure 9.14. flux and salinity distribution in 8 elements PV.

Typical diagram of SPSP system with flow control valves on permeate lines can be seen on Figure 9.15. The control valves allow continuous adjustment of front and back permeate flow according to actual membrane performance, feed water salinity and temperature. It also gives the possibility of running the system in single pass mode via a back permeate by-pass valve, which might be the case if membranes are new and feed water temperature is low. In such cases the final permeate quality could be obtained from a first pass RO operation only. Changing the permeate split ratio on a first pass RO will require the second pass high pressure pump motor to be equipped with variable frequency drive in order to operate the second pass at a different flux/capacity depending on actual conditions.

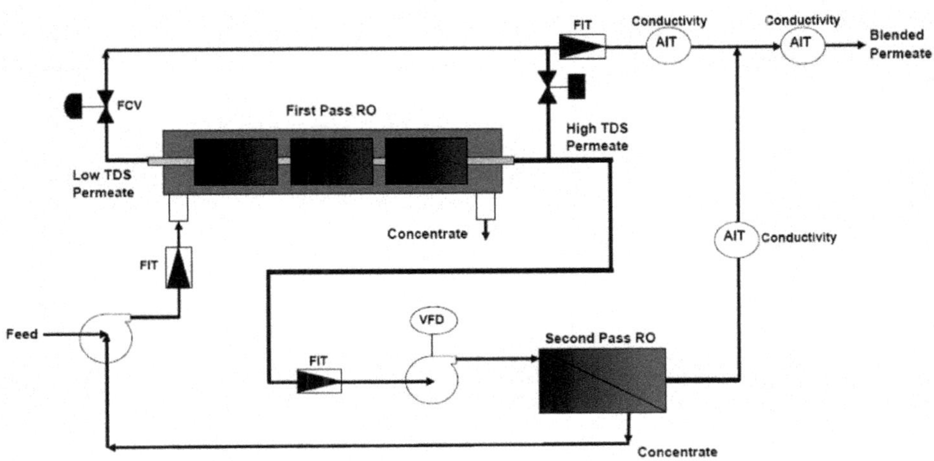

Figure 9.15. Typical diagram of SPSP with variable split ratio.

Table 9.7 compares partial two pass system with split partial two pass configuration for 10,000 m^3/day of capacity.

Table 9.7. Comparison of partial two pass and two pass configuration

Configuration	Partial two pass	Slit partial two pass
First pass capacity (recovery rate)	11,500 m^3/day (50%)	10,700 m^3/day (50%)
Second pass capacity (Recovery rate)	6800 m^3/day (85%)	3000 m^3/dy (85%)
By pass flow	3500 m^3/day	7170 m^3/day
TDS (boron) ppm	111	118
First pass array	129 PV (8M)	120 PV (8M)
Second pass array	18:9 PV (8M)	8:4 PV (8M)
Power Consumption	3.6 kWh/m^3	3.33 kWh/m^3
Total number of elements	1.248	1.056

9.8.7 Hybrid-membrane inter stage design.

The individual element permeate flow decreases uniformly throughout the series configuration from the lead element of the first stage to last element.

Permeate flow decreases because the net driving pressure, ΔP-$\Delta\pi$, is uniformly declining. (ΔP is the pressure difference between the feed side and the permeate side of the membrane; $\Delta\pi$ is the osmotic pressure difference between both sides).

A Hybrid membrane Inter-stage Design (HID) combines membranes of different nominal flux and salt rejection (different models of membranes in the same pressure vessel of seawater reverse osmosis racks) is the alternative (high rejection elements in the first positions and elements with high productivity and ultra low energy in the final positions).

The projected permeate quality and energy consumption of various hybrid and non-hybrid designs is shown in 9.16.

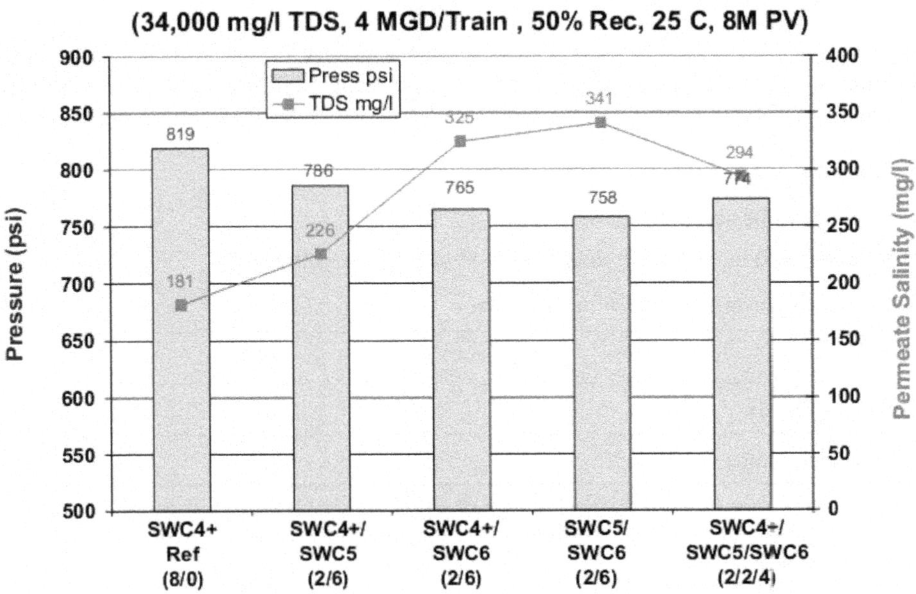

Figure 9.16. Analysis of different hybrid configurations

Different combinations with seven (7) membranes in a vessel are shown below:

- 2 membranes SWC4+Max +5 membranes SWC5+Max
- 3 membranes SWC4+Max +4 membranes SWC5+Max
- 2 membranes SWC4+Max +5 membranes SWC6
- 3 membranes SWC4+Max +4 membranes SWC6
- 2 membranes SW30HRLE 400i +5 membranes SW30ULE-400 i
- 3 membranes SW30HRLE 400i +4 membranes SW30ULE-400 i
- 2 membranes SW30HRLE 400i +5 membranes SW30XLE-400 i

Table 9.8 shows some of major SWRO plants where hybrid designs have been implemented during last years.

Table 9.8. Large scale SWRO plants using hybrid membranes design.

Project name location	*Membrane type*	*Start up*	*Plant capacity m³/day*
Barcelona- Spain	(3) SWC4+(4) SWC5 in 7MVP	2009	200,000
Escombreras. Spain	(3) SWC4+(5) SWC5 in 8MVP	2009	63,000
Las Palmas III. Spain	(6) SWC4+ 1st stage (6) SWC5 2nd stage	2008	80,000
Valdelentisco.Spain	(7)SWC4+Max 1st stage (7) SWC5 2nd stage	2009	60,000

9.9 Membranes of 16″

Market acceptance of diameters larger than 16 inches has been slow. However, large diameter RO modules, including 16-inch, are now under permanent study and large scale SWRO plants have started to operate. The main advantages and disadvantages of 16” membranes are:

1. **Potential space saving.**

Space saving can be about 10-15%.

Comparing a 8000 m^3/day train design with 8” and 16” membranes.

Table 9.9. Comparison of 8” and 16” membranes.

	8”	**16”**
Number of vessels	73	18
Number of elements	511	72
Footprint	7.85x8.125 m	5.53x 4.8 m
Volumetric space capacity	63.78 m^2	26.54 m^2

2. **Potential initial investment saving.**

Potential saving can be up to 5-6 %. To grow the 16 inch membranes market, further open discussion of the factors governing the maximum train size, the magnitude of the savings from building fewer trains, and the overall economy of scale enabled by 16-inch elements is needed.

3. **Charge and discharge of membranes.**

While a number of options for 16-inch handling have been devised and implemented the size and weight of 16-inch elements remains a perceived obstacle in some cases. Continued engineering development of loading and unloading tools is expected, and should ultimately result in options that are more efficient than manual 8-inch handling.

4. **Bad flow distribution.**

Bad flow distribution is one of the major problems of 16” membranes. Intensive research is being done in this area including the design of special flow distributors to provide an even distribution of flow within RO membranes or electromagnetic coil wounds round the RO pressure vessels. (Graham Tek patent).

9.10 Vertical installation.

As indicated in previous sections, most large scale RO systems are comprised of horizontal arrays of multi element vessels with six, seven or even eight elements of 8” diameter mounted in a series. However, a vertical arrangement of membranes is being considered and analyzed by some companies for racks with elements of 16” diameters. The thought behind this is that usually the rack support requirements to house the much heavier 16” elements are considerately higher, given the additional weight. The normal expense of trying to support these heavier membranes in a rack would be reduced due to a vertical arrangement.

This adds much greater burden on the elements and vessels themselves, and significant strength is required in the element core tube to ensure that the element can support its own weight as it is loaded into the vessel. Therefore, the core tube of a 16" element has a much greater load bearing duty than an 8" element and the first generation of these elements have used GRP rather than plastics such as PVC or ABS. Cost is therefore transferred from the system builder to the element maker.

Figure 9.17. 16" vessels. Photography Courtesy of EDI.

9.11 Membrane configurations of SWRO now operational.

Table 9.10. Configuration of racks, type of membranes, flux of large scale SWRO fully operational.

Plant characteristics	Pretreatment	1st pass	2nd pass
Ashdod 319,000 m³/d, provides two pass SWRO Permeate requirements Bo< 0,3 Cl<20	open intake, Ring filters UF	45% recovery (1st pass), 8 elements PV No of trains:12 Vessels per train: 154 Model of membrane: TORAY TM820R-440 Split variable 10-30%	Split variable 10-30% 90% recovery 7 elements PV No of trains:3 Number of stages: 3 Vessels per train: 88/29/13 Model of membrane: TORAY 720 TM-720D-440 & TM820R-440 in last stage

Plant characteristics	Pretreatment	1st pass	2nd pass
Jeddah III 240,000 m^3/day Product TDS 95 ppm	gravity DMF,	42% recovery (1st pass), 7 elements PV No of trains:16	Split variable 65% 92% recovery 7 elements PV No of trains:8 Number of stages: 2
Beni Saif 200,000 m^3/day, Open intake Single pass	Two stages 1.-Pressurized DMF (sand) 2.-Pressurized DMF (anthracite)	45% recovery (1st pass), 7 elements PV No of trains:10 Vessels per train: 288 Model of membrane: HYDRANAUTICS: SWC5	Not applicable
Barcelona, Spain commissioned in 2009 200,000 m^3/day,	Off shore sub-merged intake (2.2 km-25 m), 3 stages: 1. flocculation, DAF 2. gravity DMF 3. pressurized DMF	45% recovery (1st pass), 7 elements PV No of trains:10 Vessels per train: 230 Model of membrane: SWC4+(3)+SWC5(4) Split variable 10-30%	Split variable 10-30% 85% recovery 7 elements PV No of trains:2 Number of stages: 2 Vessels per train: 48/30 Model of membrane: ESPA-B
Aguilas, Spain 180,000 m^3/day Global conversion: 43.5% Permeate. Bo< 0.5	3 stages: 1. flocculation, DAF 2. gravity DMF 3. pressurized DMF	45% recovery (1st pass), 7 elements PV No of trains:12 Vessels per train: 190 Model of membrane: SW30HRLE-400I Split variable 60-80%	90% recovery 7 elements PV No of trains:5 Number of stages: 2 Vessels per train: 88/30 Model of membrane: FILMTEC: LE-400I
Fujairah, UAE Gulf of Oman, commissioned 2003, 170,500 m^3/d, recovery: 43% (1st pass) 90% (2nd pass), 41% (total)	Open Intake Gravity DMF	43% recovery (1st pass), 7 elements PV No of trains:17 Vessels per train: 136 Model of membrane: SWC3 By-pass:20-30%	90% recovery 7 elements PV No of trains:8 Number of stages: 2 Vessels per train: 50/23 Model of membrane: ESPA1
Perth, Australia commissioned in 2006, 144,000 m^3/day, Open intake TDS:100 mg/L	pressurized DMF+DMF	45% recovery (1st pass), 7 elements PV No of trains:12 Vessels per train: 162 Model of membrane: SW30HRLE-400 By-pass:20-30%	90% recovery 7 elements PV No of trains:6 Number of stages: 2 Vessels per train: 42/19 Model of membrane: FILMTEC:LE-400I

Plant characteristics	Pretreatment	1st pass	2nd pass
Hanriyah.Persian gulf 91,000 m^3/day Global conversion: 43.5% Permeate. Bo< 1 mgl TDS< 450mg/L Open intake	2 stages: 1. flocculation, DAF 2. UF	40% recovery (1st pass), 7 elements PV No of trains:8 Vessels per train: 110 Model of membrane: SWC5 Flux: 16,6 lmh	Not applicable
Jorf Lasfar, Morocco 90,000 m^3/day Global conversion: 43,5% Permeate. Bo< 1 Chlorides < 200 mg/L	2 stages: 1. flocculation, DAF 2. UF	45% recovery (1st pass), 7 elements PV No of trains:6 Vessels per train: 146 Model of membrane: SW30HRLE-440 Split variable 65%	87% recovery 7 elements PV No of trains:6 Number of stages: 2 Vessels per train: 16/5 Model of membrane: FILMTEC: BWHRLE-440
Valdelentisco, Spain commissioned in 2006, 70,000 m^3/day, Open intake TDS: 600 mg/L Two stages	pressurized DMF+DMF	50% recovery 7 elements PV No of trains:6 Number of stages: 2 Vessels per train: 85/57 Model of membrane: SW30HRLE-400i	Not applicable
Alicante II, Spain 65,000 m^3/day, recovery: 43%	mixed open and well intake (Neodren), each providing 50% of feed DMF, CF; SDI 2.2, NTU 0.01, TOC 1 mg/L	45% recovery 7 elements PV No of trains:7 Number of stages: 1 Vessels per train: 122 Model of membrane: SW30HRLE-400i (6)+ SWxLE-400i (1)	Not applicable
Escombreras 63,000 m^3/day Open Intake	4 stages: 1. flocculation, DAF 2. Activated carbon filters 3. Ring filters 4. UF	45% recovery 8 elements PV No of trains:7 Number of stages: 1 Vessels per train: 96 Model of membrane: SWC4+(3)+SWC5(5)	Not applicable

Plant characteristics	Pretreatment	1st pass	2nd pass
Marina Baja, Spain 50,000 m^3/day Open intake Two pass Permeate boron< 0,5 ppm	open intake, Two stages 1.gravity DMF, 2.pressurized DMF	45% recovery 7 elements PV No of trains:3 Number of stages: 1 Vessels per train: 188 Model of membrane: SW30XHR-440i (3)+ SWHRLE-440i (4)	90% recovery Split variable: 25-50% 7 elements PV No of trains:2 Number of stages: 2 Vessels per train: 37/13 Model of membrane: FILMTEC: BWHRLE-440

10

PRESSURE VESSELS AND RACKS.

10.1 Pressure Vessels.

10.1.1 Characteristics.

Pressure vessels are available in different diameters, lengths, and pressure ratings.

- 4" of diameter from 1 to 6 membranes.
- 8" of diameter from 1 to 8 membranes.
- 16" diameter.

Seven (7) membranes per vessel is the common practice.

When selecting a pressure vessel, the chosen pressure rating must be high enough to allow for a pressure increase to compensate for irreversible fouling (typically 10% more than needed in a three year design).

The design pressures of vessels are 300, 450, 600,1000, 1200 and 140 psig. Design pressure must be analyzed carefully, especially vessels installed in the second stage of seawater reverse osmosis installations. Permeate outlet is always centered on both sides.

Brine ports can be either end or side type. Multiport and side port types are becoming the preferred alternative.

Mounting design must allow for the vessels expansion, both axially and radially. Although the expansion under pressure is slight, undue restriction can result in damage to the vessel and to the system components. Typically an eight-element vessel, for example, will expand to approximately 0.20 inch (5.08 mm) in length and 0.015 inch (0.4 mm) in diameter. The following suggestions will help to ensure that the vessel is allowed to expand and will ease in servicing:

1. Mount the vessel on the urethane support pads furnished. Do not mount directly to any rigid structure.
2. Use the stainless steel straps furnished. Straps should be tightened sufficiently to hold the vessel on the urethane support pads, but not so tightly so as to restrict expansion. (A torque of 25-50 lbs./in. is sufficient.). Two (2) or three (3) supports should be used. For vessels with more than four (4) membranes, three (3) supports will be the best option.

Protection grating will be installed (approximately 2 meters from the ground floor) in the front and back side of the rack for safety reasons (feed pressure is about 55-70 kg/cm^2).

Figure 10.1. Protection grating.

1. **Manufacturers.**

The most important manufacturers are:

- Codeline-Pentair.
- Bekaert-Protec.
- Bel composite .

2. **Pressure vessel elements**

The main parts of a pressure vessel are:

- **Vessel, supports and straps.**
- **Endcaps.**
- **Head interconnectors.** The first and last elements in the pressure vessels are connected through adaptors to the pressure vessel outlets.
- **Membrane interconnectors.** The permeate tubes of each element are connected to adjacent elements through permeate interconnectors, forming a common permeate tube.

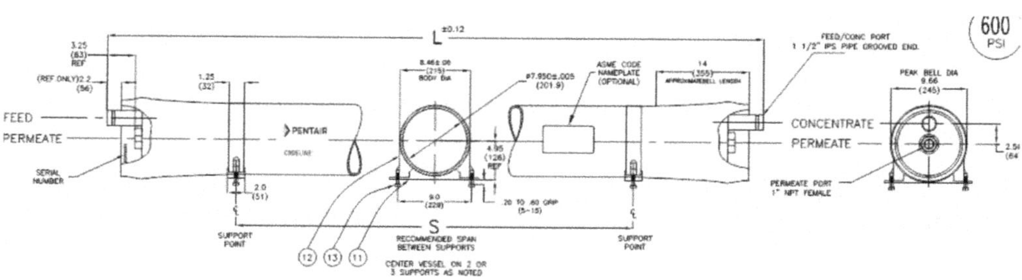

Figure 10.2. Cross section of a pressure vessel. Coursey of Bekaert

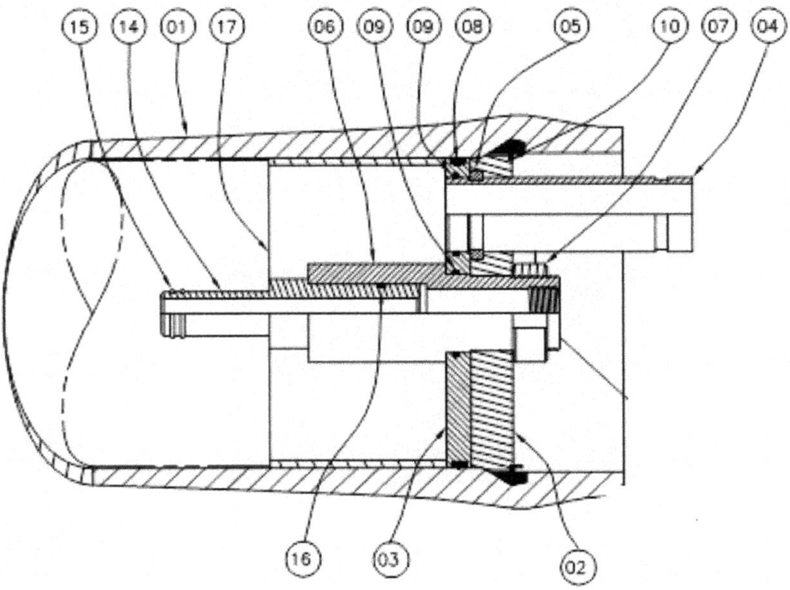

Figure 10.3. Detail of pressure vessel head. Courtesy of Bekaert.

Parts : 1.- Shell (GRP); 2.- Bearing plate (Aluminum alloy hard anodized) 3.- Sealing plate (Plastic) 4.- Feed / concentrate port (Stainless steel) 6. - Permeate port (Plastic); 14.- Adapter (Plastic)

3. **Materials:**

The materials used for the pressure vessels are:

- **Shell** Filament Would, Epoxy FRP. pH between 2 to 12
- **End Plugs/Sealing Plates:** PVC Thermoplastic
- **Retaining Rings**: Stainless Steel
- **Membrane interconnectors:** plastic for a pH between 12 and 2

10.2 Type of vessels and rack configurations.

In general, pressure vessels can have an end port, side port, or multiple port connection system. However, a multiport connection system has become the standard for large scale desalination plants. In the following images, end port and side port configurations are shown.

Figure 10.4. End port connection photograph.

Figure 10.5. a multiport connected system with seven (7) membranes in line.

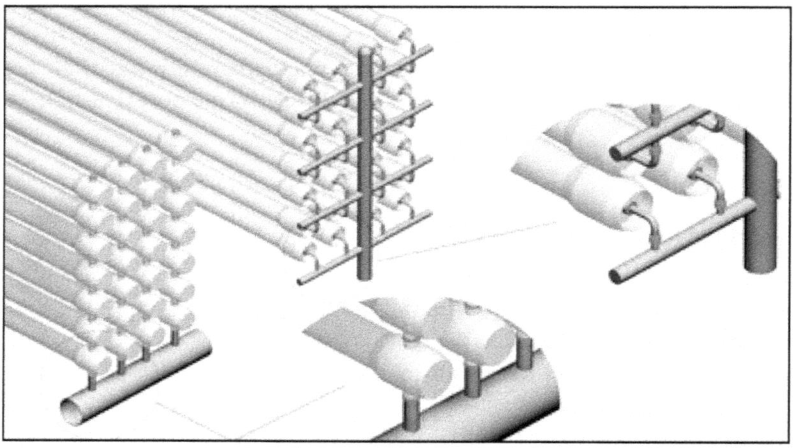

Figure 10.5. Comparison of end port and lateral multiport connection.

There are well known advantages to the multiport design that can be summarized in the following terms:

- Reduction in number of pipes and accessories needed to make the assembly. Interconnecting multiple pressure vessels eliminate the need for more costly headers, manifolds, accessories, and additional piping.
- Reduction in number of weldings. There are two advantage of this reduction: technical and economicar. Welding is performed with special stainless steel and by highly qualified approved welders. Consequently, such welders and the work they perform can be a cosy increase to any project. Interconnecting pressure vessels eliminate the need for additional pipe costs and welding costs.
- Reduction of the physical space needed for installation assembly.

The disadvantages of "multiport" connections are that the pressure vessels are more expensive and the installation tolerances are lower.

There are several different guidelines that must be followed in order to have a good long term system design. As water passes through each vessel, a pressure drop occurs. This will result in the last vessel in a connected pass receiving less pressure than in first. By flow balancing, the concentrate pressure of the last vessel will also be the lowest of any vessel. This tends to keep the pressure drop across all vessels as close as possible.

Increasing port size reduces this effect. The maximum number of vessels connected for each file will depend on port diameter:

- 3 vessels for ports of 2".
- 4 vessels for ports of 2 ½".
- 7 vessels for ports of 3".

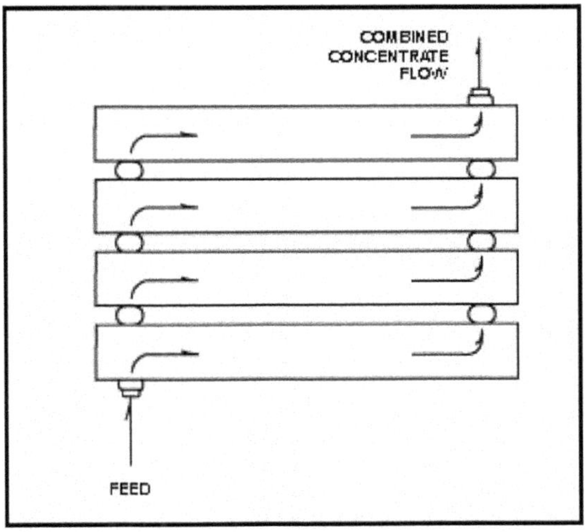

Figure 10.6. Drawing of three vessels connected. Multiport configuration.

Consider feeding them from both sides or from the center of a pass if the differential pressure when feeding from one side would be excessive. By splitting the feed flow the velocity will be reduced by one half and the pressure drop by an even greater amount since the pressure drop is proportional to the square of the flow. Feeding from both sides may be the most economically feasible when the pressure is low enough to use plastic.

Different multiport configurations are possible, including vertical or horizontal feed with a different number of vessels interconnected.

Below, the front and rear views of different alternatives.

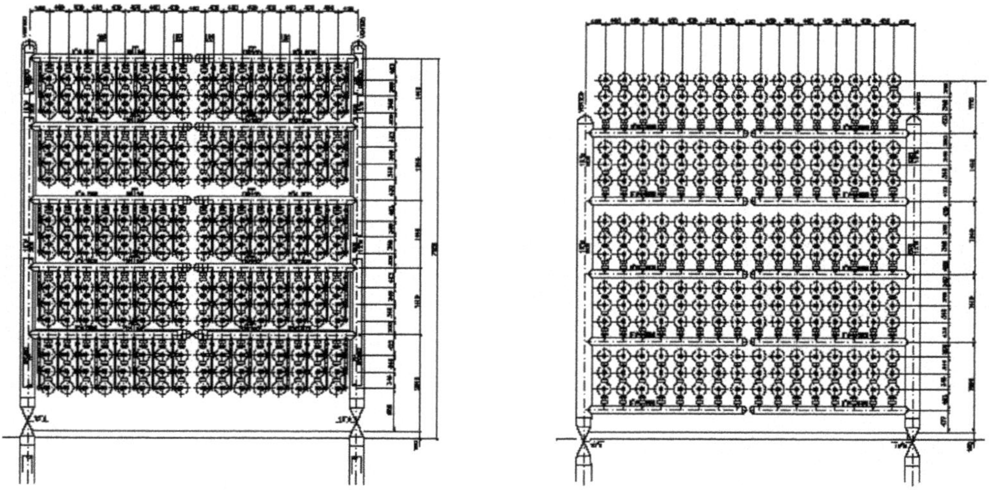

Figure 10.7. Lateral and vertical feed with (3) vessels interconnected

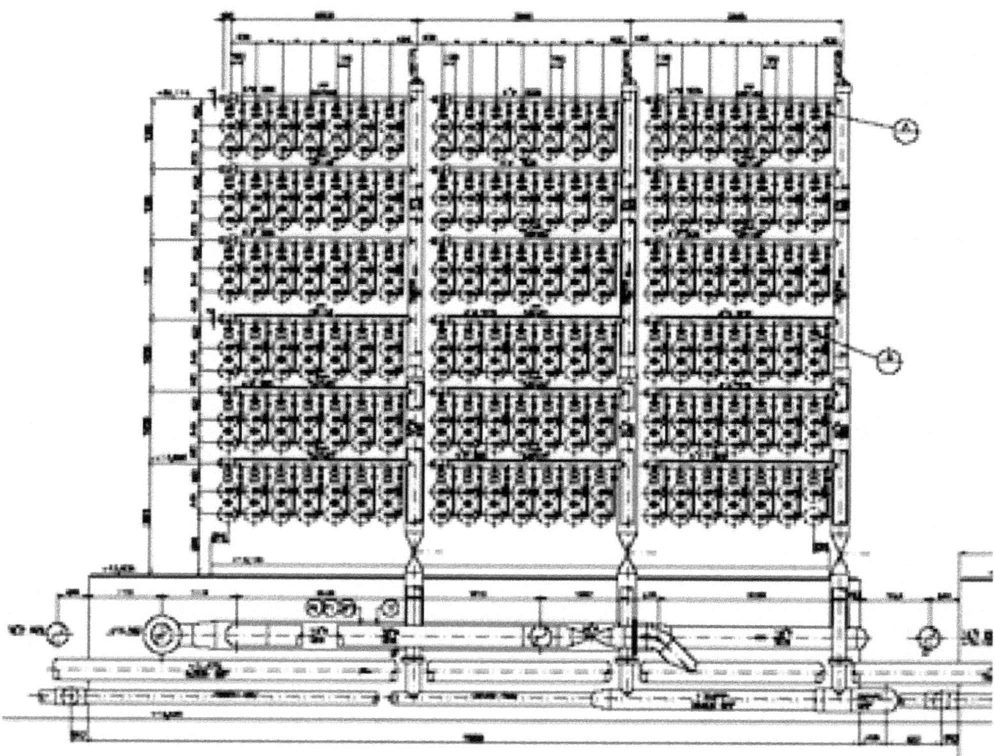

Figure 10.8. Central & vertical feed with (3) vessels interconnected.

Figure 10.9. Lateral vertical feed (3) and (4) vessels

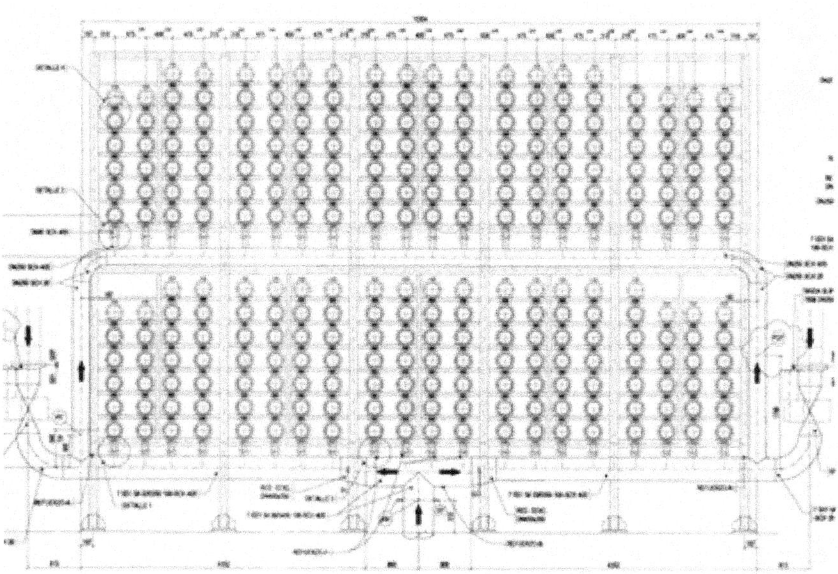

Figure 10.10. Horizontal feed (7) vessels interconnected.

10.3 Vessels with 7 or 8 membranes.

The installation of eight (8) membranes per vessel is not common practice. Some examples are Tampa, Lamarca and Ashkelton desalination plants. The advantages and disadvantages of using eight membranes instead of seven membranes are:

1. Lower investment.

There is capital cost saving of 7-8% due to fewer racks and vessels required.

2. Higher recovery.

A higher recovery rate with the same concentration profile can be obtained with longer configurations (approximately 2 points more, depending on the salinity of water).

Figure 10.11 shows that the recovery rate for the front elements is higher for the shorter configurations, while it decreases more steeply along the pressure vessel. A more homogeneous distribution is observed in the longer configurations.

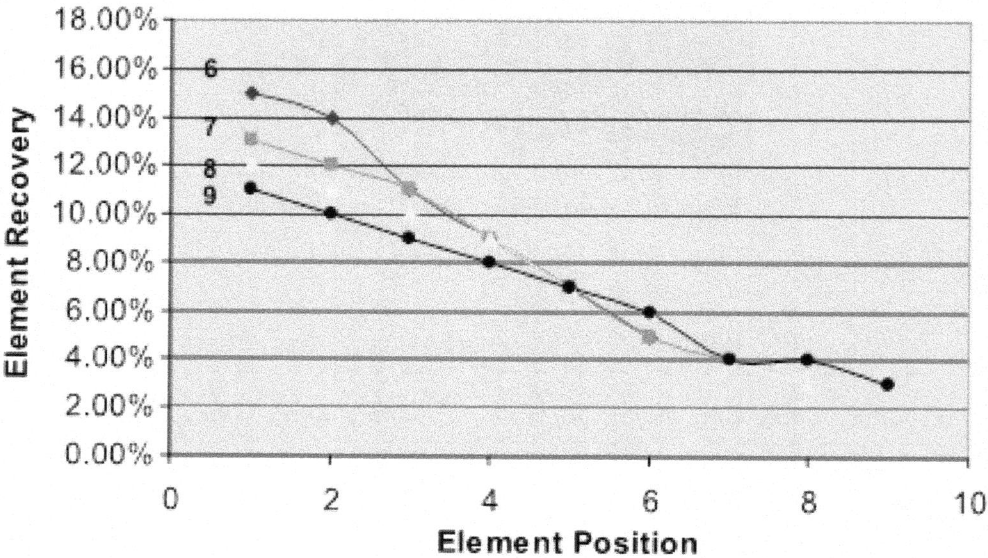

Figure 10.11. Element recovery versus element position.

3. Poorer washing.

The pressure drop can significantly increase due to fouling during normal operation. During cleaning, the feed/brine flow rate required for cleaning is significantly higher than that of normal operation resulting in an additional increase of pressure drop. The longer the vessel the higher the pressure drop during cleaning conditions.

4. Power consumption.

The longer the pressure vessel the higher the pressure drop through it, so energy consumption is slightly higher with 8 membranes per vessel.

A high differential pressure could cause a high force in flow direction on the feed side of the element. When the limit is exceeded, the element may become telescoped and damage.

10.4 Vessel installation.

Vessels are grouped in racks which are subdivided into sub-racks. A rack is a metallic or plastic structure which supports pressure vessels. A Sub-rack is the group of vessels which are washed at the same time. Sub-racks must be independent of each other with manual valves on the feed and concentrate side.

Various methods of connecting side port vessel to manifolds are possible. The recommendation method is to connect each vessel side port to the manifold using two flexible Victaulic and an intermediate piping section. A Two joint method is preferred over a single connection because it does not require as much care in vessel alignment and manifold welding accuracy. Figure 10.12. shows the different alternatives

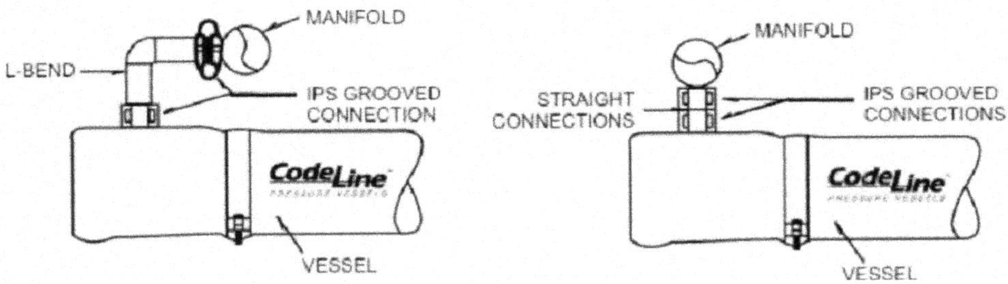

Figure 10.12. Methods of connecting side port vessel to manifold.

If one single victaulic joint is used, the manifold should be manufactured using close tolerances to help ensure a correct alignment. Installation should follow a sequence of assembly steps that ensure correct shell to manifold alignment. Firstly, the vessel should be set into the rack and secured loosely into position for the best alignment possible. At this time set the side port clearance approximately 3,1 mm. After proper alignment is achieved, secure the shell and fix the manifolds into position.

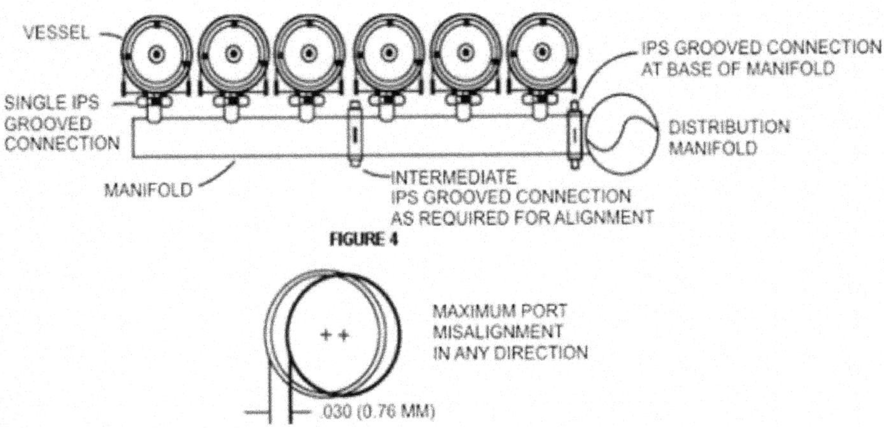

Figure 10.13. Detail of vessels connections to manifold. Courtesy of BEKAERT.

As indicated in previous section there are different alternatives to connect pressure vessels. However, configurations with vertical distribution manifolds and horizontal small manifolds are usually the most economical solutions.

Table 10.1. give an estimation of rack dimensions in function of number of vessels.

Nº vessels	flow m³/h	Gpm	velocity ft/s	Velocity ft/s	Rack size	velocity ft/s
125	937,5	4125	11,7	11,7	8x5x3+1x8	10,5
120	900	3960	11,2	11,2	8x5x3	10,1
110	825	3630	10,3	10,3	8x5x3	9,3
100	750	3300	13,5	9,4	8x5x3	8,4
90	675	2970	12,1	11,2	8x4x3	9,5
80	600	2640	10,8	10,0	6x5x3	6,7
70	525	2310	9,4	8,7	6x4x3	7,4
80	560	2464	10,1	9,0	6x5x3	6,3

10.5 Rack sizes

A Rack is defined by the permeate production. Standard permeate productions are 7500,10000,15000, 20000, 25000 and 30000 m^3/day

The number of vessels will depend on recovery, number of membranes per vessel and design flux rate.

In large desalination plants, racks are subdivided by partial-skids to reduce the size of cleaning equipment. The number of vessels per skid can range between 80 to 120 vessels. It is necessary to take into consideration that subdividing the racks for cleaning will increase cleaning periods and the number of isolation valves.

Washing pump flows usually range from 500-600 m^3/h working one or two pumps at the same type, and the required washing flow per vessel is about 9 m^3/h, so for racks with more than 110-130 vessels it will be necessary to subdivide the racks into more than one part. When the rack is divided in three parts, the complexity of the isolation valve arrangement also increases.

Table 10.2 shows approx. number of vessels per racks and sub-racks in function of rack permeate production.

Table 10.2. Production versus no of vessels

Production	no vessels/ 7 membranes	no vessels/ 8 membranes
7500 m³/day	94	82
10,000 m³/day	125	110
15,000 m³/day	2x94	2x82
20,000 m³/day	2x125	2x110
25,000 m³/day	3x102	3x92

Cost analysis summary for optimum size definition of RO banks is shown in figure 10.14. It can be concluded that 90 to 120 vessels per rack is an optimum number.

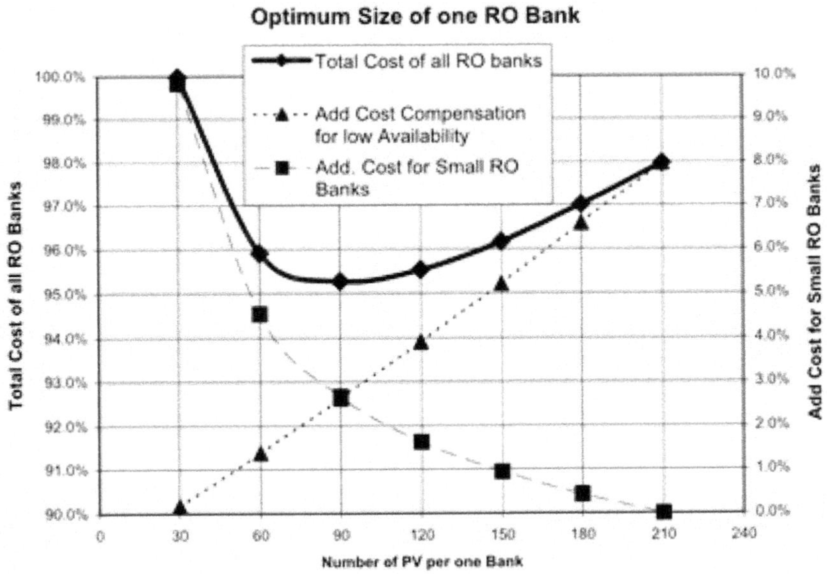

Figure 10.14. Number of PVC per one rack versus Total cost of RO bank.

10.6 Materials for construction.

Racks are workshop manufactured in carbon Steel+galvanized+painting. They are installed on site.

EPC firms have started to use GRP material at competitive prices.

Prefabrication of sub racks at a factory or other manufacturing site, then transporting the sub-assemblies to the construction site where complete assembly of the frame is carried out should be common practice in order to minimize costs and installation period.

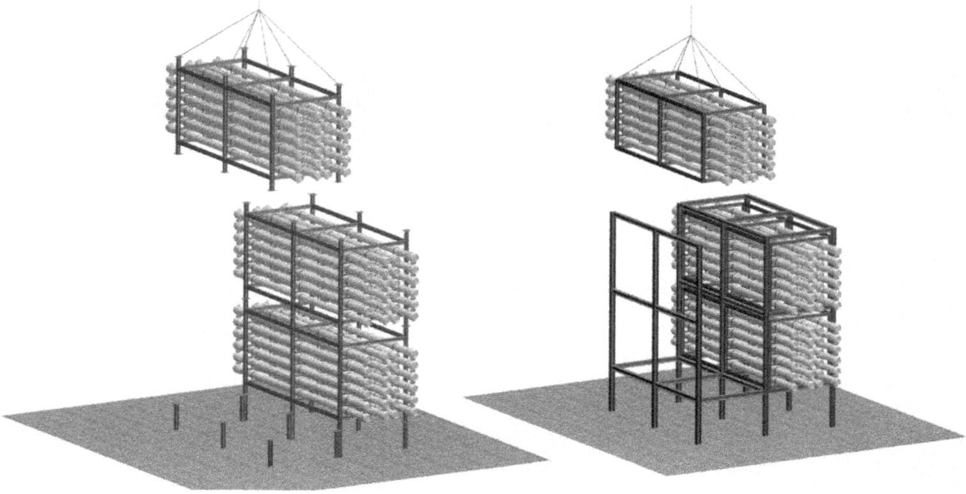

Figure 10.15. Racks installation.

11 REVERSE OSMOSIS PUMPS.

Two main services are typical for pumps in a seawater desalination plant.

1. High pressure pumps.

Before entering the seawater reverse osmosis membranes, clarified seawater is pressurized by a High Pressure Pump typically between 55 and 85 bars, depending on the temperature and the salinity of the water. Suction pressure is low (requires more than 2 bars). Consequently, they are high flow and high pressure pumps.

2. Booster pumps.

Booster pumps are used to feed reverse osmosis skids from recovery systems or to raise the pressure between first pass and second pass of a two pass RO system. Flowserve high-pressure booster pumps are designed to operate efficiently under SWRO system pressure where suction conditions can exceed 60 bar (870 psi). Differential pressure will be low (between 5-25 bars).

11.1 Types of pumps.

For each of the above indicated services, two different type of pumps are offered in the market by the pump manufacturers.

1. High pressure pumps.

- Multistage axially split casing pumps.
- Ring section high pressure pumps.

2. Booster pumps.

- Horizontal split-case pumps.
- Vertical Split-case pumps (API 610).

11.2 High pressure pumps.

11.2.1 Multistage axially split casing pumps.

Thrust loads in horizontally split case pumps can be minimized by orienting half of the impeller in one direction and the other half on the other direction. The flow passes through half of the stages (or nearly half in the case of odd numbers of stages), then crosses over to the opposite of the impeller and goes through the remaining stages. The net effect, if there is an even number of stages, and if all the impellers are trimmed to the same diameter, is that thrust loads are balanced.

Figure 11.1. Photography of multistage Split casing pump

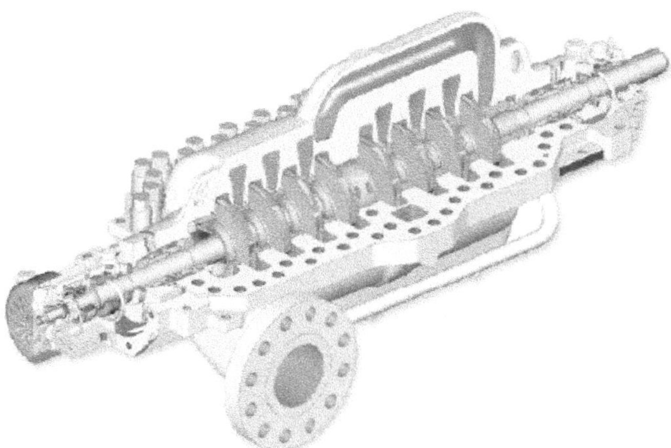

Figure 11.2. Cross section of multistage axially split case pump. Courtesy of Flowserve.

Double suction, first-stage available on most sizes for reduced Net Positive Suction Head (NPSH).

11.2.2 Ring section high pressure pumps.

More compact than multistage axially split casing pumps. The span between bearings is shorter, increasing stiffness. Minimum shaft deflection allows smaller clearance gaps for high efficiencies.

Figure 11.3. Photography of ring section high pressure pumps. Courtesy of SETA.

11.2.3 Main pumps manufacturers.

The most important seawater desalination high pressure pumps manufacturers are SULZER, FLOWSERVE, KSB, CLYDE and TORISHIMA.

11.2.4 Differences between multistage axially split case and ring section pumps.

The advantages of multistage axially Split-case pumps compared to ring section pumps are:

- The casing of an axially split pump is split along its center line. The top half can be removed and the rotor extracted as a single unit for rapid maintenance. Therefore, the maintenance of the pumps is less complex.
- Opposed impellers for axial hydraulic balance of the rotor and center balance bushing to provide rotor-dynamic stability and a high degree of dampening at critical speeds, even when the parts are worn.
- Double volute for radial balance of the rotor, which delivers both high and sustained efficiency.
- Flow capacity of multistage axially split case pumps is higher.
- A ring section pump is not hydraulically balanced. The energy is transferred to bearings and shafts are shorter.

The main disadvantage of split case pumps compared to ring section pumps is the cost.

11.3 Materials.

Super-duplex stainless steels are normally used for high pressure and booster pumps.

The most used super-duplex cast iron steels are:

- A890 Gr 5A.
- IR-885.
- Feralium 255-3SC. PREn>40.
- ZERON 100.
- 1.4469.

If duplex material is used instead of super duplex, cathodic protection should be installed as an additional corrosion protection.

Please, consider that 1.4517, 1.4593 and 1.4468 are duplex materials.

Materials selected by the different manufacturers for high pressure pumps are the following ones:

SULZER

- Casing.................................. A 890 Gr 5A
- Impeller .... A 890 Gr 5A
- Shaft .... A 276 S32760
- Shaft sleeve.......................... A 276 S32760
- Balancing device................. PEEK / A 276 S32760
- Bearing housing .... A 216 Gr WCA
- Wear ring............................. PEEK

FLOWSERVE

- Casing.................................. IR-885
- Impeller IR-885
- Shaft .... NITRONIC 50
- Shaft sleeve.......................... IR-885 /STELLITE
- Balancing device................. PEEK / A 276 S32760
- Bearing housing .... CARBON STEEL

11.4 Pump sizings.

The most important parameters that define a high pressure pump in a seawater desalination plant are:

- Flow.
- Specific speed.
- Suction pressure.
- Differential pressure.
- Efficiency.
- Salinity.
- Temperature of design.
- Velocity.

- Accessories.
- Tolerance.
- Materials.

Some important aspects to be taken into consideration are:

11.4.1 Flow.

It is a common practice to oversize the pump with a 2% excess over the theoretical water production. A minimum and a maximum operation range must be specified.

Once duty point is fixed, the pump and impeller will be selected. It is good practice to select (when possible) a pump with an impeller that can be increased in size, permitting a future increase in head and capacity. Or alternatively, an impeller, which can be reduced in size. As a guide, select a pump with an impeller size no greater than between *1/3* and *2/3* of the impeller range for the casing with an operating point in the high efficiency area. It is also important not to go too far right or left from the B.E.P. A guideline is to locate the operating point between 110% and 80% of the B.E.P. flow rate with an operating point within the desirable impeller selection area (see Figure 11.4).

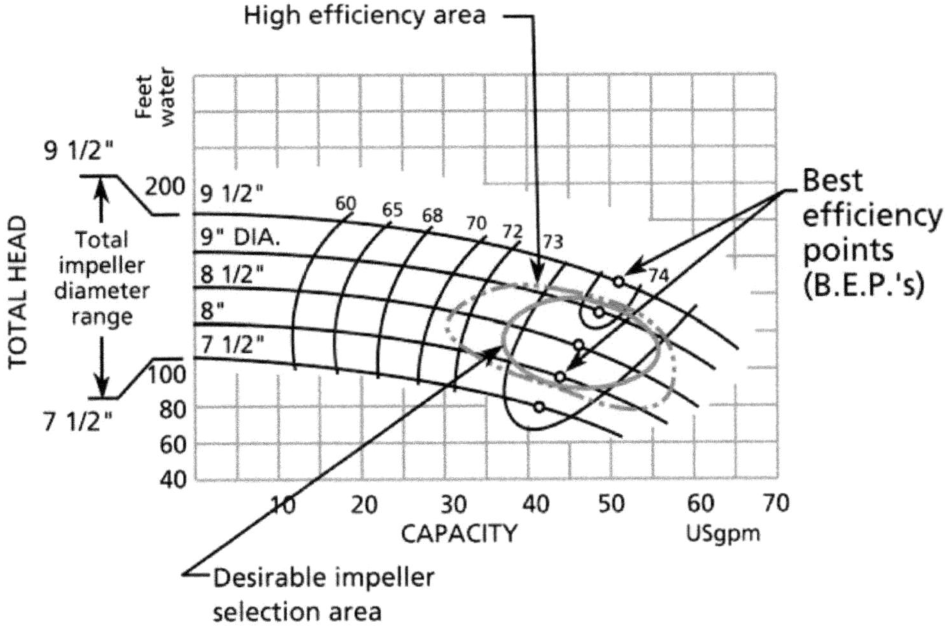

Figure 11.4. Operation point.

11.4.2 Specific speed.

Specific speed is an index number correlating to pump flow, head and the speed at the optimum efficiency point. It classifies pump impellers with respect to their geometric similarity. Two impellers are geometrically similar when the ratios of corresponding dimensions are the same for both impellers. Usually, high head impellers have low specific speeds and low head impellers have high specific speeds.

There is often an advantage in using pumps with high specific speeds. For a given set of conditions, operating speed is higher. As a result the selected pump can generally be smaller and less expensive. However, there is also a trade-off since pumps operating at higher speeds will wear out faster.

Optimum efficiency of the pump is directly related to the specific speed.

The maximum efficiency of radial flux pump is obtained for specific speeds of 34.

The flow of high and low pressure pumps installed in large scale seawater desalination plants range from 500 to 2500 m^3/hours per rack (0.14 to 0.69 m^3/sec) depending on the size of the plant, and rack and if they operate in independent lines or with a common pressure center.

The optimum efficiency at this specific speed ranges from 82% (0.14 m^3/sec) to 88% (0.69 m^3/sec).

11.4.3 Suction pressure.

Maximum and minimum suction pressure must be fixed. This point is important from an energy point of view. The usual design value of suction pressure is 2-3 kg/cm^2.

11.4.4 Differential pressure.

The pump head is measured in pressure difference between the discharge and suction sides of the pump. Discharge pressure is equal to the osmotic pressure (minimum temperature and fouling factor) plus the hydraulic head loss across the membranes.

11.4.5 Efficiency.

The minimum required efficiency must be indicated by the pump manufacturer. It will range from 85% to 87%. Depending on salinity, washing frequency, membranes characteristics, filters etc, in countries where frequency is 50 Hz, the high pressure pumps produce optimum flow, the number of stages and diameter of discharge flanges would be :

Flow	Efficiency	No of stages	Flange diam.
500-550 m^3/h	82 % at BAP	6 Stages	6”
650-750 m^3/h	85 % at BAP	5 Stages	8”
950-1050 m^3/h	86 % at BAP	4 Stages	10”
1200-1300 m^3/h	86.5 % at BAP	4 Stages	12”
1600-1700 m^3/h	87 % at BAP	2 OR 3 Stages	12”
2200-2400 m^3/h	88 % at BAP	2 Stages	14”

11.4.6 Salinity.

Salinity of feed water is a critical factor for a pump manufacturer, as it will determine the selection of materials.

11.4.7 Temperature.

Design temperature is a critical factor to evaluate if cooling of bearings are required.

11.4.8 Velocity.

Pumps can work at 3000 or 3600 rpm.

11.4.9 NPSH.

The required net positive suction head (NPSHr) is the amount of energy required to prevent the formation of vapor-filled cavities of fluid within the eye of an impeller. The formation and subsequent collapse of these vapor-filled cavities is called cavitation and is destructive to the impeller.

Cavitation can produce the following effects

- Loss of head.
- Loss of efficiency.
- Noise.
- Vibration.
- Complete collapse at full cavitation.

11.4.10 Accessories.

- **Mechanical seals.** Must be manufactured in duplex, cartridge type. Split case pumps must have two seals.
- **PT-100** (thermoresitances) will be furnished by measure bearing temperatures. High pressure pumps must have three PT-100.
- **Vibration sensors.** Bearing housing vibration can be monitored on pumps furnished with ball bearings. Pumps supplied with sleeve bearings can be furnished with non-contacting vibration probes to measure actual rotor vibration. It is not very common.
- **Coupling protection**: Is required to specify if material is stainless steel or not. Manufacturer usually does not include piping, drainage and vent valves.

11.4.11 Tolerances.

The actual performance of a new centrifugal pump is usually close to its published performance curve. However, testing such pumps requires precision instrumentation and exacting procedures detailed in the Hydraulic Institute Standard ANSI/HI 1.6 Centrifugal Pump Tests, which are usually difficult to achieve in the field. For this reason, the manufacturers guarantees of performance are usually predicated on the acceptance of a factory test. The above standard also includes performance Tolerances which must be specified.

- Flow: ± 2%.
- Pressure: ±2%.
- Efficiency: 0 % minimum.
- Power: ±1%.

Other standards are : DIN-1944, ISO-9906, API-610. API 610 accepts 0.5% of deviation in the efficiency and +4% in power.

11.4.12 High pressure pump design. Optimum design combined with low pressure piping.

Once a recovery system has been selected (isobaric chamber or pelton turbine) and has been chosen, the process of optimum selection must follow the steps indicated below:

1) Selection of number of pumps depending on:

- Flexibility required for SWRO.
- Selected configuration (separate lines or common pressure center).

- Racks backwashing philosophy.
- Availability of frequency converters for low pressure pumps.

2) For the selected flow, with a Nss =9500, it must be calculated at NPSHR 3%. NPSHR 40000 h will tell us about the suction pressure of the pump in permanent operation.
3) NSPHR 40000 h plus head loss in piping and filters will be the minimum static head of low pressure pumps; adding the osmotic pressure variation between low and high temperature, low and high salinity and new and old membranes so the maximum feed pressure can be obtained.
4) Once the operating range of the low pressure pump is selected, the rest of the operating pressure to complete the required pressure must be provided by the high pressure pump.
5) For a specific velocity (nq=34) with the selected number of stages, the differential pressure the pump can provide will be obtained for the given flow and total pressure.

11.5 Required instrumentation.

Minimum instrumentation to be installed in high pressure and booster pumps is:

- Suction .... Manometer and pressure switch. If pressure is lower than 1.5-2 kg/cm², the pump will stop to prevent cavitation.
- Suction .... Flow transmitter.
- Discharge .... Manometer and pressure transmitter.

A pressure transmitter could be installed on the suction side of HPP to regulate seawater pumps. On the suction side of booster pumps, the pressure is also controlled. Therefore, in both cases, power consumption can be monitored based on flow and pressure measurements.

11.6 Foundations.

A good foundation will be strong enough to provide stability and give rigid support. It will be able to absorb the normal strains and shocks that may be transferred to it while operating in service.

The foundation must also minimize the vibration that the pump and driver create, and from external sources, such as piping, building structures or other operating equipment. The vibrations can produce forces which can be 4-5 times the weight of the turbo group. VDI norms will be used for vibration analysis, which will be measured in mm²/dseg. Maximum value of vibrations will be 7 mm²/second, but normal values will round between 2-3 mm²/second.

General industry practice provides a foundation that is 3 to 6 inches (7.6 cm to 15.2 cm) longer and wider that the baseplate that will be installed. Practice in the industry provides a design that insures the foundation is wide enough to provide a rigid support. This practice draws imaginary lines downward 30° on either side of the vertical of the pump, through the shaft centerline.

Hardware used must be stainless steel to prevent corrosion.

11.7 Motors.

The main manufacturers of high pressure pumps motors are:

- SIEMENS. Used by IDP.
- ABB. Used by Sulzer.

Premium efficiency motors are preferred.

11.8 Speed converters.

The cost of equipment limits the use of frequency converters (VFD) in high pressure pumps However, important advantages have made it more usual practical for the installation of VFD.

- Power consumption saving.
- There is the possibility of installing a large range of membranes, both currently from different suppliers and of future generations, increasing the rack production in specific cases by the increase of the specific flow per membrane in the event of increased demand, etc.
- Limited motor start-up current (gentle start-up), is a factor which is highly appreciated by the electricity company and in many cases is demanded.
- Imperceptible voltage drops, whereby the effect on the other plant receivers is zero.
- A low level of harmonics to the system.
- Low reactive energy consumption.

It is possible to get some of the advantages above by installing speed converters from the feed pumps to the high pressure pump (HPP). However, when flow /temperature and salinity variations are high, a more detailed technical analysis will be recommended and the alternative of installing a speed converter in the HPP should be considered to optimize power consumption.

A speed converter is always installed in booster pumps, but special care to the variation range must be taken when the booster pump feeds two sub-racks .

It is important to inform the pump supplier of maximum and minimum pressure and flow when a FVD is going to be installed.

11.9 Configuration. Pressure center design.

Conventional SWRO plants design contains independent modular units, where each unit is the combination of the three main reverse osmosis components: high pressure pump, reverse osmosis membrane rack and the energy recovery system. Each of these modules runs independently. However, pressure center design with larger high pressure pumps forming a common pumping center, which supplies seawater common feed lines to all RO racks and energy recovery devices forming a common energy recovery center, which collects pressurized brine from all RO racks is becoming a common practice for large scale plants due to these important advantages:

- Installing larger capacity equipment, such as larger membrane racks, multi energy recovery skids and higher flow rate high pressure pumps (higher flow rate results in higher pump efficiency), the specific energy cost per cubic meter of water produced decreases.
- Stopping one unit, for example, one high pressure pump, does not affect the other components which can continue to operate.
- The pressure center concept reduces equipment and instrument wear, shortens startup time, minimizes array shutdown and shortens the time required to stabilize the process and quality of water after production changes.

The main disadvantages are:

- Increase of motor power and impact on frequency converters at startup.
- Forced lubrication.
- The problem of equal distribution of feed flow and proportional collecting of brine. The feed flow coming from the energy recovery systems is a little saltier than the feed coming from the high-pressure pumps. Therefore the RO banks located near a DWEER/ERI center are exposed to higher salinity. On the other hand, the drop pressure in distribution and collection lines causes an unevenness in flow sharing between RO banks. In each of these cases a combination of them has to be treated differently.

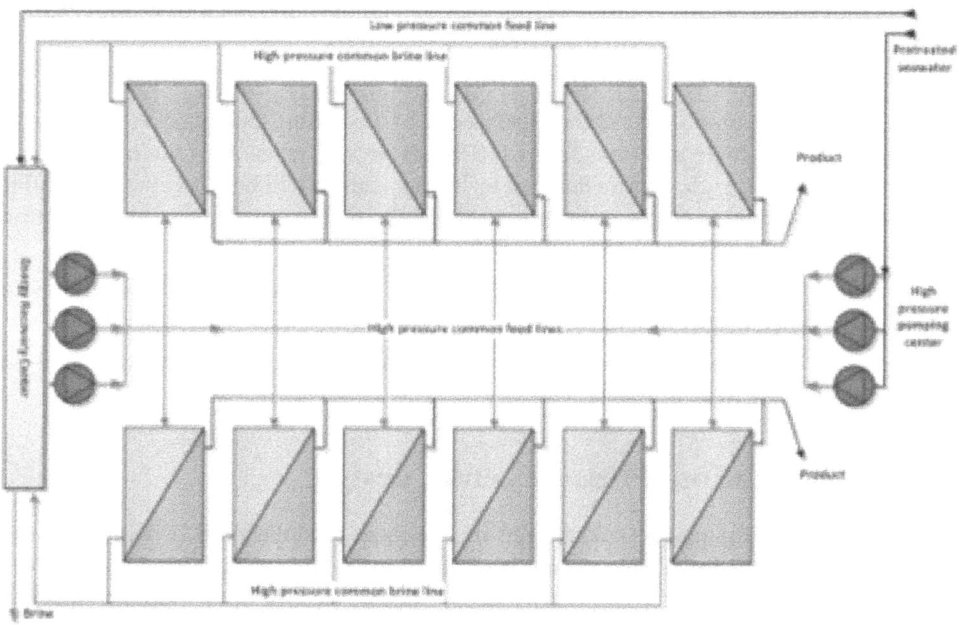

Figure 11.6. Three center design.

A summary of stages and efficiency of high pressure pumps comparing independent of center concept philosophy are shown in table below.

Plant production (m^3/day)	Number of lines/ flow per line (m^3/day)	Concept	Number of HP pumps	HP flow (m^3/h)	HP efficiency
60,000	6/10000	Independent	6	500	82%
60,000	4/15000	Independent	4	750	85%
60,000	--	Pressure center	3	1000	86%
100,000	8/12500	Independent	8	625	85%
100,000	5/20000	Independent	5	1000	86%
100,000	--	Pressure center	2	2500	88%

12 RECOVERY SYSTEMS.

Brine pressure is higher than 60 bars at membranes discharge. This energy is recovered in recovery systems.

In reverse osmosis, (3) three type of recovery systems are used:

1. **Pelton Turbine.**
2. **Isobaric chambers.**
3. **Turbocharger.**

12.1 Pelton turbine.

Transforms the potential energy of brine into kinetic energy through nozzles. Nozzles direct forceful streams of water against a series of spoon-shaped buckets mounted around the edge of a wheel. As water flows into the bucket, the direction of the water velocity changes to follow the contour of the bucket. When the water-jet contacts the bucket, the water exerts pressure and the water is decelerated as it does a "u-turn" and flows out the other side of the bucket at low velocity. In the process, the water's momentum is transferred to the turbine. This concept allows a smaller, less costly motor to be utilized and saves a very considerable proportion of the power and cost necessary to drive the high pressure feed pump.

Pelton turbine has one (1) or two (2) wheels and each nozzle will have installed (1) or (2) nozzles.

Figure 12.1. Pelton turbine

12.1.1 Recovered power (P_{rt})

The power recovered in a turbine can be obtained from the following equation:

$$P_{rt} = Q \times H \times d \times \frac{r}{(270 \times 1.36)}\ kw$$

Q = flow m^3/h

H = height in mcw (meters of column of water)

d = density 1.04 kg/l

r = efficiency 86-88%

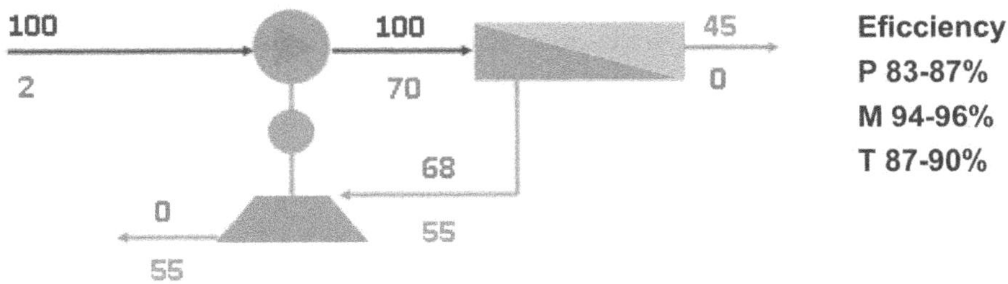

Figure 12.2. Pressure & flow in a pelton recovery system.

12.1.2 Typical plant example.

Train capacity:	10000 m^3/day.
Product flow:	417 m^3/h.
Recovery:	50%
Membrane configuration:	single stage.
Req. membrane pressure:	68 bars.
HP pump flow:	417/0.5 = 834 m^3/h
Suction pressure:	2 bars.
Discharge pressure:	68 bars.
Pump efficiency:	87%
Energy recovery turbine flow:	417 m^3/h.
Brine pressure:	67 bars.
Turbine efficiency:	90%

Calculation of energy consumption:

Required pump power:	834 m^3/hr x (68-2) bars/0.87/36 = 1757.5 kW
Turbine recovered energy:	417 m^3/hr x 67 bar x 0,9/36 = 698.5 kW
Power absorbed by motor:	1757 kW-695.5 kW = 1095 kW
Specific energy consumption:	1059 kW/0.965/417 m^3/h = 2.63 kWh/m^3

12.1.3 Motor power.

Usually, motor power for the group pump-turbine is calculated by the manufacturer of the pump. However, the basic design criteria is the following:

- Motor must have enough power to start the pump with turbine nozzle valve closed (typically, the consumption is 60-70%) of pump power.
- An additional margin of 15-20% must be added to the power consumption of motor (pump consumption-recovered power in turbine/motor efficiency)

12.1.4 Restrictions

Discharge of brine is by gravity, so the drainage pipe must be located 2 meters above turbine discharge flange. The drain collector must be sized with a minimum slope of 1% and filled at 50%, as the rest will be filled with the air content of brine. The collectors size will depend on the turbine but it will be a minimum of 800 mm.

12.1.5 Turbine wheel.

It is the key element of turbine as it transforms the Kinetic energy on the jet into mechanical energy. A series of spoon-shaped buckets are mounted around the edge of the wheel. Each bucket reverses the flow of water, leaving it with diminished energy and the resulting impulse spins the turbine.

The buckets are mounted in pairs to keep the forces on the wheel balanced as well as to ensure smooth, efficient momentum transfer of the fluid jet to the wheel. The wheel is mounted on the high-pressure shaft, which together with a motor drive pressurizes the SWRO system.

The diameter of Wheel (D_1), usually fixes the flow that can be treated by the turbine, the rotation speed and the bucket capacity (D_1/B_2) which must be higher than a certain value.

Bucket size is very important to optimize turbine efficiency and avoid cavitation

$$D_1 = \frac{K_{u1} \times 84.6 \times (H)^{1/2}}{n}$$

$D_{1:}$ Diameter in mm

H: in m

n: in min^{-1}

K_{u1}: between 0.455 – 0.51

12.1.6 Type of arrangements and installations.

The pump manufacturer also supplies the turbine. The motor will be supplied independently by the motor manufacturer. The complete arrangement will be assembled by the pump manufacturer. All the elements will be mounted on a common baseplate. The coupling with motor will be flexible.

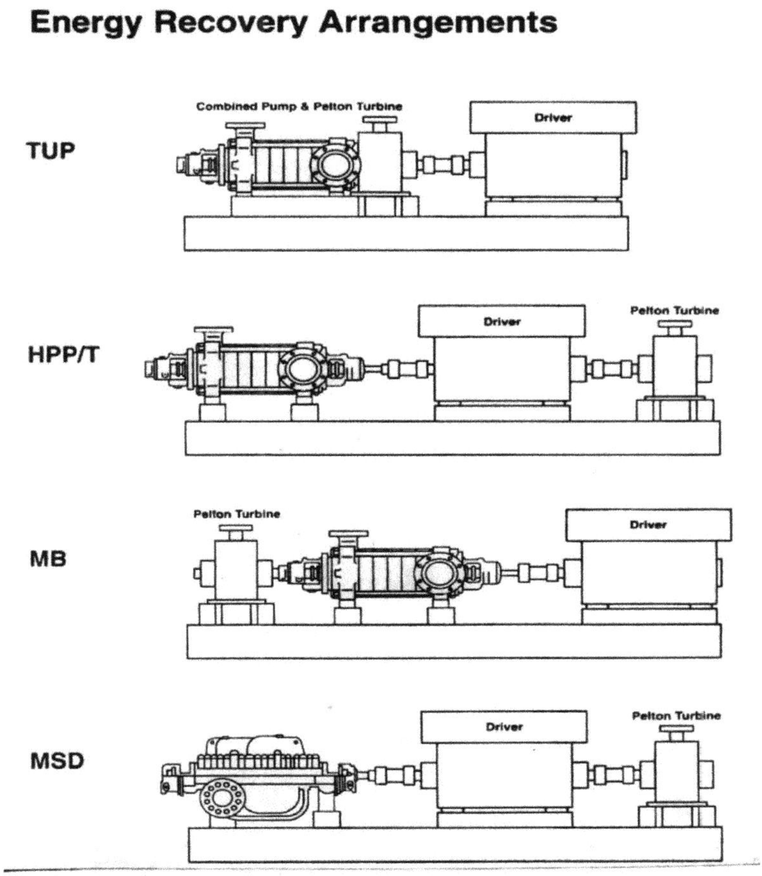

Figure 12.3. Different type of configurations Pelton Pump Motor.

12.1.7 Turbine manufacturers.

There are two (2) manufacturers CALDER and VATECH though CALDER is the largest.

12.1.7.1 Differences between turbines.

There are not important differences between VATECH and CALDER.

12.1.8 Size of turbines manufactured

The size of turbines is directly linked to the rack production

Rack (m^3/day)	Flow (m^3/h)	Type of turbine*
5000	250	RO-350-100-1
7500	330	RO-350-80-2
10000	550	RO-350-100-2
15000	650	RO-350-100-2
20000	1020	RO-350-100-4

*Calder

When turbines work at 60 Hz 3600 rpm, the diameter of wheels will be 310 mm instead of 350 mm.

12.1.9 Materials.

Brine is in contact with turbine materials for a short period of time. For that reason, duplex (not superduplex) will be enough to guarantee there is no corrosion. The materials used are:

- Body..1.4462 or 1.4468
- Wheel..1.4468
- Shaft 1.4462
- Injector 1.4468
- Seals ....Teflon

12.1.10 Sizing.

The most important parameters to be defined are:

1. Maximum and minimum design flow.
2. Design brine pressure.
3. Salinity.
4. Temperature.
5. Velocity.
6. Efficiency.

1. Flow.

Turbine flow is determined by high pressure pump flow minus product flow (Fluctuations can range between +-2%)

It must be specified not only the design, but also by the minimum and maximum flow.

2. Pressure.

The operating pressure of the turbine depends on the membranes age, temperature and membranes fouling. It will be the maximum and minimum discharge pressure of the pump minus piping head loss, and membranes head loss.

It must be specified not only in the design, but also in the minimum and maximum pressure.

3. Salinity and temperature.

Important for the design of buckets in order to prevent cavitation.

4. Efficiency

Ranges from 86% to 88% for the same velocity of 3000 rpm. It is recommended for the turbine to work at a fixed velocity. Velocity variations will affect negatively the efficiency of the system and could potentially produce cavitation, as the lower the velocity, the higher the risk of cavitation.

5. Accessories.

Usually, PT-100 are specified to be installed in turbine bearings.

12.1.11 Cavitation velocity variation,

Turbines, as centrifugal pumps, have cavitation problems.

There is a formula to obtain critical cavitation, $C_{C=}N_S^{(1.6/3400)}$, specific velocity of a turbine is defined by the velocity, wheel diameter and nozzle diameter.

Operation constant (sigma) is obtained from the equation:

$$C_f = \frac{(H_a - H_s)}{H}$$

H_a = Atmospheric pressure.

H_s = Vapor pressure at operation temperature.

H = Liquid height at the turbine inlet.

12.1.12 Instrumentation to be installed in a turbine.

A pressure transmitter is usually installed in the suction side of turbine. The flow is measured by the difference between inlet flow and product flow.

12.1.13 Type of diagrams turbo-pumps and rack.

- <u>One pump for each turbine.</u>

Pump is attached to the rack so that if one of them stops, the other will also stop. It is the most common operation mode. The spare group is attached to each rack with valves and piping. The problem is that It is expensive. If a spare element is not installed, a manual valve will not be required in the pump.

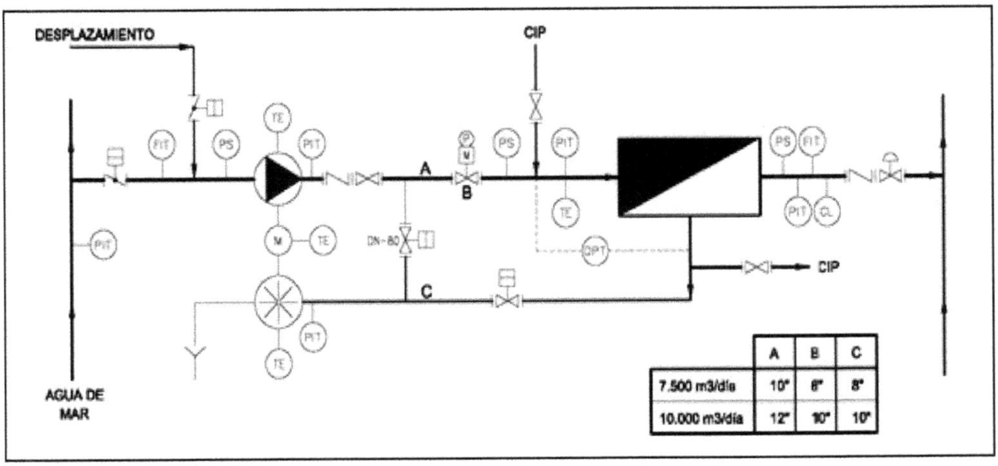

Figure 12.4. P&ID. Pelton energy recovery system.

- <u>Racks and turbo pumps.</u>

Turbo pumps and racks with common collectors. Isolation valves must be installed to start and stop independently each rack. The main disadvantage is that the stand-by turbo pump can only work with adjacent racks and not with all racks.

- Racks with independent start/stop systems

Instead of installing valves to isolate racks, an independent control valve will regulate the recovery during start/stop of racks. The collector will be common for all racks, with an automatic and a control valve for each rack which connects the reject and start -up rack. A motorized control valve will be installed in the feed to rack.

12.2 Isobaric chambers.

Isobaric chambers convert the hydraulic energy of brine to mechanical energy and then back to hydraulic energy in cylinders with pistons.

12.2.1 Type of isobaric chambers.

Isobaric chambers can be divided between slow and fast. Slow models are:

- **DWEER** supplied by CALDER.
- **Sal TEC** supplied by KSB.
- **RO-Kinematic** manufactured in Canary islands for small installation.
  As fast isobaric chambers:
- **PX** supplied by ERI.

12.2.2 Systems descriptions.

The most typical systems are **DWEER** and **PX,** so this book will focus its attention on both systems, which have some points in common:

- Both systems need a high pressure pump and a booster pump to boost the sea water back to the rack inlet.
- High pressure pump flow is equal to the permeate flow plus losses. Booster flow is equal to brine flow plus losses.
- Seawater consumption is the sum of booster plus high pressure pump flow and losses.
- The concept of losses in both systems are similar.

12.2.3 Definitions.

- **Brine outlet membranes.** $\mathbf{Q}_{se}$ = Brine flow; $\mathbf{P}_{se}$ = Brine pressure; $\mathbf{S}_{se}$ = salinity.
- **Seawater inlet.** $\mathbf{Q}_{ame}$ = Seawater flow; $\mathbf{P}_{ame}$ = seawater pressure; $\mathbf{S}_{ame}$ = seawater salinity
- **Seawater outlet.** $\mathbf{Q}_{ams}$ = seawater IC outlet flow; $\mathbf{P}_{ams}$ = Seawater IC outlet pressure; $\mathbf{S}_{ams}$ = Seawater IC outlet salinity.
- **Brine IC outlet.** $\mathbf{Q}_{ss}$ = Brine IC outlet flow; $\mathbf{P}_{ss}$ = Brine IC outlet pressure; $\mathbf{S}_{ss}$ = Brine IC outlet salinity
- **Product water flow** = Q_{ap}
- **High pressure pump flow** = Q_{ah}; P_{ah} = High pressure pump pressure.
- **Mixing** Seawater is contaminated with brine. This increases the total dissolved solids in feed water, and can result in an increase in the required pressure of the feed. Because

of the increase in pressure requirement, the HPP has to work harder to achieve the desired pressure. Therefore, it consumes more energy, leading to a lower efficiency and higher power consumption.

$$M = \frac{S_{ams} - S_{ame}}{S_{se} - S_{ame}} \times 100$$

- **Leakage** This is also called "lubrication flow" and occurs when reject at high pressure leaks into the feed that is at low pressure. To compensate for this, the high pressure pump has to work harder to provide more flow and will consume larger amounts of energy, resulting in a decreased efficiency of the plant.

$$L = \frac{Q_{se} - Q_{ams}}{Q_{se}} \times 100$$

$$Q_{ams} = \left(1 - \left(\frac{L}{100}\right)\right) \times Q_{se}$$

- **Overflush** A small flow of sea water is transferred to the brine LP side. It is directly related to mixing.

$$O = \left(\left(\frac{Q_{ame}}{Q_{ams}}\right) - 1\right) \times 100$$

- **HP differential pressure.**
- **LP differential pressure.**

  This pressure differential occurs when there is a high difference in pressure between the feed inlet and the brine outlet, because of which, the feed pump consumes more energy to supply higher pressure.

High pressure pumps supply permeate flow plus leakage.

A frequency converter is installed in the booster pumps, which is designed to boost the water to compensate head losses in piping, membranes and HP pump.

12.2.4 Energy consumption.

Isobaric chambers have a high energy recovery > 90%, and the most important point is that the efficiency is better at lower recoveries. However, it is necessary to analyze each scenario and take into consideration the impact of reducing the recovery on the pretreatment size. (lower recovery means bigger pretreatment).

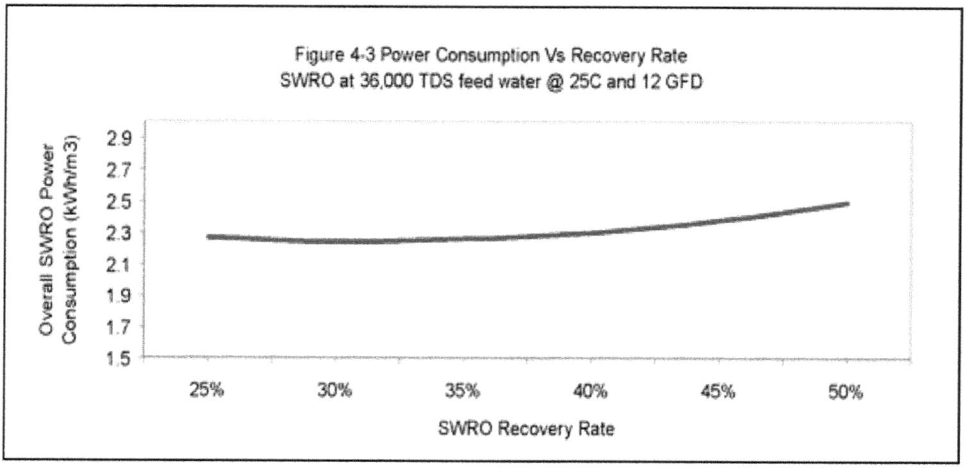

Figure 12.5. Curve energy-recovery.

- *Efficiency calculation*

Performance can be quantified by its energy transfer efficiency. Efficiency of the PX device is the ratio of the total energy output by the device to the total energy input to the unit expressed as percentage. In this case the total energy is obtained as the product of flow and pressure which is given in equation.

$$Efficiency = \frac{(Q_{ams} \times P_{ams} - Q_{ame} \times P_{ame} - (Q_{ah} - Q_{aP}) \times P_{ah})}{(Q_{se} \times P_{se})}$$

The following losses must be taken into consideration (Not usually considered by manufacturers):

- There is more energy consumption due to leakage.
- Discharge at higher pressure than atmospheric is an equipment requirement.

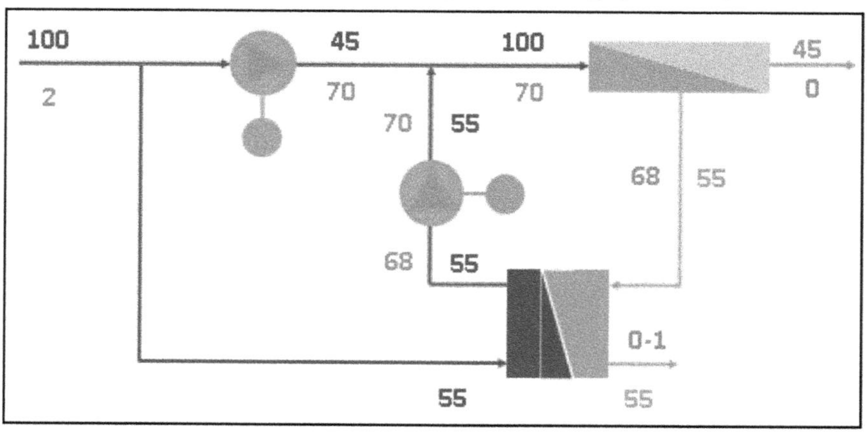

Figure 12.6. Efficiency study of isobaric chambers.

12.2.5 DWEER.

1. Description.

The DWEER™ belongs to the third class of energy recovery devices (ERD) which are hydraulically driven pumps in parallel. These are designed to separate the reject and the feed by a piston for diminished intermixing between them (Figure 12.7). By separating the feed and the reject it can be ensured that that the increase in salinity and pressure of the feed is always at its lowest.

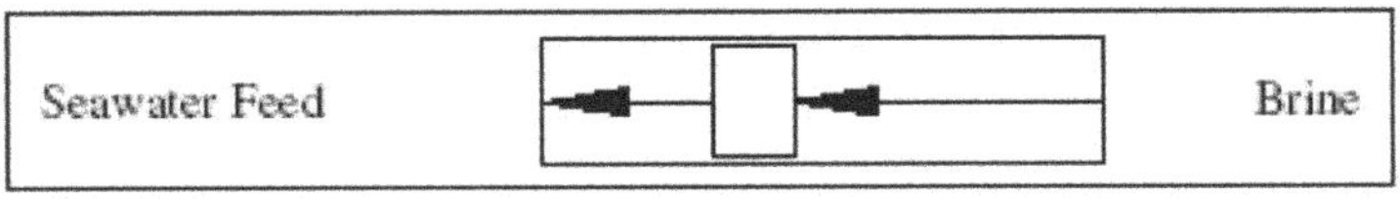

Figure 12.7. Piston to Ensure Minimum Mixing of Brine and Seawater Feed.

The DWEER™ has three main subassemblies: the pressure vessels made in superduplex in 14", the LinX™ valve at one side and a check valve nested on the other side. (a T is installed at the discharge of the vessel. T outlets are connected to four (4) check valves of 6").

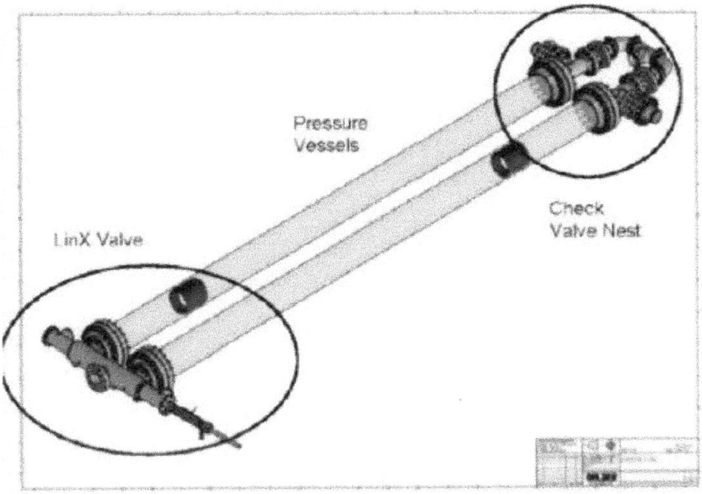

Figure 12.8. DWEER system. Courtesy of KSB.

Inside of the vessel, as shown above, a ceramic piston will transfer the pressure from seawater brine and will separate the feed and reject.

This equipment works slowly (4 cycles per minute) and has an operating range between 150-280 m^3/h. It is considered a heavy piece of equipment (5700 kgs full of water).

Figure 12.9 illustrates the integration of a work exchanger into the plant. In contrast to the conventional integration with a Pelton wheel, the feed stream is split into two, one of which is driven into the high pressure pump (HPP) and the other to the ERD. Since the HPP can handle a flow, which is equal to or just a little higher than the product flow, its size has to be reduced. Therefore, a recovery of 40% is achieved with a high pressure pump. A booster is added to compensate for the dip in pressure. However, this pump consumes less energy.

In an RO, there is a sharp decrease in pressure between the feed and the reject, which is entering the DWEER™. Since the piston and the membrane are at a similar pressure, it is placed in a pressure vessel. Owing to the limited volume of this vessel, a valve is used to enable the vessels with exchanging their functions. A booster pump enables the flow pressure in the effluent exiting from the DWEER™ to equal the discharge pressure of the high pressure pump, thereby making the system able to function in a loop. The rate of flow from the booster pump will be equal to the flow rate from the membrane, unless there is a leakage present in the membrane.

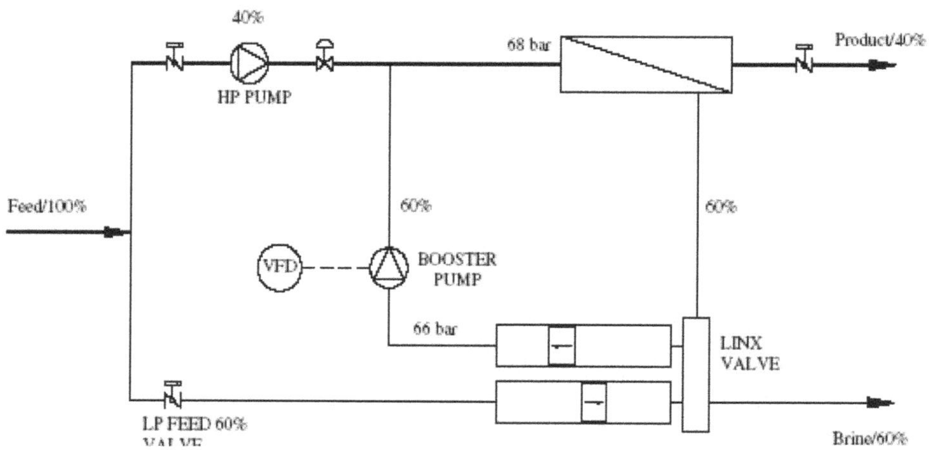

Figure 12.9. P&ID DWEER system.

2. Operation.

The equipment operates with a PLC which controls LINX valve. A timer signals when the valve needs to start its cycle. At normal conditions of operation, there is no variation in the cycling time. Each DWEER™ unit is designed so that there is utilization of 90% of the vessel so that about 10% addition in brine flow will not cause much difference in its functioning.

The DWEER™ units are also supplied with special sensors that detect overflush as well as under flush. If any of these are detected, the information is sent to the Programmable Logic Controllers (PLC) of the plant. When there is more than one dweer, PLC controls the operation in such a way that it will be sequential and uniform.

A LINX valve is hydraulic, and a hydraulic system will be supplied by the manufacturer per train.

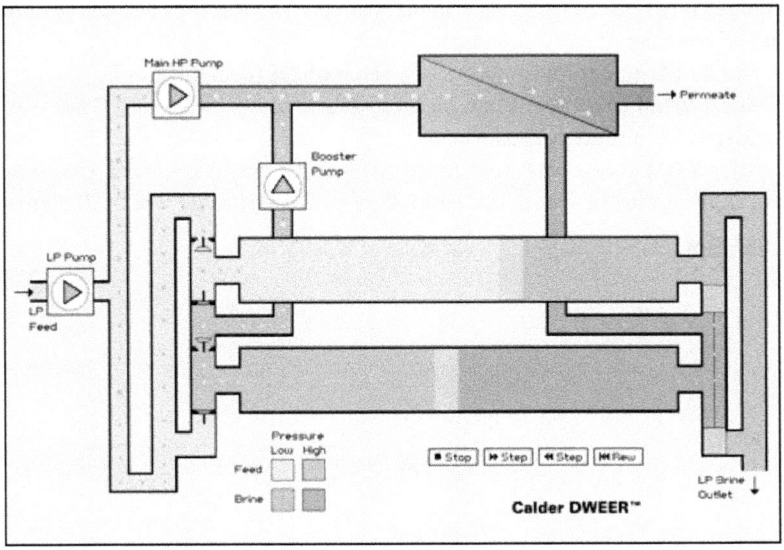

Figure 12.10. DWEER System. Courtesy of KSB.

3. Flow graphic.

DWEER works as an alternative pump. Flow is pulsed (4 pulses per minute).

When various DWEER systems operate together and are perfectly synchronized, the duration of pulses that the membranes receive are lower and the flow will be more uniform.

This is an aspect which needs to be taken into consideration when sizing piping.

4. DWEER losses.

- **Mixing**: It depends on overflush. For an over flush of 2%, mixing will be 1.5%. For design purposes, 2% will be considered. See attached graph for over flush/mixing relationship.

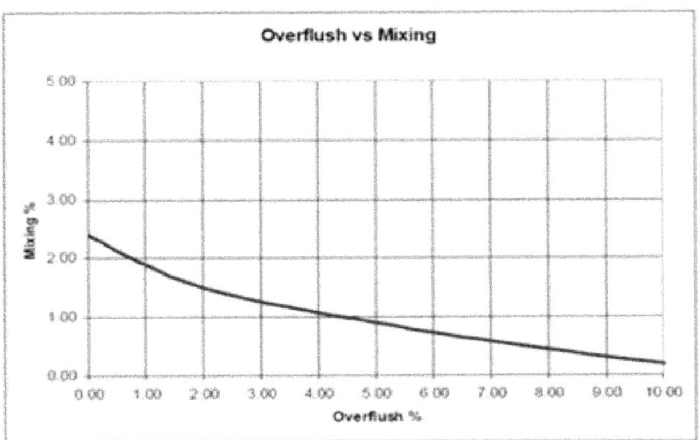

Figure 12.11. Oveflush and Mixing DWEER system.

- **Leakage:** DWEER guarantees a value between 1.8% to 2%. For design purposes, 2% must be considered.
- **Overflush:** Manufacturer guarantees a value of 2% (design value).
- **HP differential pressure:** Manufacturer guarantees a value of 1.2 bars (it should be considered for design purposes).
- **LP differential pressure:** A minimum value of 2.5 bars is required. Head loss value is 2 bars. The rest may be lost in the control valve. Design value will be 2.5 bars.

5. **Flux diagram.**

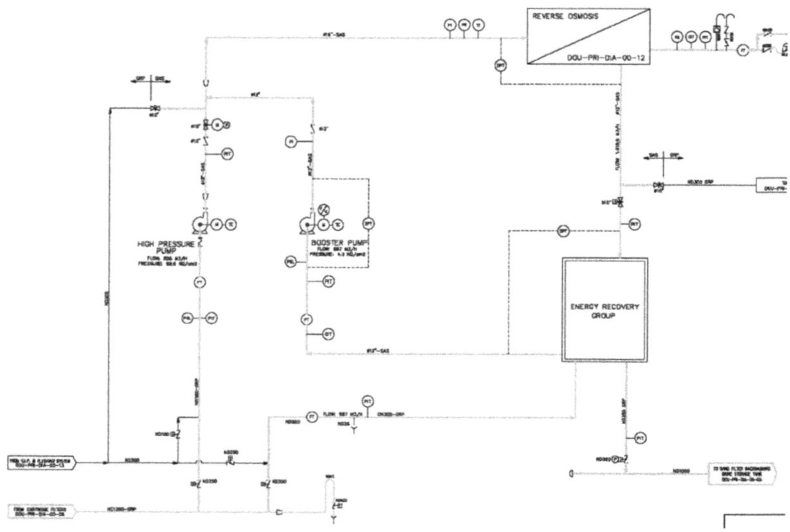

Figure 12.12. P&ID DWEER.

Sometimes a flow transmitter is installed in brine inlet. It must be clarified during contract signature.

- **Advantages.**
- Brine and feed are separated by a piston to ensure minimum mixing.
- For a piston designed for a minimum drag, the transfer of energy is 100%.
- All the materials are superduplex.
- It is quieter (80 dB) compared to PX.
- Less fouling.

- **Disadvantages.**
- High capital cost.
- Booster pump is needed.
- Mixing, lubrication, overflush, high pressure differential, low pressure differential.
- Equipment must be controlled with a PLC.
- Bigger maintenance costs (seals, springs, check valves, LINX valves).
- Less experienced operators than with ERI equipment .

6. Configuration .

DWEER systems can be mounted in different ways

Figure 12.13. Horizontal installation. Installation under membranes.

12.2.6 PX.

PX device works on positive displacement and is an isobaric energy recovery device.

It contains a ceramic cartridge which is at the heart of the device The cartridge has a feed water end cover, rotor, sleeve and concentrate end cover as shown in figure 12.14 & figure 12.15.

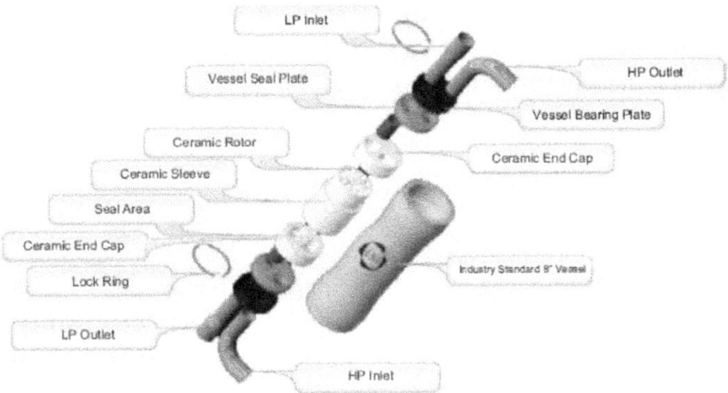

Figure 12.14. ERI Cartridge filter. Section I. Courtesy of ERI Inc.

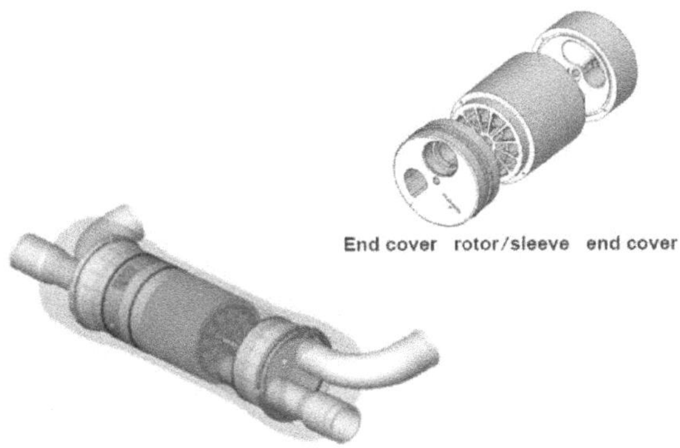

Figure 12.15. ERI Cartridge filter. Section II Courtesy of ERI Inc.

The design of the rotor consists of axial ducts, which are arranged circularly around a central tension rod. It rotates in the sleeve and both of the end covers enclose the rotor. Sealing and bearing surfaces are also made of ceramic and the lubrication of the brushing surface is accomplished by the high pressure feed water. The end cover at the concentrate end consists of High Pressure HP Concentrate port for allowing in the HP IN (HP concentrate) and also a low pressure concentrate port for out flowing of concentrate (LP OUT). At the feed water end cover side, there are two feed water ports termed LP IN and HP OUT.

For the design of this element, ceramic a natural choice because of its toughness, corrosion resisting properties and its dimensional stability. The high-pressure concentrate is used to direct pressurization of the feed water in the PX device. In order to lubricate the rotor, the flow rate of concentrate is set at a somewhat higher flow rate than the feed water so that a small amount of feed water helps the lubrication of the rotor. So, the PX device supplies the membranes a portion of the feed flow (portion of concentrate) and the high pressure pump supplies the permeate portion of feed water flow and also accommodates a small flow for the lubrication.

Figure 12.16 shows the operating cycle of the device. As explained above, a rotor duct is exposed to the LP IN and LP out end-cover ports. Shown in Step 1, is the LP feed water rotor duct. The positioning and the placement of the rotor ports is such that it provides tangential flow that exerts turning force on the rotor. The rotor ducts terminate in a Seal Zone where the flow is stopped and duct is sealed from the HP and LP flow circuit as seen in step No 2. The duct rotates further to expose the incoming HP concentrate that pressurizes the feed water and pushes it toward the membranes as shown in Step 3. The rotor continues to advance until the duct is sealed as seen in step 4 and the whole cycle begins again.

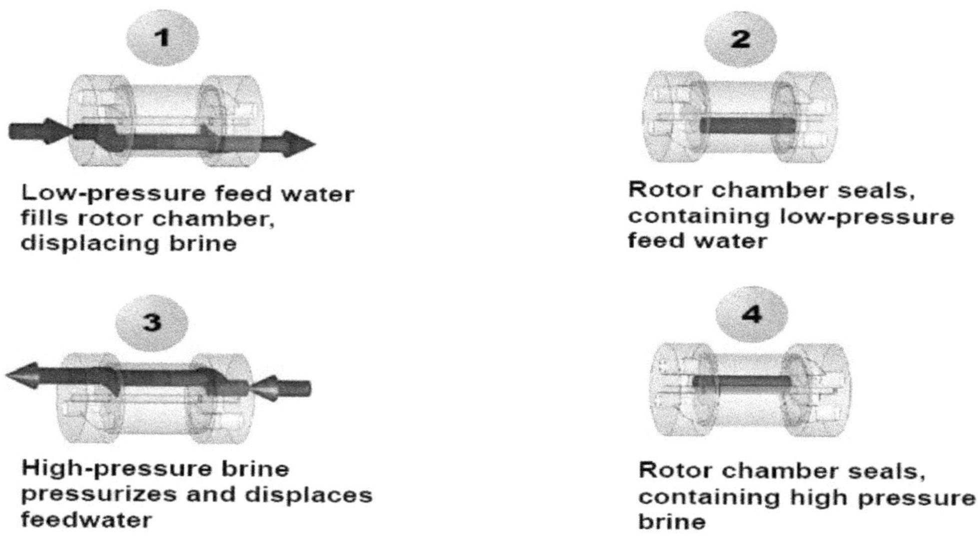

Figure 12.16. Operation cycle of ERI. Courtesy of ERI.

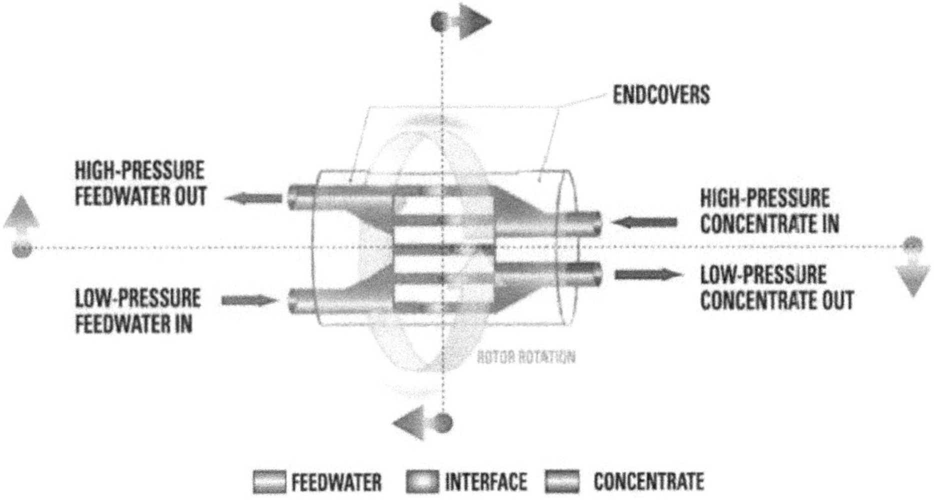

Figure 12.17. Operation principles of ERI. Courtesy of ERI.

Though the feed water and the concentrate come into direct contact inside the rotor, due to the water barrier that exists inside the duct, there will be limited mixing between the feed water and the concentrate.

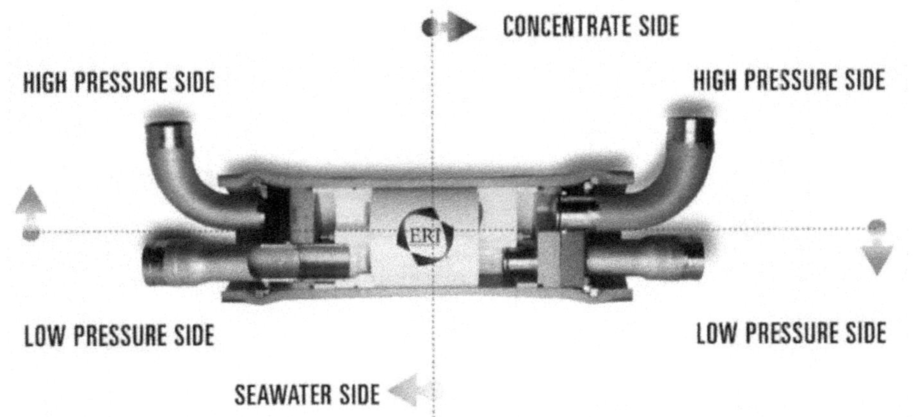

Figure 12.18. Section of ERI element. Courtesy of ERI.

The design of the ceramic components is such that the interfacing feed water and concentrate does not reach the end of the duct before it is sealed and so retains separation. If there is any water remaining from the previous cycles, it acts as a barrier against the mixing of these flows. The ducts are designed with an aspect ratio to be long and narrow so that the flow is stabilized and there is no breakup of the barrier due to turbulence.

There is a continuous evolution and increase of capacity and treatment flow of equipment with the positive impact on the number of required components (saving on material and installation). The three models being used in large desalination plants are shown in table 12.1:

Table 12.1 Models of PX.

	Maximum flow	Guaranteed efficiency	Sound level	Maximum lubrication flow
PX-260	59 m^3/h	96.8%	94 dBA	0,78 m^3/h
PX-300	68 m^3/h	96.8%	93 dBA	1 m^3/h
PX-Q300	68 m^3/h	97.2%	<85 dBA	

1. Operation.

This equipment operates in an automatic mode. As the liquid circulates, the rotor will start to rotate.

The sequence the fluids will follow to start is the following:

- Seawater.
- Booster pumps.
- High.

No control is required for flux distribution between different PX units.

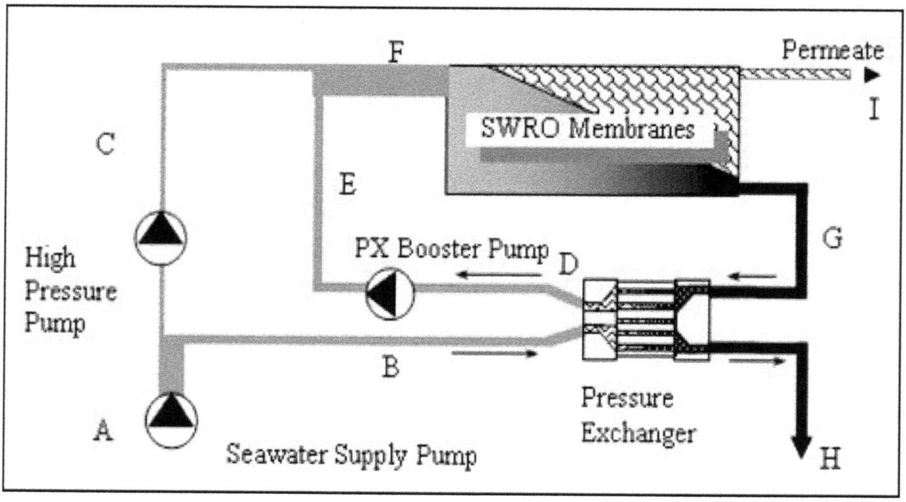

Figure 12.19. Flux distribution of PX unit.

2. Installation

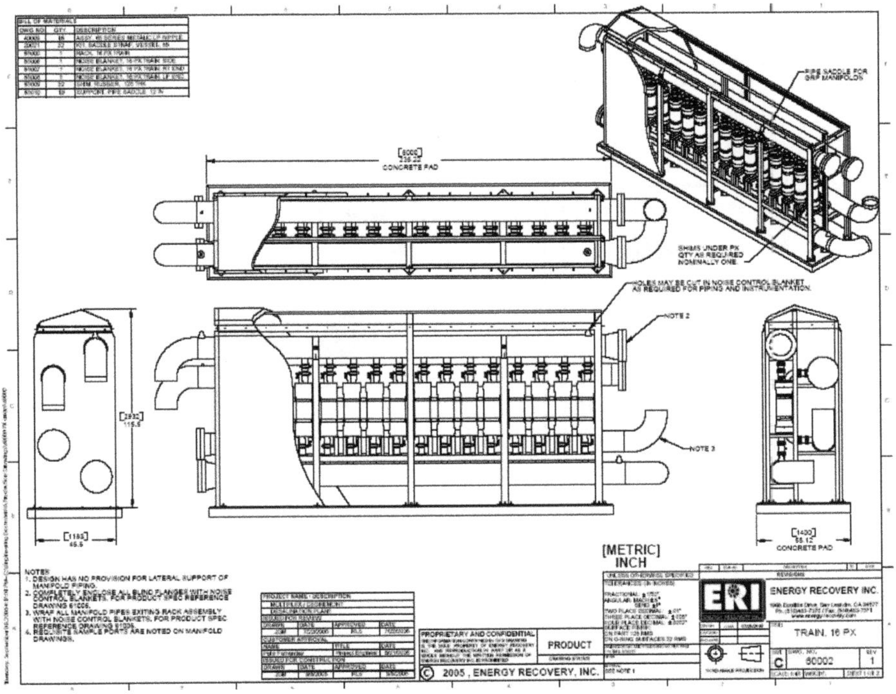

Figure 12.20. Views of a PX units skid. (I) Courtesy of ERI.

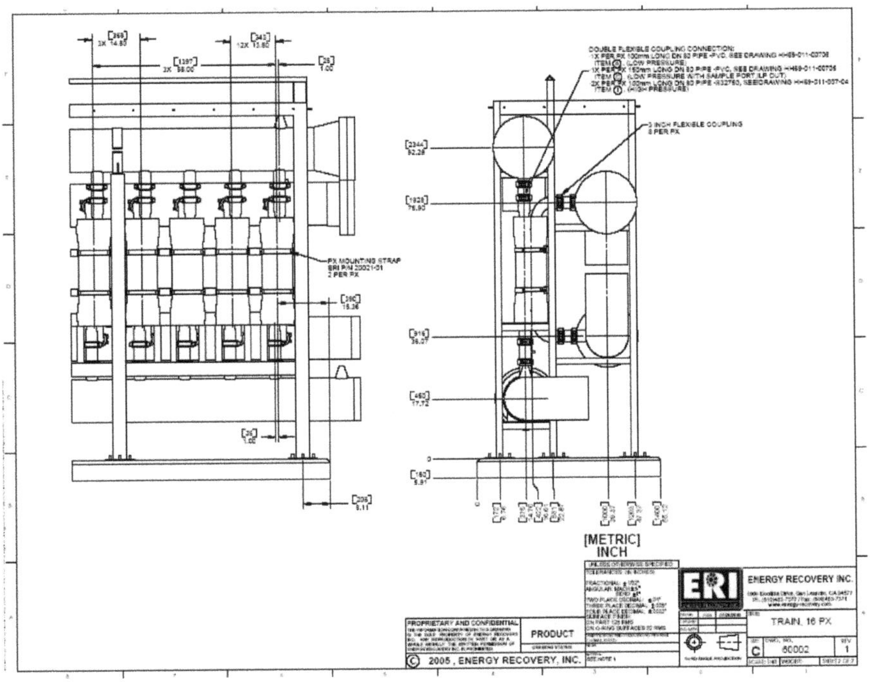

Figure 12.21. Views of a PX units skid.(II) Courtesy of ERI.

Find below the details of victaulic connections between exchangers and collectors.

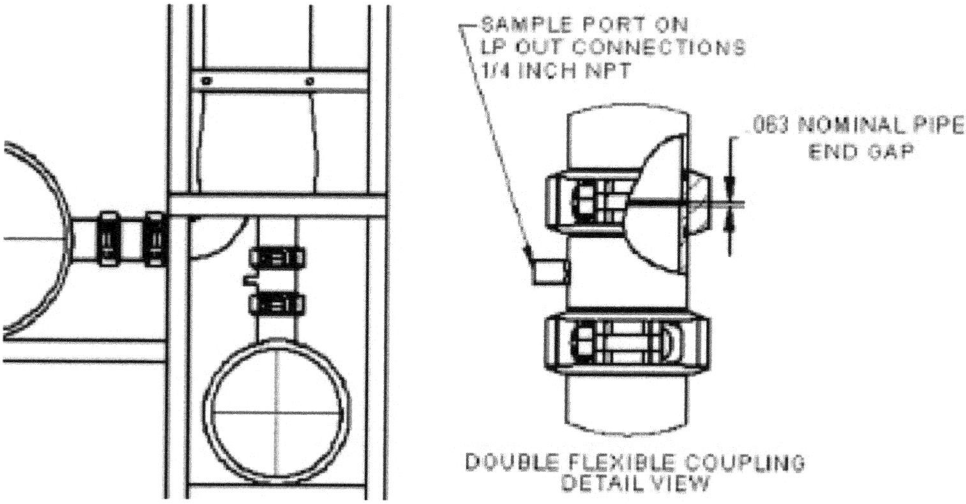

Figure 12.22. Connection victaulic ERI.

3. PX losses.

- **Mixing.**

Mixing is approx. 6% with an over flux of 0%.

If a balanced flow is maintained (and high and low pressure flow rates are equal), then the mixing is calculated by using the following equation and is approximately fixed at above indicated at 6%.

$$Volumetric\ mixing = \frac{(HPOUT_{salinity} - LPIN_{salinity})}{(HPIN_{salinity} - LPIN_{salinity})} \times 100$$

For the design it an increase of salinity of 6% must be considered at the PX outlet and 3% in the membranes. This 3% increase in the salinity at the membrane region results in increase of operating pressure by approximately 1.3 bar. If the PX performance is used to express the efficiency, the decrease caused by mixing with balanced flow is calculated to be 1%.

- **Leakage.**

Lubrication flow will calculated according to the formula:

$$Lubrication(gpm) = 3 \times 10^{-6} \times Q_{se} \times P_{se} + 1.4 \times 10^{-3} \times P_{se} + 0.059 \times T_e - 0.7$$

Q_{se} = gpm ; P_{se} = psig ; T_e = ° F.

It will be the difference between the low pressure brine from the PX unit minus the low pressure feed water to the PX unit or the high pressure brine to the PX unit minus the high pressure feed water from the PX unit.

For design it must be considered at a minimum of 1%.

- **Overflush.**

It must be considered 1% for design.

- **HP differential pressure.**

Design HP differential pressure will be 1 bar.

- **LP differential pressure.**

Design LP differential pressure will be approximately 1.8 bars.

4. **Advantages and disadvantages.**

- **Advantages.**
  - Core built of ceramic selected to be the ideal material for its toughness, corrosion resistance and dimensional stability for withstanding the harshest saline environments.
  - Unlike turbines no transformational losses occur in a PX device.
  - Stable efficiency over wide range of recoveries.
  - Lack of traditional seals and bearings.
  - No control required.

- **Disadvantages.**
  - Very noisy equipment, and it is necessary to place in sound proof enclosures and install them underground if possible.
  - All the parts in contact with fluid are made of plastic or ceramic materials, and based on the experience when installed into open intake plants, fouling can be excessive.

5. **Flow diagram.**

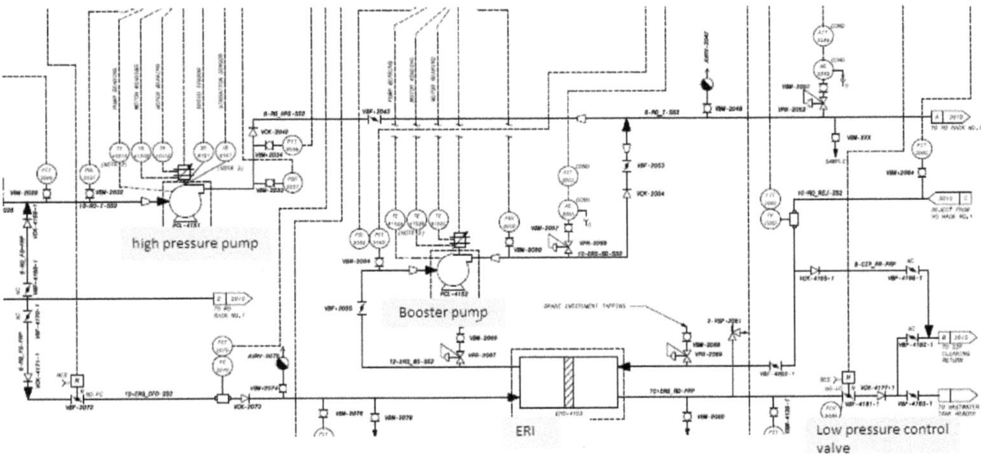

Figure 12.23. P&ID Energy recovery system. ERI.

6. **Isobaric recovery systems operation modes.**

In contrast to the conventional integration with a Pelton wheel, in ERI & DWEER energy recovery systems the feed stream is split into two, one of which is driven into the HPP and the other to the ERD. The decrease of the flow implies a reduction of efficiency.

To improve the efficiency there are different solutions:

- Bigger racks (15,000 m³/day or 20,000 m³/day). High pressure pumps of 7500 m³/day and 10000 m³/day would be used. Working with large racks has the disadvantage of the number of tubes needed to be installed. This will involve an increase of leakage, stops etc.
- A common pumping center, decreasing the number of pumps in operation compared to the number of racks. It allows for a significant reduction in specific power.
  However, if keeping the size of racks (7500 m³/day-10,000 m³/day), the number of booster pumps will also be high. To solve this issue, there are two alternatives which will be listed below.

First alternative:

Big racks of 15000, 20000 or 25000 m³/day with three (3) sub racks of 5000, 7000 or 8400 m³/day respectively, so that if one of the sub-racks stops, the other two can continue working. The booster pumps must be capable of operating at 67% of flow. In this case, when one of the sub racks starts, the flow of the rest of the sub racks must be adjusted to a minimum production percentage to avoid excessive recovery.

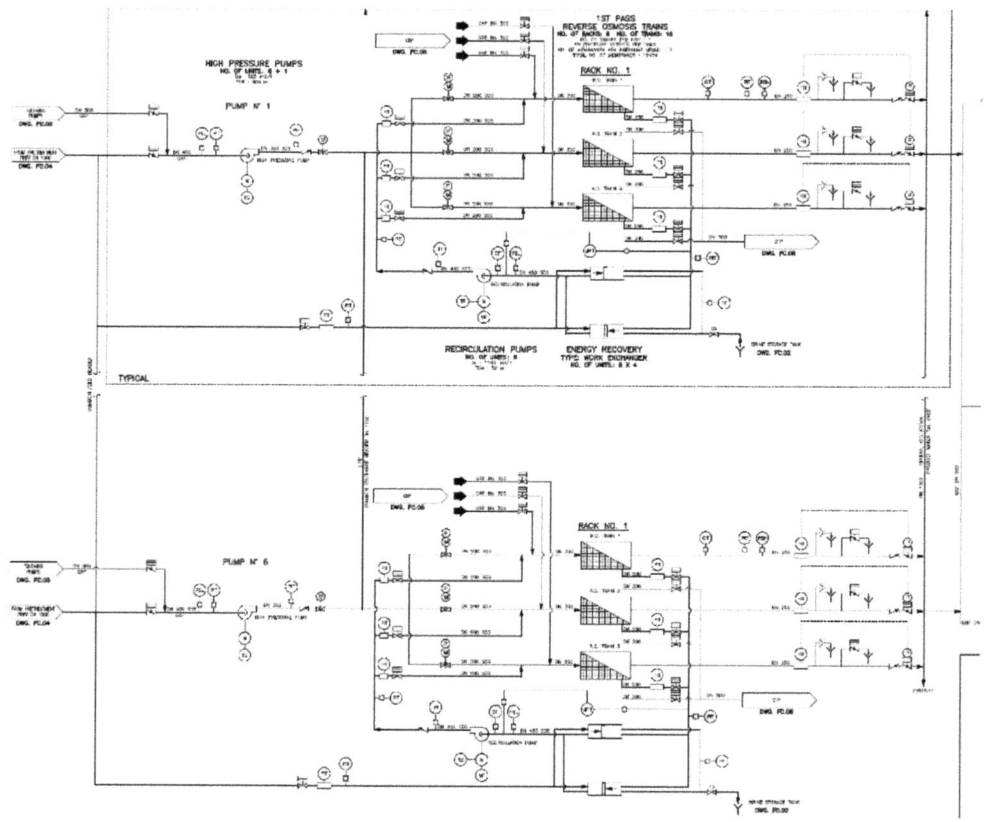

Figure 12.24. P&ID of first alternative.

Second alternative:

Racks of 7000, 8000, 9000 or 10000 m³/day with the booster pumps working in a common collector. The number of pumps decreases, but two start-up collectors (feed and outlet) are required which will put all of the racks on line. The number of booster pumps and high pressure pumps can be the same or different. If spare booster pumps are required, one could install two pumps of 50% flow, which would give more flexibility in case of variations in the of racks operating.

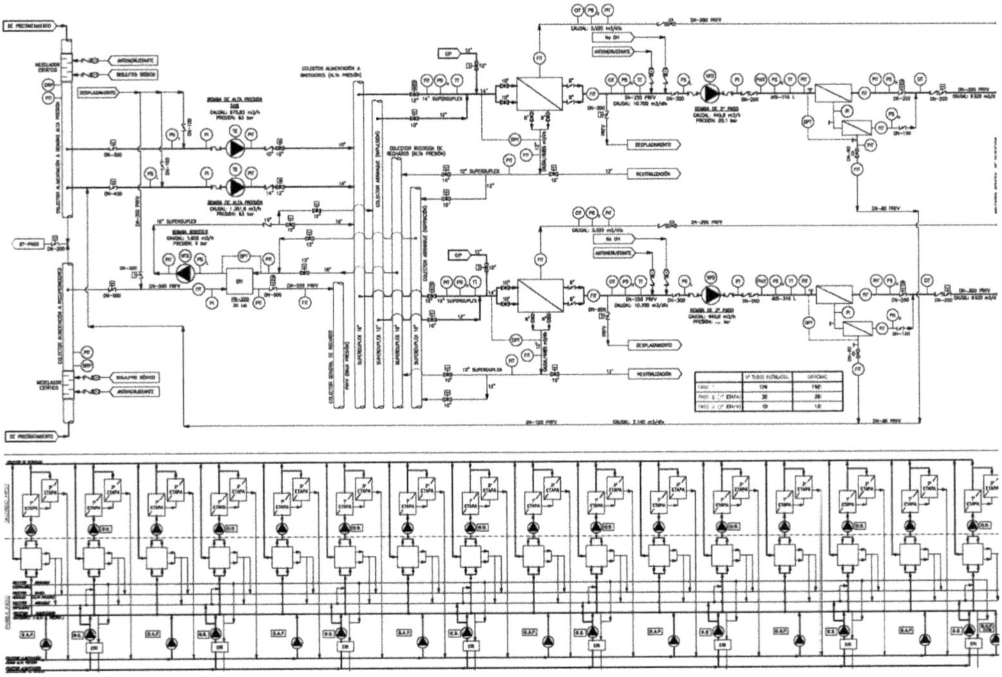

Figure 12.25. P&ID of second alternative.

o **Valves.**

Each recovery system has four (4) collectors and the number of isolation valves to be installed will depend on the operation mode (one recovery system for each rack or operating system with sub-racks).

An on-off butterfly automatic valve is installed in the low pressure sea water inlet.

Flushing of recovery systems is necessary so membrane flushing is usually carried out with booster pumps by installing flushing automatic valves in parallel with the sea water.

An automatic valve with a positioner will be installed in the low pressure brine discharge to keep a constant back pressure in the collector. A butterfly valve is a standard option.

To isolate the recovery systems, when chemical cleaning of membranes is carried out, automatic valves in the suction side of the booster pumps for each sub-rack are installed.

Isolation valves will not be installed in booster pumps with high pressure discharge unless the system works with a common collector. Then, automatic valves will be installed to isolate the system.

7. Instrumentation.

The following instrumentation is required in an isobaric recovery system.

- Flow transmitter (FIT) in the booster pump suction side.

If there are various racks associated to an ERI system, one FT will be installed in discharge side of booster pump plus one (1) FT per rack.

- FIT in feed water inlet.

It measures seawater flow that the recovery system needs.

- Differential pressure transmitter (DPT) in the high pressure side of the recovery system.

It measures the head loss in the recovery system.

- DPT in the booster pump.

It measures the differential pressure of booster pump.

- PIT in the suction side of the booster pump.

It measures the pressure in the recovery system outlet side.

- PIT in the feed water inlet.

It measures the pressure of the seawater.

- Conductivity analyzer (CIT) in the suction side of the booster pump.

To measure the conductivity at the outlet of the energy recovery system and to control salinity increases generated by the system.

- PIT in the brine discharge.

To control the back pressure needed by the recovery system and to prevent cavitation.

EXAMPLE

An example of the design of Energy recovery system-ERI.

P: Rack permeate production m3/h 1000

PS: Permeate salinity, ppm..............................200

R: Recovery, %...40%

FW: Feed water salinity ppm ....40,000 ppm

For the design it must be considered that an increase in salinity of 2-3% in membranes. This increase in salinity will affect the RO projection. With a 2.1% salinity increase, the membranes feed salinity will be:

MEMBRANE FEED

MS: Membrane salinity ppm........................... 40,853 ppm

Membrane feed pressure...... ....62 bars*

Membranes feed flow.......................................PS/R = 2500 m^3/h

HP IN

Salinity ppm..67,955 ppm*

*Membrane concentrate pressure *..........................60 bars*

Flow..1500 m³/h*

** These values are provided by the membrane manufacturer using membrane projection software.*

LP IN

Salinity ppm...... ....40,000 ppm (same as feed water salinity)

Pressure ....2 bars (selected by the designer)

Flow..1487 m³/h (with lubrication factor of 0.8%).

HP OUT

Salinity ppm........... .... 39,362 ppm (Obtained from salinity mass balance)

40,853 ppm x 2500 m³/h = 1012,3 m³/hx40,000 ppm +1487 m³/h x 39,362 ppm

Pressure ....HP IN pressure –differential pressure = 60-1 = 59 bars

Flow...................................... 1487 m³/h (Same value as LP in flow in balanced system)

Booster pump characteristics: 3 bars and 1487,5 m³/h

HPP FEED

Salinity ppm........... .... 40,000 ppm (same as feed water salinity)

Pressure ....2 bars (selected by the designer)

Flow......................................Membrane feed flow-HP out flow = 2500 -1487 = 1012,4 m³/h

If the energy recovery model is PX-Q300, the minimum number of PX units should be 23.

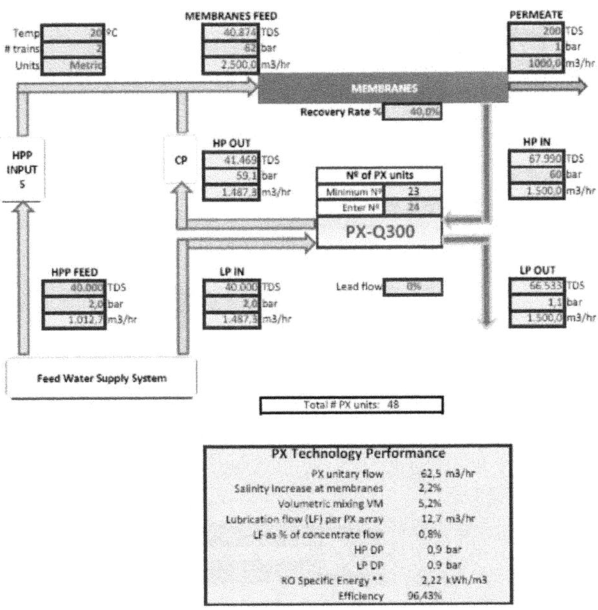

Figure 12.26. Energy recovery system Example.

12.2.7 Retrofitting of existing SWRO .

The pressure exchangers (ERI, DWEER) make it possible to significantly increase the capacity of the existing systems by adding little or no additional power and/or to reduce the power consumption of the existing systems by as much as 75%. There are several proven approaches to retro-fitting existing SWRO system using pressure exchanger technology.

We are going to focus on the multiple train retrofitting of SWRO with Pelton systems to be substituted by ERI/DWEER .

The multiple train power reduction takes advantage of the fact that the main HP pump flow equals the permeate flow and combines multiple trains of existing RO systems while eliminating their associated high pressure pumps resulting in significant power savings. For example, the plant in Figure 12.27 is operating with Pelton wheel energy recovery turbines. Each plant is producing approximately 3200 m³/day at 49% recovery and 60 bars feed pressure.

After the retro-fit, the PX system had replaced one of the main HP pumps and both of the Pelton turbines. The original motor of the main HP pump had sufficient horsepower to handle the full load of the HP pump without the assistance of the Pelton turbine.

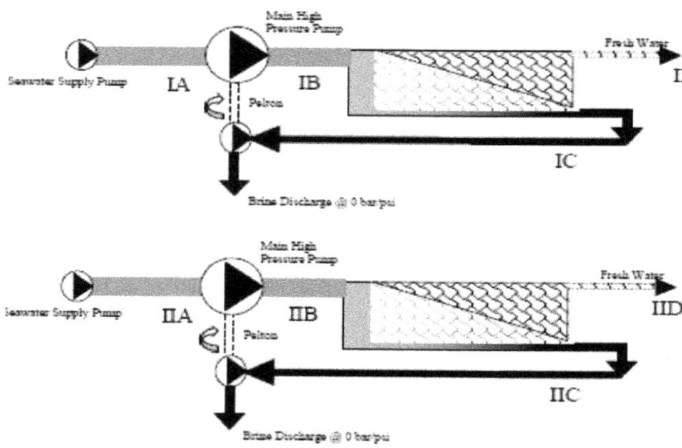

Figure 12.27. Rack operating with Pelton turbine.

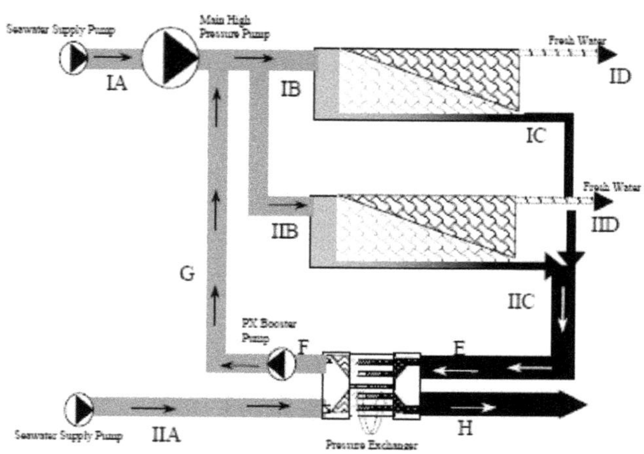

Figure 12.28. Two rack operating with same high pressure pump after retrofitting.

Table 11. Multiple Train Power Reduction Flows and Pressures Before and After

Stream	Description	Before PX m^3/hr (gpm)	After PX m^3/hr (gpm)	Before PX Bar (psi)	After PX Bar (psi)
IA	Seawater Feed supply Train I	277.3 (1220)	277.3 (1220)	1.8 (26)	1.8 (26)
IB	RO Feed Stream Train I	277.3 (1220)	309.8 (1363)	59.6 (865)	59.6 (865)
IC	Reject Stream Train I / PX HP Inlet	141.4 (622)	164.3 (723)	58.3 (845)	58.3 (845)
ID	RO Product Flow Train I	135.9 (598)	145.4 (640)	0 (0)	0 (0)
IIA	Seawater Feed Supply Train II	270.4 (1190)	294.1 (1294)	1.6 (24)	1.4 (20)
IIB	RO Feed Stream Train II	270.4 (1190)	261.8 (1152)	60.3 (875)	59.6 (865)
IIC	Reject Stream Train II / PX HP Inlet	135.2 (595)	132.9 (585)	59.3 (860)	58.3 (845)
IID	RO Product Flow Train II	135.2 (595)	128.9 (567)	0 (0)	0 (0)
E	PX HP Inlet / Reject	n/a	297.3 (1308)	n/a	58.3 (845)
F	PX HP Outlet	n/a	294.1 (1294)	n/a	57.2 (830)
G	Booster Pump Outlet / Seawater	n/a	294.1 (1294)	n/a	59.6 (865)
H	PX LP Outlet / Reject	n/a	297.3 (1308)	n/a	0.3 (5)

In many larger installations where multiple trains typically exist, the main HP pumps are producing full system pressure. This approach can be used in conjunction with expansion schemes in order to optimize the match between the high pressure pump flows and the fresh water production rates while simultaneously achieving significant power savings.

12.2.8 Turbo Charger/Hydraulic Pressure Booster (HPB).

A nozzle drives the reject into a turbine, which is connected to a centrifugal spinning impeller. The use of the turbocharger is to boost the pressure of the feed that has been pressurized by the high pressure pump (HPP) for reaching the required feed pressure.

A turbocharger unit consists of a hydraulic turbine and an HPP. The turbine is similar to a reverse running pump. The HPP and turbocharger are not connected directly to overcome the disadvantages that are observed in FTs and Pelton wheels, thereby allowing operational flexibility. In addition, turbochargers are easy to install and are energy efficient.

The impeller of the HPP is made of the same material as the impeller of the turbocharger. Both the impeller and the turbine of the turbocharger are a centrifugal "close-coupled mixed-type" with both axial and radial flows. The transfer of energy by the turbocharger increases the pressure of the feed, and the rotating element of the device is balanced. It has a by-pass that allows manual control of the flow. This is required when the flow rate of the brine at the second stage is higher than needed. This is required even more when the feed experiences large variations in temperature or when the membrane is old. The energy recovery by the turbocharger is achieved when the discharge pressure required by the HPP are decreased.

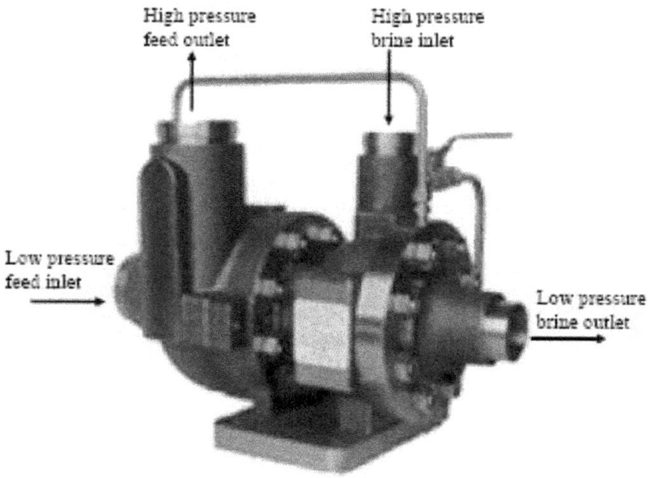

Figure 12.29. Photography of a turbocharger.

The highest transfer efficiency that can be achieved by a hydraulic turbocharger is calculated by multiplying the efficiency of the impellers, nozzles and turbine, ie: 90% x 90% x 99% = 80%. Each of these three factors influences the efficiency of the device. The efficiency depends on the pump flow, but will be between 60 to 80% with flows ranging from 70 to 1200 m^3/h. Control valves and nozzles can help in adjusting the performance.

- Materials

They are manufactured in the following materials.

- Body: AISI-316 or DUPLEX (2205).
- Impeller: AL6XN.
- Bearings: Resin or graphite/carbon.
- Bolts: AISI 304.

- Diagrams/operation modes

There are three working modes:

1. When a double stage is installed, it is necessary to consider that 1st stage pump pressure must be increased.
2. Double pass.
3. With one single stage, it is necessary a support pump.

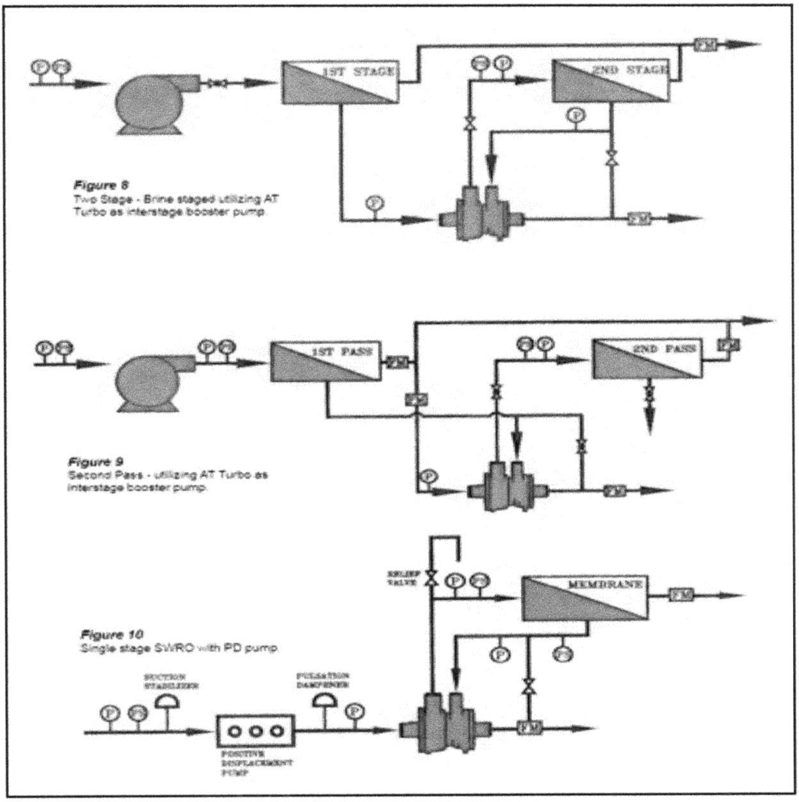

Figure 12.30. Diagram of turbo charger operation modes.

- Pumping pressure calculation:

The following formula is used.

$$P_1 = N_{te} \times R_r \times (P_{br} - P_{be})$$

P_1 = Pressure increase in centrifugal pump.
N_{te} = Equipment efficiency.
R_r = Relationship between turbine/pump flow.
P_{br} = Equipment inlet brine pressure at the equipment inlet.
P_e = Minimum discharge pressure 5 psig.

Two types of equipment are manufactured.

- LPT model is a low pressure equipment (lower than 24 bars)

From the smallest model LPT-125 with a flow of 40 m^3/h to the LPT-2000 of 650 m^3/h.

- HTC model can work with a pressure as high as 82 bars. The models available range from HTC AT-50 with connections of 1" to the model HTC AT-9600 with connections of 12 inches.

Specifically designed for seawater RO, relatively small plants and brackish water plants.

12.2.9 SalTec® N System.

The SalTec System is comprised of a high-pressure pump, a booster pump, the pressure exchanger SalTec DT, and the control unit for the system, SalTec System-Control.

It is recommended to operate the booster pump at variable speed by means of a frequency inverter. The Pressure Exchanger operates trouble free when controlled by the speeds of the recirculation pump and the step motors controlled rotating valve.

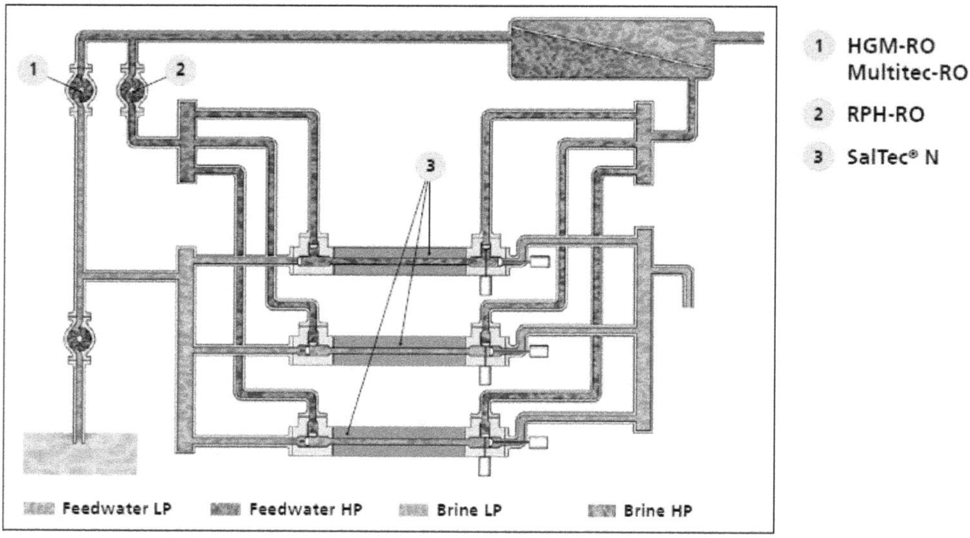

Figure 12.31. Saltec N system arrangement. Courtesy of KSB.

13

REVERSE OSMOSIS RACKS CONTROL.

13.1 Single Stage/One rack.

The following instrumentation should be installed

- Sea water inlet:...................... PS, PT, TE
- Brine outlet: .... DPT
- Product outlet:..................... PS, FIT, PIT, CI

PT: Pressure transmitter PS: Pressure switch; DPT: Differential pressure transmitter; FIT: Flow transmitter; CI: Conductivity analyzer; TE: Temperature analyzer.

A motorized + positioner plug valve will be installed in the high pressure pump discharge side to increase the feed pressure slowly (1-2 minutes). It is important not to use this valve as a control valve!

An automatic isolation valve shall be installed in the brine discharge, usually in the energy recovery system suction side to isolate the rack when the system stops. To prevent water leakage of the rack due to valve malfunction, the brine pipe will discharge through the upper part of rack (an automatic vent will be installed).

A control valve will be installed in the permeate outlet to regulate the permeate production. Depending on the operation range, it will be necessary to select the valve. See below typical one stage P&ID.

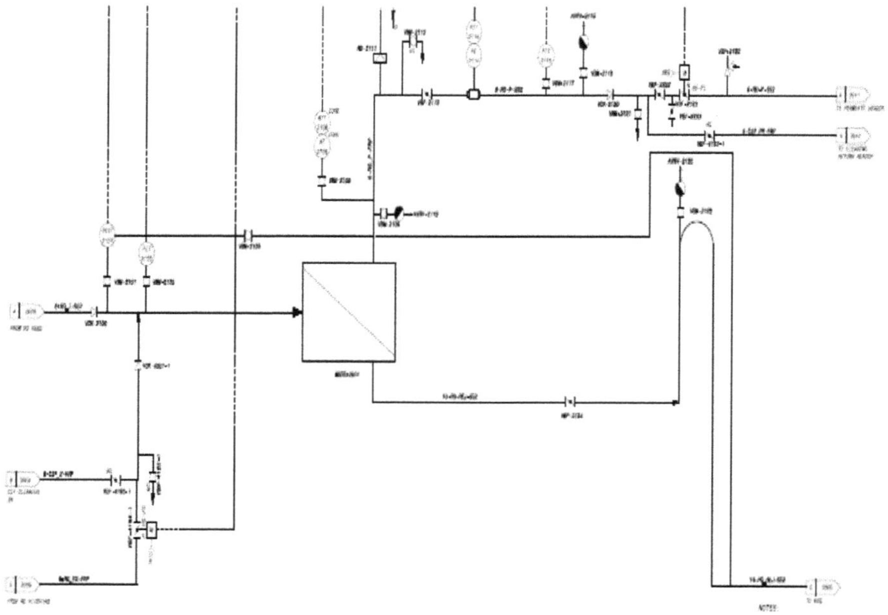

Figure 13.1. P&ID single stage/one rack.

The typical control configurations of a single stage RO rack with ERI/DWEER systems are shown below.

13.1.1 Alternative I.

A permeate control valve will control production. It will compare the set point for production (Q_p) with the difference between the raw water flow (Q_a) and the brine flow (Q_s), while opening or closing to keep the production at a constant.

Control valve in the discharge side of the booster pump will control recovery. It will compare the set point for recovery (FC) with the value calculated from $(1\text{-}Q_s/Q_a)*100$, opening or closing to keep the recovery at a constant.

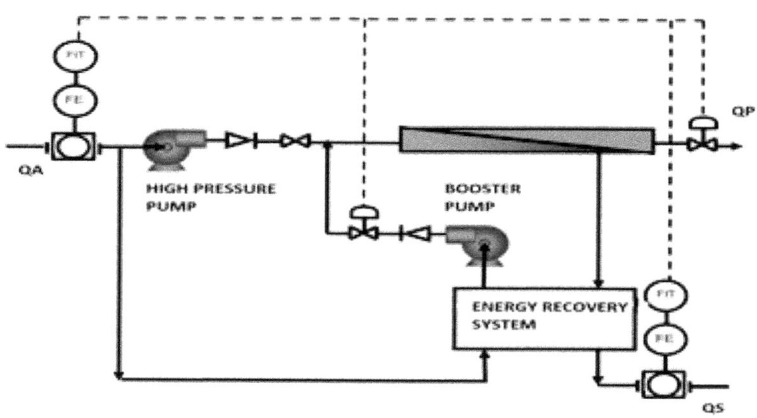

Figure 13.2. Alternative I. Control RO single stage rack.

13.1.2 Alternative II.

The operation philosophy of the permeate control valve will be the same as indicated in first alternative. A frequency converter will be installed in the booster pump, modifying the pump curve to keep the recovery constant at the fixed set point. This is the preferred alternative for most of the plants in terms of energy consumption and investment. The booster pumps are low voltage pumps so the cost of speed converters is not excessive and the installation of speed converters is usually a better alternative to optimize the energy consumption compared to the installation of a control valve in the discharge side of the booster pump when the efficiency of speed converters is above 92-93%.

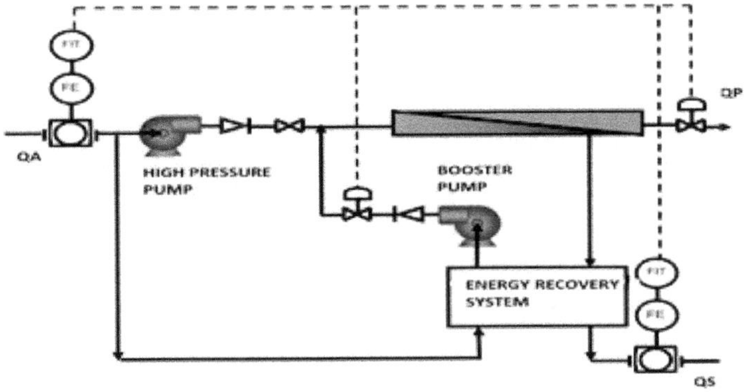

Figure 13.3. Alternative II. Control RO single stage rack.

13.1.3 Alternative III.

Operation of the booster pump control valve will be the same as alternative I.

A frequency converter (VFD) will be installed in high pressure side for the control of production, then modifying the pump curve to keep the production constant at the fixed set point. It must be taken into consideration that speed converters are expensive and complex equipment that need specialized maintenance. The efficiency of VFD must be above 95-96% to be a solution from an energy point of view.

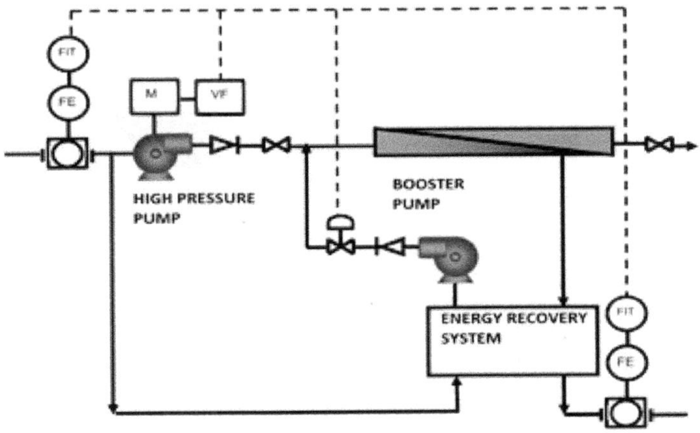

Figure 13.4. Alternative III. Control RO single stage rack.

13.1.4 Alternative IV.

A frequency converter will be installed in the booster pump, modifying the pump curve to keep the recovery at the fixed set point constant.

A frequency converter will be installed in the high pressure pump for control of production, modifying the pump curve to keep the production at the fixed set point constant. This is the most complex and expensive solution, and it will only be considered when the operating temperature and salinity range is high. A detailed CAPEX and OPEX study should be required. The continuous improvement in the efficiency of speed converters is turning this alternative into a more attractive solution for control.

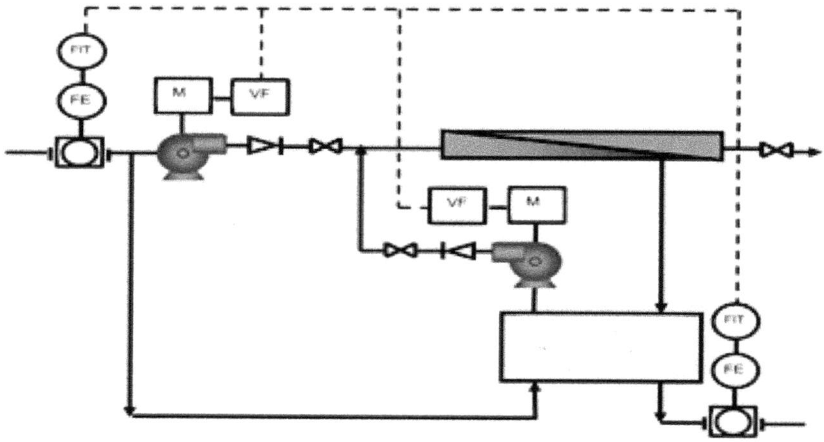

Figure 13.5. Alternative IV. Control RO single stage rack.

13.2 Rack with turbine.

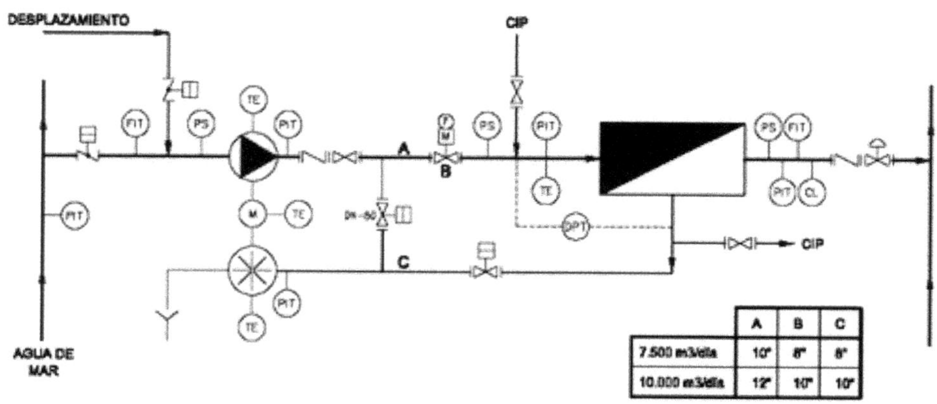

Figure 13.6. P&ID Rack with turbine.

13.3 Two (2) Stages and two (2) passes to increase recovery.

This arrangement is used to increase global recovery to 55-60%.

The instrumentation installed for the 2nd stage will be the same as indicated for one the stage system.

Between stages, a centrifugal pump (booster) will be installed. This pump will have a frequency converter so that a control valve is not required. In the permeate outlet, a control valve is required in the 1st stage. It is not common to install a control valve in the permeate line of the second stage.

Pressure vessels in the second stage (1st pass) should be designed for 1400 psig. The same will be applicable to membranes. Problems in the 2nd stage are:

- 2nd stage is operated at higher pressure.
- It is necessary to check the schedule of piping.
- The flushing procedure will be critical as brine concentration is higher.

Below, the diagrams of both the 1st pass with double stage system and the 2nd pass with a double stage system are shown. A control valve to start the 1st pass is necessary.

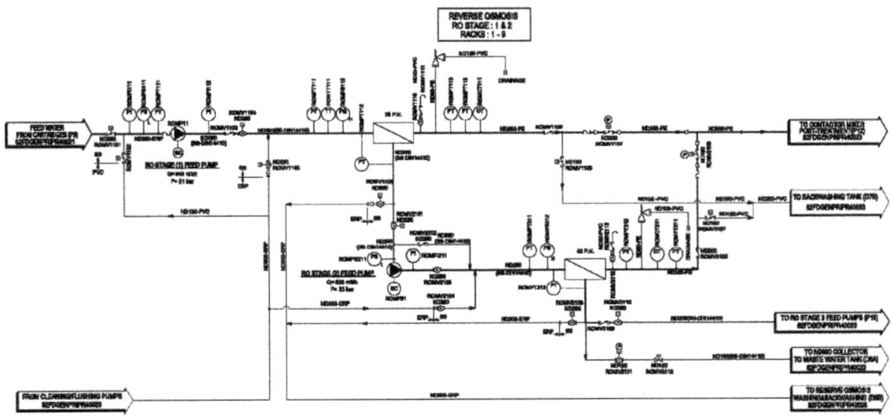

Figure 13.7. p&id 1st pass (two stages).

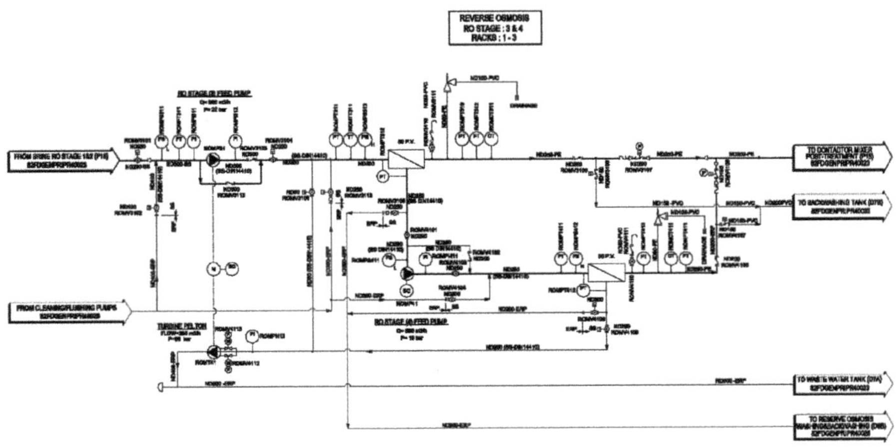

Figure 13.8. P&ID. 2nd pass (2 stages).

13.4 Two stages in the same rack.

When a second pass is installed to improve the permeate quality, a two stage rack is the alternative used to improve the global recovery of the system.

The instrumentation will be the same as the instrumentation installed when there is only one pass, but with a double permeate outlet. A pipe with a control valve that connects both product lines is required.

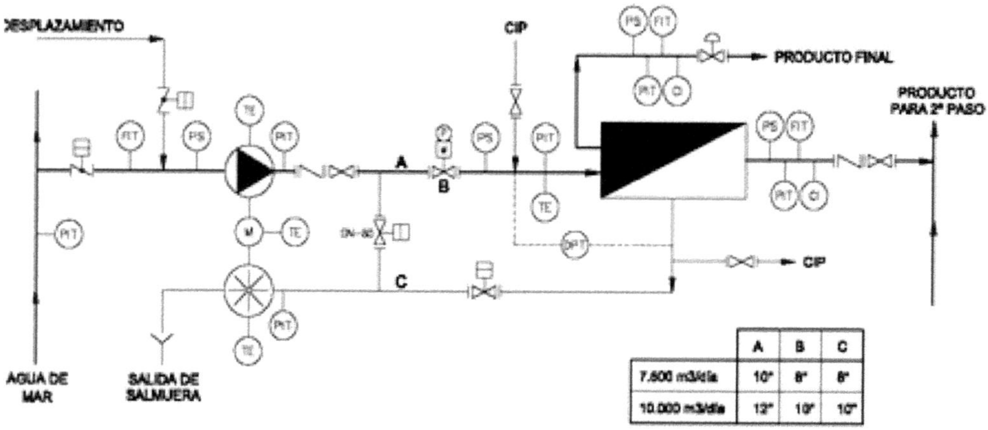

Figure 13.9. P&ID. Two stages in the same rack.

13.5 Double pass with two racks to improve water quality.

100% or a portion of the permeate flow from the first pass RO is treated by a second pass RO to obtain the required product quality and quantity.

Typical recovery of a two pass SWRO system: First pass – 50%, Second pass – 90%, Total – 47% with the second pass brine recycle.

- The first pass configuration is similar (valves and instrumentation) to a first pass rack which operates alone.
- The second pass is arranged in two stages as it operates at a recovery of 90%. Some differences compared to a two stage system which treats seawater must be identified:

- No washing connections are required. Fouling potential is very low (water is the permeate from 1st pass.
- A booster pump speed converter will feed the second pass.
- A FIT, PS and a manual valve will be installed.

- *First and second pass diagram.*

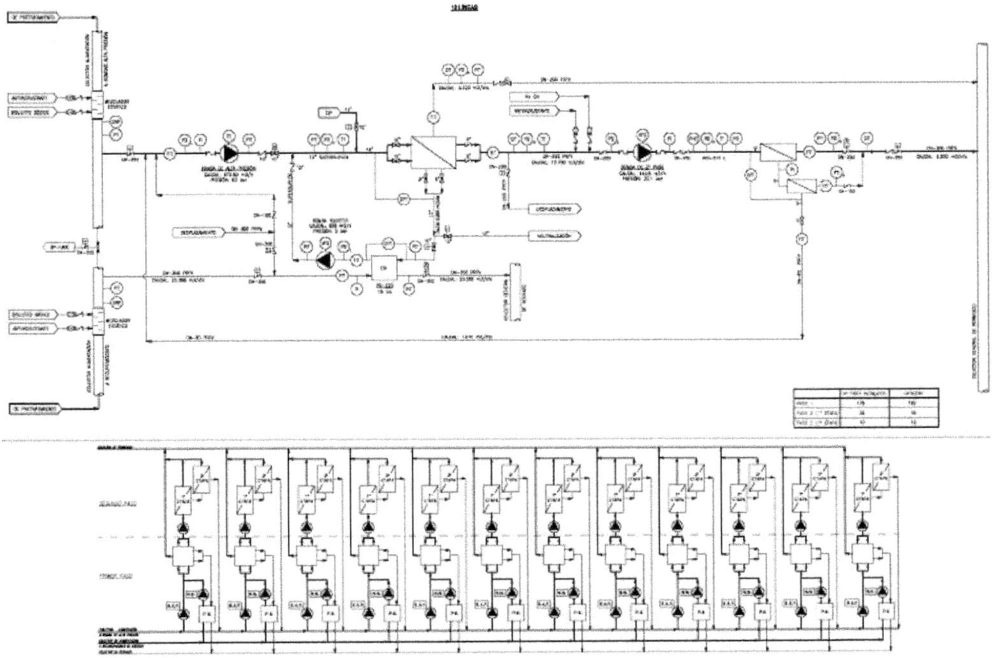

Figure 13.10. 1st Pass/one stage–P&ID

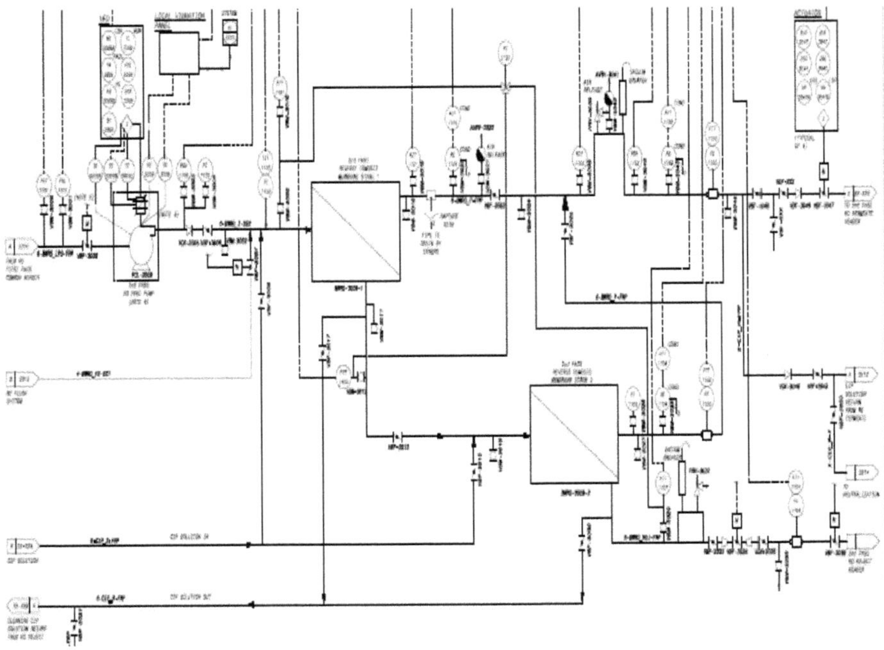

Figure 13.11. 2nd Pass/two stages–P&ID

14

REVERSE OSMOSIS RACKS EQUIPMENT.

14.1 High pressure piping.

Seamless stainless steel piping will be used for diameters less than 4". Welded stainless steel pipe will be used for larger diameters. The ANSI/ASME norms cover the standardization of dimensions for welded and seamless wrought stainless steel pipe. ASTM norms cover the fabrication and the quality of the different components of piping.

The norm applicable for seamless pipe is B677 ASTM.

The norm applicable for welded pipe is SB677 ASTM.

There are other norms which need to be taken into consideration for high pressure piping designs such as A-312 for seamless piping, and UNS for austenitic stainless steel.

The table shows the dimensions for Sch 40 piping.

(Kg/cm^2)	NOMINAL (inch)	NOMINAL (mm)	MINIMUM (mm)	MINIMUM	SCHEDULE (mm)
70	½	12.70	0.16	10S	2.11
70	¾	19.05	0.24	10S	2.11
70	1	25.40	0.32	10S	2.77
70	1 ¼	31.75	0.40	10S	2.77
70	1 ½	38.10	0.49	10S	2.77
70	2	50.80	0.65	10S	2.77
70	2 ½	63.50	0.81	10S	3.05
70	3	76.20	0.97	10S	3.05
70	3 ½	88.90	1.13	10S	3.05
70	4	101.60	1.29	10S	3.05
70	5	127.00	1.62	10S	3.4
70	6	152.40	1.94	10S	3.4
70	8	203.20	2.59	10S	3.76
70	10	254.00	3.23	10S	4.19
70	12	304.80	3.88	20	6.35
70	14	355.60	4.53	20	7.92
70	16	406.40	5.17	20	7.92
70	18	457.20	5.82	20	7.92

The piping system connected to Victaulic connectors should be Sch 40. (up to 70 kg/cm^2).

14.2 Accessories.

The piping system must be connected to the ports using flexible connectors in order to allow relative movement of the vessels and movement of the piping system. (Victaulic or equivalent connections are recommended. The Victaulic flexible couplings were supplied in the past in galvanized carbon steel. However, nowadays the materials employed are Super Duplex as it provides crevice corrosion due to oxygen absence.

The most important manufacturers are VICTAULIC and PIEDMONT.

Figure 14.1. Victaulic connections photograph.

14.3 High pressure valves.

14.3.1 Plug valves.

Plug valves are typically used in reverse osmosis desalination plants for on/off services except in the discharge side of high pressure pumps where plug valves with electric actuators and positioners are installed to guarantee a controlled opening (1 to 2 minutes). Once the high pressure pump has been started, the feed pressure control valve will slowly open increasing the feed pressure and feed flow rate to the membrane elements until the design concentrate flow is reached. The feed pressure increase to the elements should be less than 10 psi per second. If the feed pressure is ramped up too quickly, the housing of elements may be damaged by excessive forces in flow/radial direction.

Double acting pneumatic actuators are preferred for on/off service except as commented above where a positioner is installed.

The valve installed for the discharge of high pressure pumps should not be used for control, as energy consumption will increase.

Figure 14.2. Plug valve installed in high pressure pump discharge.

Plug valves are also installed in the backwash water inlet and outlet piping from rack.

For a rack of 10.000 m^3/day, with a feed flow of 930 m^3/h, a pipe of 12 inches will be acceptable. A plug valve of 10" will give a head loss at 100% of 0.27 bars.

The standard materials for these valves for body, bonnet and plug are super duplex stainless steel ASTM A995 Gr5A, 904 L stainless steel UNS N08904 or 254SMO ASTM A351/A744 Gr CK3MCuN.

The seats and seals are manufactured in PTFE.

The valves will be provided with butt weld full rated ASME Cl 600.

The suppliers of plug valves are Xomox, Victaulic (MTS), Flowserve and E-Control.

Ball valves can be used instead of plug valves (It is a more expensive option). Ball valves will be flanged and complete valve must be dismantled to replace it.

14.3.2 Check valves.

Dual check valves mounted between flanges for a pressure of 600 psig will be used.

14.3.3 Butterfly valves.

Butterfly valves of 600 psig mounted between flanges are available which can be used in the discharge of pumps and collectors isolation. It must be considered that butterfly valves require two (2) flanges in superduplex.

14.3.4 Others.

In the upper part of the brine collector a depressurization valve should be installed. It will open when the rack stops to release residual pressure.

Vent valves should also be installed at the top of the brine collector.

14.4 Low pressure valves.

Butterfly valves are installed in the low-pressure areas of a desalination plant, usually in the early stages of production upstream of the membranes and then again at the end on the permeate lines. They are similar to the valves installed in the pretreatment area but with the disk manufactured in AISI 316L.

A sample port is located in the permeate stream from each vessel. The 3-way ball valves ensure that the permeate flow will never be closed. This is a very important security function: Thus, the risk of faulty manipulation is completely eliminated. At the same time the permeate can be directed to the receiver and/or for sample taking. In this case it is recommended to use a 3-way ball valve with limitation of 180° with T-bore. By directing the water to a receiver, it is possible to know the production flow of each vessel. If one pressure vessel shows a significantly higher permeate concentration than the other vessels of the same stage, then this vessel could be probed. The installation of three-way valves allow for locating a problem within a pressure vessel while online without unloading elements. Probing involves the insertion of a plastic tube (approx. 6 mm in diameter) into the full length of the permeate tube in order to measure the permeate conductivity at the different locations inside the pressure vessel.

Figure 14.3. Low pressure PVC valves.

Butterfly valves with a positioner can be installed the in the permeate discharge to regulate back pressure if the required head loss is lower than 3 bars. (In the past on/off valves have been installed).

For higher headloss, it is necessary to install "glove-rotary" control valves (MASONEILAN or SAMSON) with electro pneumatic positioners for a signal, 4-20 mA, with two limit switches and an auxiliary hand wheel (MASONEILAN or SAMSON).

A check valve is installed in the permeate discharge to prevent water backflow due to the osmotic effect when the rack stops working.

Automatic air vents should be located in the upper part of the feed and reject piping.

14.5 Low pressure piping.

Treated water piping. Material will be UPVC (DWI approved) or polypropilene. Design pressure will be 10 bars.

14.6 Sample panel.

Sampling panel will be installed for each rack. Probe tubing is used to connect the plugged permeate port of rack vessels with the quick connections installed in panel. A conductivity analyzer will measure the conductivity.

Figure 14.4. Sample panel photograph.

14.7 Instrumentation.

- Flow transmitters.

Electro magnetic flow transmitters for the low pressure side is the right selection. In the suction side of booster pumps, the design pressure of transmitters will be NP64. Hastelloy C will be the selected material for electrode if the fluid is brine or sea water.

If the flow transmitter is mounted in plastic pipe a reference electrode should incorporated.

- Pressure transmitters.

Enclosure and wetted parts are made of stainless steel (AISI 316L). Hastelloy C will be used for diaphragm.

- Temperature transmitters.

Thermowell material will be superduplex.

- Conductivty and pH analyzer.

pH and conductivity instrumentation will be installed in a by-pass line. Flow cells must withstand at least (1) bar to send treated water to services deposit.

Figure 14.5. Analyzers simple panel.

15

RACKS CLEANING SYSTEM and FLUSHING.

Over time, membrane systems can become fouled with any number of foulants such as colloids, organic matter, metallic scales, and biological constituents. These materials can build up on the membrane surface and in the feed brine channel. If left uncorrected, the accumulation of these foulants can cause a severe loss of performance in the system: such as pressure requirement increases to maintain flow, pressure drop increases, and salt rejection can suffer. If the system is not cleaned and the system continues to build up foulants, the elements may "telescope," or shear internally, causing the integrity of the membrane surface to be compromised and rendering the membrane irreversibly damaged.

Fouling characteristics that signal you need cleaning are a 10-15% decrease in normalized permeate flow, 10-15% decrease in normalized permeate quality, or a 10-15% increase in normalized pressure drop as measured between the feed and concentrate headers.

What you clean for can vary site by site depending on the foulant. Complicating the situation is that more than one foulant can be present. Typical foulants are:

- Calcium carbonate scale.
- Sulfate scale of calcium, barium or strontium.
- Metal oxides of iron, manganese, aluminum, etc.
- Silica scale.
- Colloidal deposits (inorganic or mixed inorganic/organic).
- Organic material of natural origin or man-made origin.
- Biological (bioslime, mold, or fungi).

Figure 15.1. Washing pump.

15.1 Products used for cleaning.

There are a number of factors involved in the selection of a suitable cleaning chemical (or chemicals) and proper cleaning protocol.

These chemicals can be generic and are available from a number of suppliers or can be private-labeled proprietary cleaning solutions. The proprietary solutions can be more expensive but may be easier to use and you cannot rule out the advantage of the intellectual knowledge supplied by these companies.

Low and high pH cleaning are the most common, with liquid and solid products:

- For CO_3Ca, SO_4Ca, SO_4Ba, SO_4Sr, F_2Ca scaling . Weak acids are employed.

  - o Citric Acid at 2%.
  - o Phosphoric acid at 0.5%.
  - o Clorhidric acid at 0.5%.

- For metallic oxides, weak acids are used as:

  - o Phosporic acid at 0.5%.
  - o Sulphamic acid at 0.2%.

- To clean inorganic colloids and bio fouling, use detergents from HENKEL. P3-ULTRASIL 10 al 0.7-1.5% with sodium hydroxide to raise the pH between 11 to 12 are employed.

15.2 Membrane cleaning procedure.

Reverse osmosis cleaning procedures may vary dependent on the situation. The time required to clean a stage can take from 4 to 8 hours. The basic steps for cleaning are:

- Perform a low pressure flush with feed or permeate water to remove service concentrate and foulants.
- Make up the cleaning solution per the manufacturers instructions.
- Introduce the cleaning solution to the first stage for 60 minutes. You may want to throttle the flow up slowly to minimize the plugging of the feed path with dislodged foulant. Send the displaced water with up to 20% of the fouled cleaning solution to drain before returning the cleaning solution back to the RO Cleaning Tank. Readjust the pH to the target when it changes more than 0.5 pH units.
- An optional soak and recirculation sequence can be used. The soak time can be from one (1) hour to overnight depending on the manufacturers recommendation, but be cautious that the proper temperature and pH be maintained and that this does increase the chemical exposure time of the membrane.
- A low pressure cleaning rinse with permeate water is required to remove all traces of chemical from the cleaning skid and from the reverse osmosis skid.
- Once all the stages of a train are cleaned, the RO can be placed back into service.

The recirculation flow at the maximum flow will be between 6 to 10 m^3/h per vessel of 8".

Major cleaning system components are:

- **RO Cleaning Tank**:

When the tank is large, it is common to install a smaller tank of 2000 liters where the product is mixed and it is then pumped to the big deposit. The cleaning pump is recirculated to mix it. Therefore, the agitator in the RO cleaning tank can be eliminated.

- **RO Cleaning Pump**:

This pump needs to be sized to develop the proper crossflow velocity to scrub the membrane clean.

- **RO Cleaning Cartridge Filter**.

Normally 5 to 10 microns and it is designed to remove foulants that have been displaced by the cleaning process.

- **RO tank mixer.**

This is recommended to get optimal mixing of chemical, though some designers rely solely on the slow introduction of chemical while maintaining a recirculation through the RO cleaning pump back into the tank.

- **Instrumentation.**

Flow transmitter in the discharge.

PH meter for cleaning control.

Conductivity analyzer for flushing control.

- **RO Tank Heater or Cooler.**

The optimal temperature for cleaning is 35°C to 45°C. One cannot forget that heat is generated and imparted by the RO cleaning pump.

- **Valves** For each rack, two (2) valves (rack inlet and rack outlet) are necessary. It is recommended that the valves to be automatic. One automatic valve will be installed in the recovery system.
- **Sample Points:** Sample valves should be located to allow pH and TDS measurements off the RO cleaning pump discharge and the concentrate side recirculation return line and permeate return line: A small amount of the cleaning solution can permeate through the membranes and so a permeate side return line is recommended.

15.3 Equipment sizing.

- **Mixing tank.**

The mixing tank should be constructed of polyethylene (PE) or fiberglass-reinforced plastic (FRP). The tank should be provided with a removable cover and a temperature gauge. A rule of thumb for sizing a cleaning tank is to use the approximate volume of the empty pressure vessels, then add the volume of the feed and return hoses or pipes and guarantee that the pumps can work without cavitation.

- **Centrifugal pumps.**

The cleaning pump should be sized for a flow of: number of vessels x 9 m^3/h (8 inches vessel). The pressure must be between 5 to 6 bars. The pump should be constructed of 316 SS or nonmetallic composite polyesters. If a frequency converter is used, a (1+1) configuration is acceptable. If not a (2+1) configuration is required. These pumps are used for flushing.

- **Cartridge filters.**

They must be designed for the maximum backwashing flow (15 – 20 m/h). The material of the filter can be glass reinforced plastic (GRP).

- **Feed and return piping.**

It must be designed for a velocity between 2.5 and 3 m/s. It is a high velocity, but the operation is not continuous so it can be considered as acceptable. Material of piping will be GRP and the design pressure 7.5 bars.

- **Plug valves.**

Plug valves will be installed in the connecting lines to the high pressure piping for the racks. It is recommended that the valves be automatic. If manual valves are selected, limit switches will be installed to not allow the high pressure pump to start if the valves are open.

- **Butterfly valves.**

They can be manual, except for the service water inlet valve (for flushing) and the (2) two valves of the chemical cleaning return (rainage valve and deposit inlet).

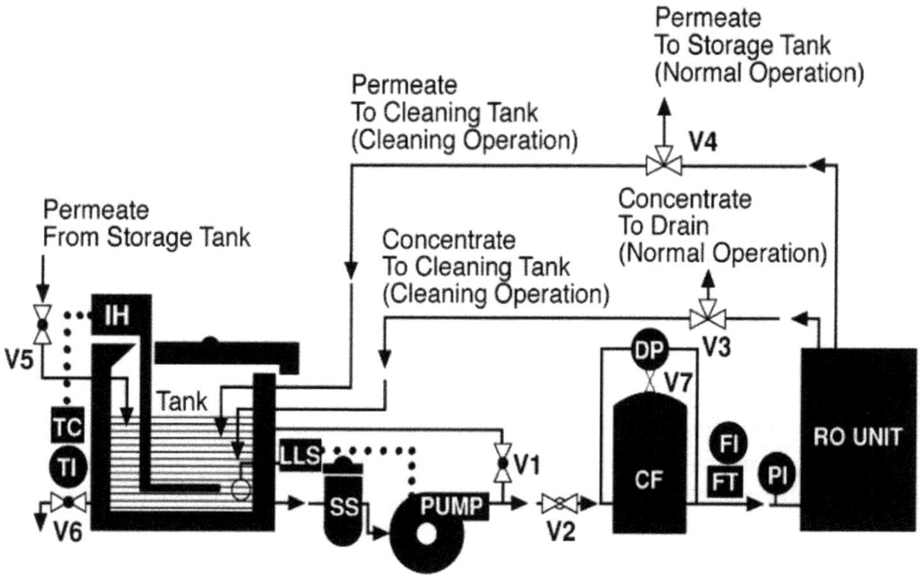

TANK Chemical Mixing Tank, polypropylene or FRP
IH Immersion Heater (may be replaced by cooling coil for some site locations)
TI Temperature Indicator
TC Temperature Control
LLS Lower Level Switch to shut off pump
SS Security Screen–100 mesh
PUMP Low-Pressure Pump, 316 SS or non-metallic composite
CF Cartridge Filter, 5-10 micron polypropylene with PVC, FRP, or SS housing

DP Differential Pressure Gauge
FI Flow Indicator
FT Flow Transmitter (optional)
PI Pressure Indicator
V1 Pump Recirculation Valve, CPVC
V2 Flow Control Valve, CPVC
V3 Concentrate Valve, CPVC 3-way valve
V4 Permeate Valve, CPVC 3-way valve
V5 Permeate Inlet Valve, CPVC
V6 Tank Drain Valve, PVC, or CPVC
V7 Purge Valve, SS, PVC, or CPVC

Figure 15.2. P&ID membranes washing.

EXAMPLE

Preparation tank

Volume of the vessels:

Considering a 50% free volume, the total volume of water in each vessel will be V = 0.228 - 7 x 0.5 x 0,032 = 0.116 m3

The number of membranes cleaned will be 46 so maximum volume will be 46 x 0.116 = 5.34 m3.

Volume of piping:

If the diameter of piping is 250 mm and the approximate total length is 105 m, the volume of piping will be 0.049 x 105 = 5.15 m3

Minimum volume of the tank will be:

5.15 + 5.34 = 10.49 m3

The tank will have a capacity of 15,000 liters and the dimensions will be:

Diameter................................ 2400 mm

Cylindrical height.................. 3500 mm

Backwashing pumps:

Backwashing flow of each vessel is 6.81 – 9.8 m^3/h

A design flow of 9 m^3/h will be adopted. Therefore, flow requirement will be 9 x 46 = 414 m^3/h.

Two backwashing pumps with a design flow each of 230 m^3/h and a pressure of 6 kg/cm^2 will be installed.

Cartridge filters:

a) Number of cartridge filters

Considering a length of 1000 mm and a flow per cartridge of 2.8 m^3/h, the number of cartridges to be installed are:

$$\frac{414}{2.8} = 148\ cartridges$$

b) General characteristics

Design pressure...................... 6 Kg/cm^2

Diameter................................ 990 mm

Cylindrical height.................. 1.350 mm

Filtrating surface .... 39 m^2

Figure 15.3. Photography of different components of a cleaning system.

15.4 Flushing

To preserve elements from biological growth and to help maintain performance over time, membranes are flushed prior to use to eliminate residual preservatives in the product stream.

Upon shutdown for seawater systems, a flush with RO permeate is recommended to remove the high concentration of salts from the membranes. The permeate valves should be opened to drain during this flush to prevent damage to the membranes. If RO permeate is temporarily unavailable, the membranes should be flushed with RO feed at low recovery (with brine valve wide open). The membranes should then be flushed with RO permeate as soon as it is available. The seawater RO/NF system should not be left unflushed with a high concentration brine on the membrane surface.

The quantity of water used in both the normal RO System Start-up and the RO System Shutdown flushing should be equal to or greater than that which is retained in the system. For standard 16-inch x 40 inch elements assume 130 liters per element. For standard 8-inch x 40 inch elements assume 37.85 liters per element. The pressure of the flushing pumps is approximate to the pressure drop in the pipes plus 2 bars in the membranes.

The high pressure pump should also be flushed. The estimated time for flushing is approximately 10 minutes.

A way of saving money is sharing the osmosis washing pumps with the flushing pumps. Usually, washing water flows can be diverted to the flushing volumes (consider time between 5 to 15 minutes).

Pressures will be a bit higher (the washing pumps will be dimensioned to 6 bars), but this energy loss through valves is balanced with the cost saving resulting from sharing pumps.

Post-treatment reagents should be avoided. In figure 15.4, the two arrangement alternatives to avoid contact with reagents during flushing are shown.

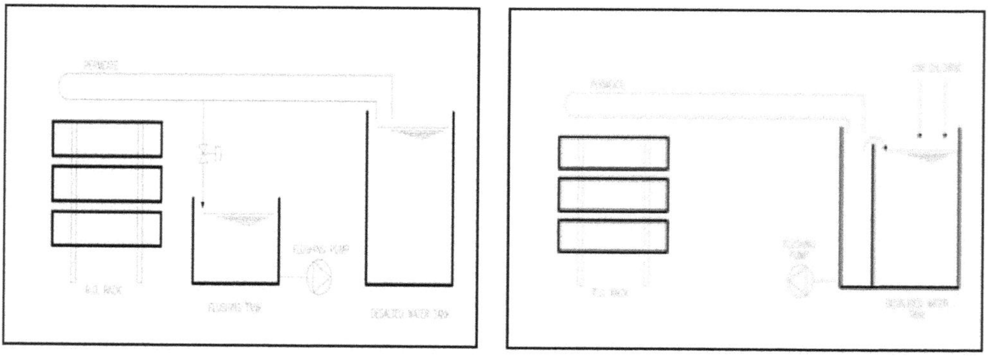

Figure 15.4. Flushing alternatives.

Flushing will be effective when conductivity of water after flushing is below 5000 μs/cm.

EXAMPLE

PUMP SIZINGS

Each rack produces 280 m³/h.

Number of vessels installed is : 92

First block .... 46 vessels

Second block .... 46 vessels

Product average flow per membrane is:

$$\frac{280}{92 \times 7} = 0{,}435 m^3 / h$$

Considering a recovery rate of 10% per membrane, flushing maximum flow is 2 – 4.5 m^3/h

This flow is for each membrane. There is no production during flushing so this will be the flow for each vessel. With a flow 4.5 m^3/h per vessel and 46 as the number of vessels being flushed

1st block .... 4.5 x 46 = 207 m^3/h

2nd block................................ 4.5 x 46 = 207 m^3/h

Flow will be 207 m^3/h.

Head loss will be calculated for each block of vessels

For the 1st block:

Feed flow .... 309 m^3/h

Production flow .... 139 m^3/h

Reject flow............................. 170 m^3/h

Average flow will be

$$\frac{309 + 139}{2} = 227 m^3 / h$$

Headloss in membranes when they are clean is 1 kg/cm^2 and 2.5 Kg/cm^2 when fouled

Two pumps (1+1) will be installed with the following characteristics:

Flow .... 230 m^3/h

Pressure................................. 6 Kg/cm^2

15.5 Service water deposit

Water is required for different services: reactives dilution, flushing, and chemical cleaning.

Service water deposits must have enough capacity to do a flushing of all racks (number of racks x water volume per rack). Deposits will be filled automatically with a automatic butterfly valve (ND 100) and controlled with a level transmitter.

In the past, permeate from an RO system was directly connected to the service water tank and the water overflowed into CO_2 chamber but the different chemical treatments to the treated water have forced operators to modify this arrangement by doing it independently.

A pipe will connect the service water tank to the flushing and chemical system. An automatic butterfly valve will be installed for the discharge of the service water tank.

16

TREATED WATER CONDITIONING

Osmotized water requires some additional post-treatment. These treatments include:

- Treated water SAR control.
- Boron reduction.
- Treated water re-mineralization.
- Treated water disinfection.
- Other treatments.

16.1 SAR control.

The average value of the SAR or sodium adsorption ration is around 9. This high value may cause damage not only to the soil but also to certain crops, due to an excess of sodium with respect to calcium and magnesium. For this reason it is better to re-mineralize or to add Ca and Mg directly to fertilizers.

SAR (Sodium Adsorption Ratio) of the water is a reflected in the following formula:

$$SAR = \frac{Na}{\left({(Ca \times Mg)}/{2} \right)^{1/2}}$$

Where Na, Ca and Mg are expressed in meq/L

And where the assessment criteria are as follows:

SAR> 12	Very bad damage.
9< SAR< 12	Moderate damage.
6< SAR<9	Growing problems.
5<SAR < 6	Initial problems.
3< SAR< 5	Low risk.
1< SAR<3	No risk.

16.2 Boron Removal.

Seawater contains approx. 5 mg/L of boron, which due to its size and change, is not well rejected by reverse osmosis. Boron has increasingly become a concern in recent years due to its adverse effects on agriculture at concentrations as low as 1 mg/L. Additionally, because the human health effects of boron are under investigation and not yet fully understood, the world health organization (WHO) has recommended a maximum concentration of 0.5 mg/L.

The typical boron rejection of seawater membranes at nominal test conditions is 92% - 93% depending on element permeability. This corresponds to about 80% - 86% boron rejection when operated in a typical commercial seawater system at a flux of 8 gfd (14 lmh).

Like other ions, the rejection of boron is influenced by various operating conditions including temperature, flux, and the ionic strength of the feed water. However, boron is unique in how its rejection is influenced by pH. The boron rejections stated above are typical at a pH of 7, when boron is present in the form of boric acid, $B\,(OH)_3$

The ionization equilibrium of boric acid, a weak acid, is represented as:

$$B\,(OH)_3 + OH^- \rightarrow B(OH)_4^-$$

The lack of charge and small size of the boric acid molecule results in poor rejection by a reverse osmosis membrane. But at high pH, Boron is present as the borate ion, which is well rejected by the reverse osmosis membrane due to its larger radius and negative charge. Figure 16.1 shows estimated nominal boron rejection of the typical sweater membranes at 24°C and 30°C as a function of pH.

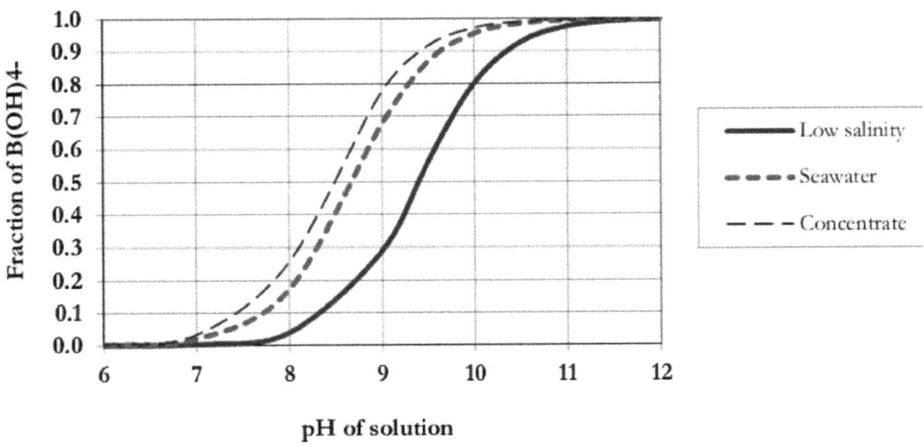

Figure 16.1. Nominal boron rejection in function of pH.

16.2.1 Design considerations.

A single pass equipped system may not be sufficient to achieve 0.5 mg/L boron levels when treating typical seawater, especially at high temperatures. For this reason, there are a number of design solutions to further reduce the permeate boron levels. These solutions include:

- Partial or full second pass reverse osmosis processing of first pass permeate with pH increase in the second pass using boron rejection brackish membranes in the second pass.

- Additional permeate polishing with boron selective ion exchange resins.
- First pass pH increase.

Two pass system.

This concept, shown in Fig. 16.2, has been widely proposed for systems where the boron requirement is between 0.4 and 0.5 mg/L, and the boron concentration of the seawater feed ranges between 4.0 and 6.3 mg/L. As explained above, high Boron retention is achieved at a pH higher than 9.5.

However, the rise in the first pass increases the membranes scaling with calcium carbonate and magnesium hydroxide. The first pass is therefore commonly equipped with high-pressure seawater RO membranes and operates at a pH below 8.2.

After significant reduction in the ion concentration in the first pass, the second pass is constructed from low-pressure brackish water RO membranes and operated at a pH of 9-10.5. Some of the first-pass permeate is by-passed to keep some minerals in the water, as well as for mixing with the second-pass permeate in order to reduce the pH in the product water. Due to the high feed pH in the second pass, special attention should be paid to calcium and magnesium concentrations, antiscalant should be added.

High boron rejection brackish water will be used in the second pass (BW30LE400, ESPAB), which will operate at pressure of 10—15 bars with a recovery of 90% in two stages and an average flux of approx. 35-40 l/h/m^2

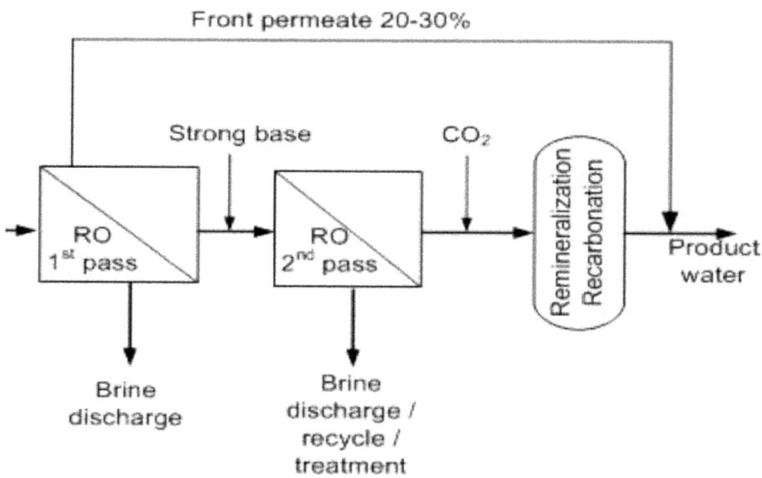

Figure 16.2. Schematic of two pass system to remove Boron.

Booster pumps for the second pass must be regulated with a frequency converter. Material used for equipment will be AISI 316. Butterfly valves of nominal pressure NP16 can be used. The control valve at the brine outlet must be a double seat valve, manufactured in AISI 316 (MASONEILAN or SAMSON).

Usually, there are very few tubes in the 2nd pass, so two design procedures can be followed:

- Install 2nd pass tubes above or beside the 1st pass rack with an independent pump for each rack.

- Install independent rack for 2nd pass with feed pumps for 2nd pass.

The disadvantages of this alternative are:

- High energy demand.
- Limited recovery (~85% at pH 10) due to chemical scaling (higher recovery requires further stages - more energy).
- Antiscalants required at relatively high concentration.

Increase in first pass pH.

Typically seawater can have 6000 mg/L $CaCO_3$ of hardness which is concentrated to more than 12000 mg/L $CaCO_3$ in the brine stream. Consequently, $CaCO_3$ scaling is inevitable at pH 9, even with antiscalants, which are considerably less effective at high pH values.

However, despite these high concentrations, real scale plant operations have demonstrated the absence of scale on the sea water membrane after operating at elevated feeds of a pH below 8.5. The high ionic strength and the use of antiscalant prevent such scale formation up to pH of 8.5.

Boron concentrations below 1 ppm will be obtained with high boron rejection membranes working at pH of 8.5. However, such membranes are typically characterized by lower permeate flux, which would imply a reduction in the rate of water production or an increase in energy demand.

Also, high boron rejection SWRO membranes could produce a boron concentration lower than 0.5 mg/L at pH less than 9. This scenario should be analyzed in detail, and probably a second pass will be required. However, using the elevated pH process will reduce the second pass.

Ion exchange.

The ion exchange system for boron reduction utilizes boron selective resin, which is regenerated with acid and caustic. The advantages of ion exchange resin are a low concentration of boron in the effluent (less than 0.1 ppm), low power consumption and low water loss. The disadvantages are the high cost of resin replacement and chemicals for regeneration. Also, because the ion exchange process does not result in any reduction of water salinity, the required level of permeate salinity must be produced by the first pass unit.

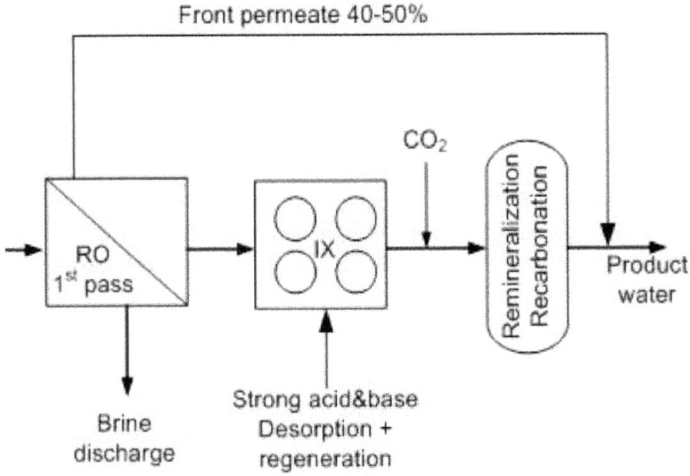

Figure 16.3. Schematic of an Ion exchange system to remove boron.

16.3 Treated water remineralisation.

Desalinated waters are a solution with a mean content of 280 mg TDS/l and approximately 72% of ClNa. They have a very low content of HCO_3 (< 4 mg HCO_3/l) and Ca (< 2.5 mg/L Ca/l) and Mg (< 4 mg/L). The Mg content is approximately twice that of Ca, both expressed in Mg/l-

The pH at the outlet of the reverse osmosis racks ranges from 5.3 to 6.7 depending on the plant. CO_2 content varies between 0.8 and 16 mg/L of CO_2

Certain waters may move outside this range as a result of specific treatments. The permeate with a high pH is related to the need to increase boron rejection. Desalinated water has an LSI of less than -4, as a result it tends to dissolve calcium carbonate. It also implies the risk of problems of corrosion due to its higher chlorine (60-200 mg/L of Cl) and sulphate content (2-6 mg SO_4) compared to its content of bicarbonates (2-4 mg/L of HCO_3).

The purpose or remineralization is generally to produce water with a Langelier level close to zero or slightly positive that remains stable in contact with the atmosphere, to increase the alkalinity. It is important accomplish this with a minimum consumption of CO_2 in order to minimize operating costs.

There are basically two methods to re-mineralize water: Either with calcium hydroxide (lime) and CO_2 or with limestone (calcite) contactors and CO_2.

The first method is the most widely used one. Over 90% of the existing desalination plants worldwide use a sequential feed of calcium hydroxide and carbon hydroxide to re-mineralize the water. The second one has been developed by the "centro canario del agua en España".

Investment of lime dosing systems is lower but the operation cost is higher due to the CO_2 consumption.

16.3.1 Remineralization with calcium hydroxide (Lime) and CO_2

The re-mineralization using lime and carbon dioxide follows the chemical reaction presented below:

$$2CO_2 + C_a(OH)_2 \rightarrow (HCO_3)_2Ca$$

There are different softwares used to calculate the lime and CO_2 dose, which will depend on the values of alkalinity and LSI. Lime oxide can also be used instead of calcium hydroxide but it needs to be converted into calcium hydroxide with water, as this is an exothermic.

Based on the equation above, 0.74 mg/L of hydrated lime (0.56 mg/L of quick lime) and of 0.88 mg/L of carbon dioxide would need to be added in order to increase the hardness and alkalinity by 1 mg/L (as $CaCO_3$) each. Therefore, for a target recommended dosage of alkalinity and hardness in the water of 100 mg/L the water produced by the desalination system will need to be treated with 74 mg/L of lime and 88 mg/L of CO_2. It has been proven that with 60 ppm of lime and 80 ppm of CO_2, corrosion will be avoided.

Once the amount of CO_2 and lime has been established, in order to mix them with the treated water, the following aspects must be considered:

- Calcium hydroxide (lime) must be prepared in the form of lime slurry (milk of lime) at 3% maximum.
- Depending on the type of installation and the necessary volume of lime, milk of lime can be dosed directly into the pipe of treated water.

- Usually, the lime slurry is blended with the fresh water produced by the process to create a saturated lime water which is then injected into the unconditioned desalinated water.
- CO_2 is dosed first. Saturated limewater is added to the system after the carbon dioxide in order to prevent calcium carbonate scaling. It is recommended to dose CO_2 in excess to not only prevent carbonate scaling but to regulate the pH between 6.5-8.

There are three (3) systems used to mix CO_2 and saturated lime water with the treated water

1. When the treated water deposit is not close the plant (at a certain distance).
2. Mix CO_2 and saturated lime water in separate chambers.
3. Mix CO_2 and saturated limewater in the pipe which feeds the treated water deposit.

- *When treated water deposit is not in the SWRO plant and is located a certain distance from plant.*

CO_2 is dosed in the suction side of pumps and lime slurry or saturated lime water in the discharge. A static mixer should be installed to guarantee proper mixing.

If the lime amount is not so large and dosing pumps can be used, this alternative will be selected instead of installing a lime saturator (the lime slurry deposit should have two or three hours of retention time and a filter must be installed to remove sands).

CO_2 dosage and lime slurry will be proportional to flow and a pH-meter will be installed in the discharge to regulate the lime slurry supply.

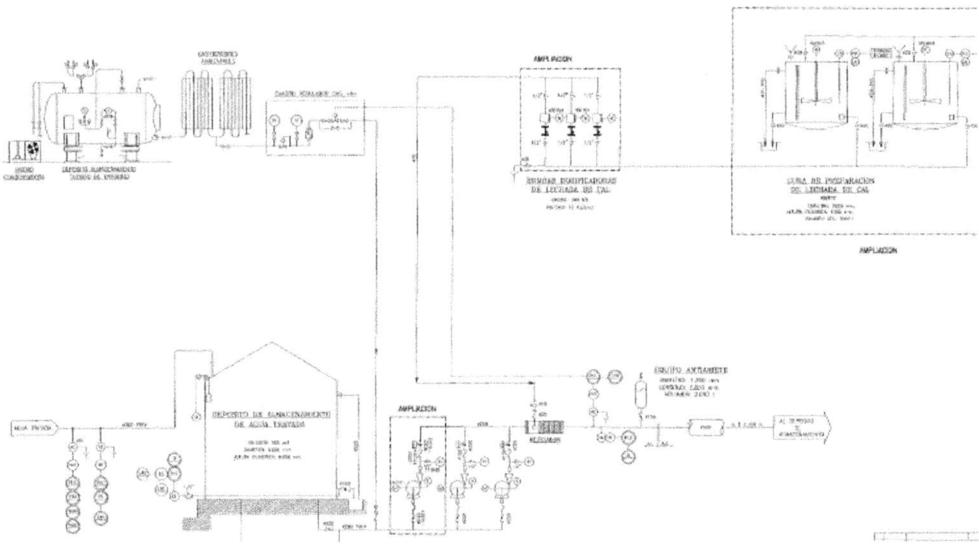

Figure 16.4. Remineralization I.

- *Mix CO_2 and saturated limewater in separate chamber next to the treated water deposit.*

To obtain a good efficiency the level of water in the chambers must be at a minimum 4.5 to 5 meters and if higher better.

CO_2 is distributed by porous tubes manufactured in AISI 316 from the manufacturer PORAL or equivalent. Required surface is calculated with the criteria of 6 kg/h of CO_2 per m^2. The number of elements is calculated through the information indicated by the manufacturer. CO_2 dosage will be proportional to the flow with a control valve installed in the CO_2 circuit and a flow transmitter (with equipment supplied by the CO_2 supplier). It is necessary to add that CO_2 injection is not a good system for obtaining a good efficiency (70-80% will be the ratio between the CO_2 absorbed by water and consumption).

Water saturated with CO_2 flows through a submerged passage into the following chamber where it is mixed with saturated limewater. To accelerate the mixing process, a low speed submerged agitator must be installed. The discharge to the treated water deposit must be done through an overflow to guarantee that the level of mixing chambers is independent of the treated water tank level.

Saturated lime water regulation will be done with a pH meter installed in the overflow of the deposit entrance and will regulate the flow of saturated limewater pumps.

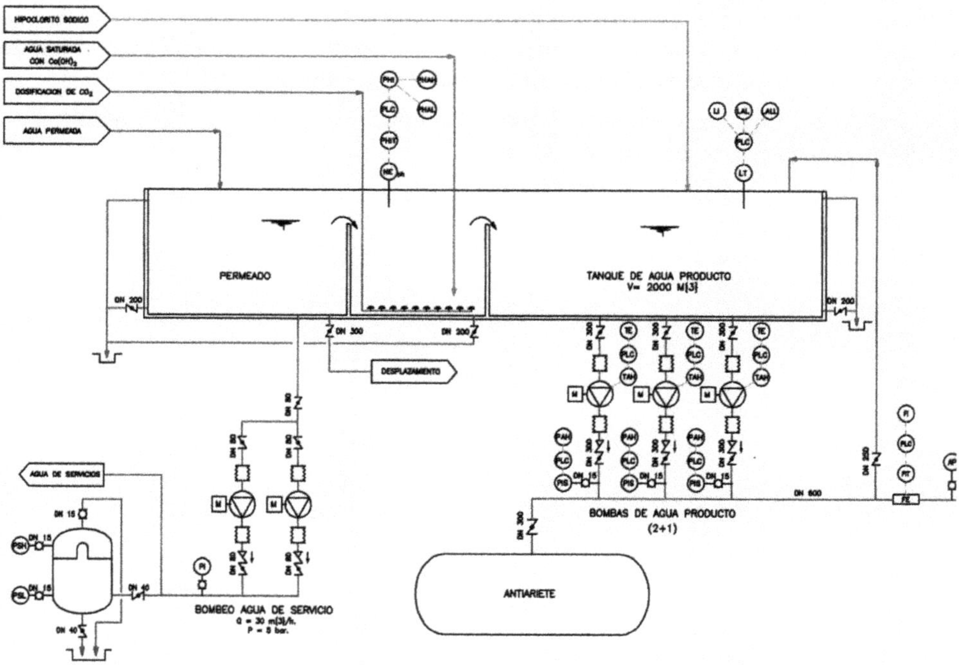

Figure 16.5. Remineralization II.

As indicated previously, to prevent calcium carbonate scaling, CO_2 must always be dosed before lime, and in order to enhance lime solubility, CO_2 is added in excess.

The deposit will consist of four (4) chambers

A	B	C	D

TREATED WATER	
LIME	CO2
NON TREATED	

A) A first chamber with non-treated water. This chamber is necessary for flushing and back-washing.

B) A second chamber for CO_2 injection.

C) A third chamber to introduce lime.

D) A final treated water deposit.

- *Mix CO_2 and saturated limewater in the pipe which feeds the treated water deposit.*

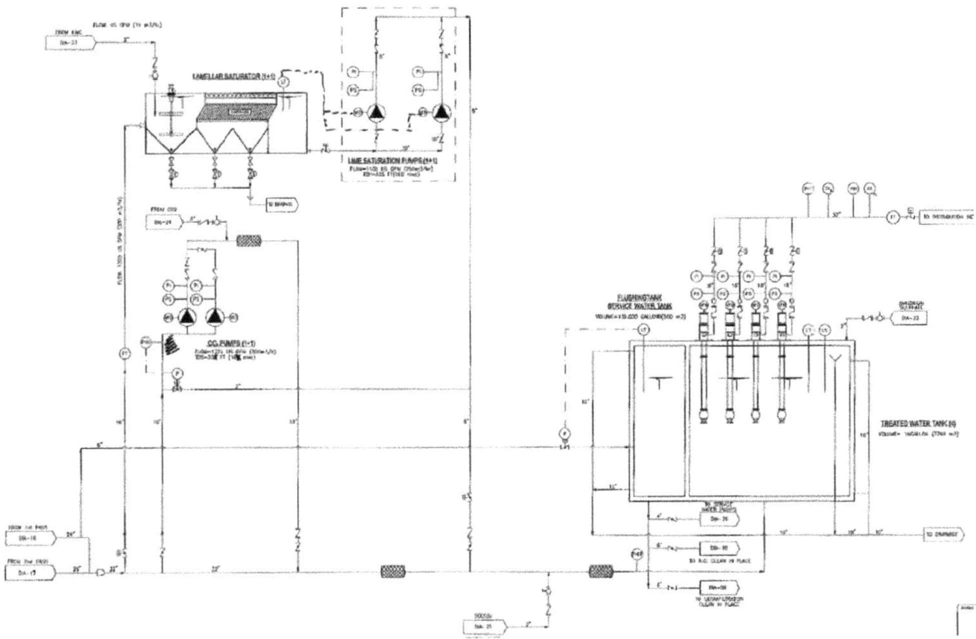

Figure 16.6. Remineralization III.

Mixing is previously carried out to the treated water storage tank with static mixers.

First, turbulent mixing of CO_2 with treated water is carried in an independent circuit, to the main collector. Centrifugal pumps (1+1) will be used with a flow of 2 m^3/h, 1 ppm of CO_2 and a pressure of 2 bars (depending on headloss of static mixer). These values are provisional, as there is not enough experience so these values are obtained from static mixers suppliers.

CO_2 dosage is proportional to the flow through a control valve installed in the CO_2 circuit and a flow transmitter (equipment supplied by the CO_2). The chore is to dissolve all the CO_2 in the water and guarantee the production of carbonic acid before mixing it with the saturated lime water.

Water mixed with CO_2 is re-circulated and mixed with the rest of the treated water in the 1st static mixer of the main line (as can be observed above in figure which shows the details of water recirculation and CO_2 injection system). After mixing the water and CO_2, the treated water will be added to the saturated limewater in the 2nd static mixer. PH is controlled In the discharge of this element and it will regulate saturated limewater flow to the main circuit.

The difference between system no 2 and system no 3 is the efficiency of the CO_2. In system no 2 an efficiency of 90-95% is achieved. In system no 3, it can be achieved an efficiency of 100%.

16.3.2 Saturated lime solution preparation

Water is saturated by lime with different amounts depending on the temperature of water and according to the values of table 16.1 below shown:

Table 16.1. Lime milk preparation.

Temp°C	10	20	30
Kg/m³	1,76	1,65	1,53

Saturated lime water is created in a saturator where lime slurry is blended with water from the desalination process and is thoroughly mixed. After a retention time, saturated limewater is then injected into the unconditioned desalinated water.

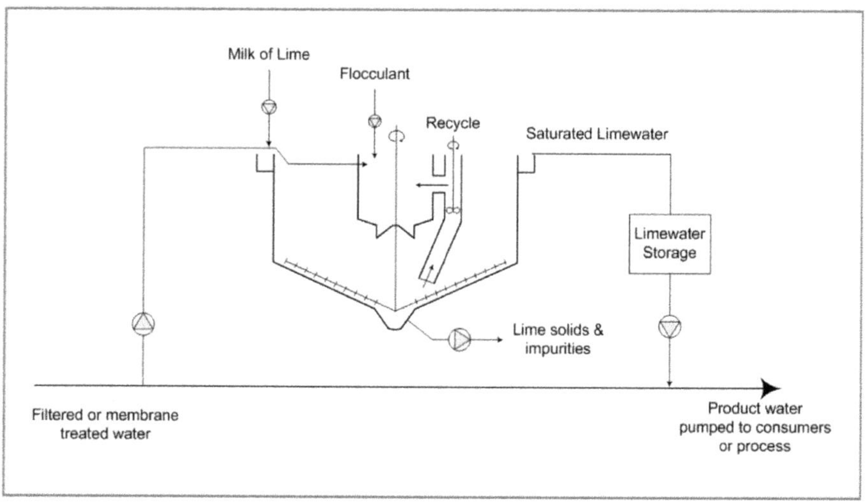

Figure 16.7. Scheme of saturator.

Speed converters will be installed in pumps. With a pH meter we will regulate the amount of saturated limewater which will be mixed with treated water.

The supply to the saturator will be regulated with a level transmitter and a butterfly valve with a positioner installed in the inlet. A flow transmitter will be installed to know what the water flow is.

Most installed saturators are circular with a turbine to mix lime slurry with the water to be saturated. Circular saturators are expensive and difficult to transport. It is possible to substitute them with lamella saturators and a flocculation chamber.

Figure 16.8. Photographs of saturators.

Design velocity of lime saturators is of 2 m/h and of 10 m/h for lamella saturators.

It is necessary to take into consideration that when the quality of water is good, there are no solids except for calcium carbonate as impurities in the water are not soluble (sand) and will be removed in the saturator.

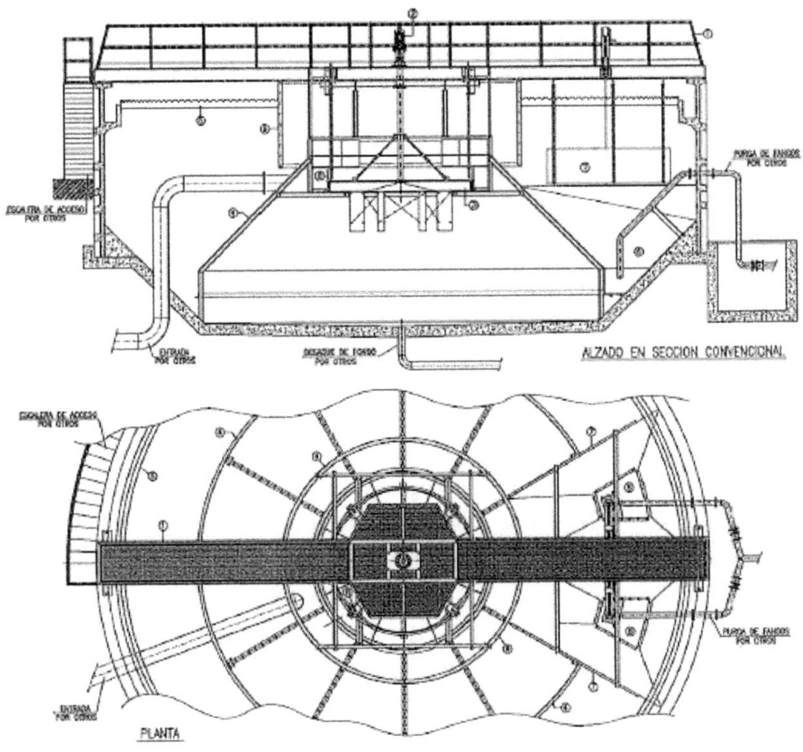

Figure 16.9. Lime saturator.

Treated water has CO_2 that when reacting with excess lime produces calcium carbonate. Most calcium carbonate will not be in soluble form and will add turbidity to the water (Usually between 100 150 NTU). This turbidity disappears when saturated limewater is mixed with CO_2, as CO_2 dissolves it and forms bicarbonates.

16.3.3 Lime slurry (lime milk) preparation

Lime is usually delivered and stored in silos as powdered hydrated lime or as pebble-lime (which is then slaked to produce hydrated lime). Specific equipment is required to prepare lime slurry at a maximum 3%.

The required equipment is indicated below:

-**Lime silo:** Usually two (2) silos are installed depending on the size of the installation. Storage silos are generally vertical cylindrical units with a straight side to diameter ratio of between 3:1 and 4:1 for best flow characteristics. A conical bottom is generally supplied with a minimum cone angle of 60 degrees.

Sizing the silo is a function of the usage rate. A 7-15 day minimum storage is typically used. The minimum storage time may be altered depending on lime availability, transportation constraints or special problems.

Silo design should account for local seismic, wind and snow loadings as well as potential mass flow of the product. An allowance should be made for some freeboard in the silo; typically 3' to account for the angle of repose of the product. This freeboard also increases the efficiency of the bin vent filter operation.

Material of construction for the silos include welded or bolted steel, concrete, or stainless steel. Steel construction is the most popular for these systems. Shop fabricated welded steel silos are economic up to 14' in diameter. Beyond this size shipping constraints preclude their use and field fabricated silos are used. Bolted steel factory coated steel silos offer an attractive option at diameters bigger than 12'. They are shipped knocked down and can be quickly erected on site with local labor.

Steel silos are generally supplied with epoxy or inorganic zinc coated exteriors. Interior product areas can be left uncoated but are more commonly coated with an epoxy primer. For difficult products the cone area can be coated with a special super slick Teflon based coating, which greatly improves flow in this area.

Silo accessories should include the following:

- Access ladder (caged) with rest platforms (if required) and perimeter guard rail.
- Vacuum pressure relief hatch.
- 4" fill line with long radius elbows typically of Sch. 40 or 80 carbon steel c/w compression couplings and grounding strips.
- High and low level probes (capacitance type).
- Knife gate shut off valve for maintenance.
- Other accessories as desired for inventory control.

With regard to silo coatings we recommend that the silo interior be epoxy primed after surface preparation to SSPC-SP6 with special anti-friction coatings applied to assist in flow promotion. The exterior and skirt interior area are usually epoxy primed with an acrylic finish coat.

-**Volumetric dosing system** with frequency converter.

-Dilution tank with mixer, These tanks are made of black HDPE (high density polyethylene) and are fitted with a mixer, a water board with a solenoid valve, a level check system and spigots needed for pumping and emptying a lime milk loop. In order to prevent any dust emission, the dilution tank is completely closed. Usually, one dilution tank is installed for each storage silo. However, if there are more silos, it is possible to install (1) mixing tank for (2) silos.

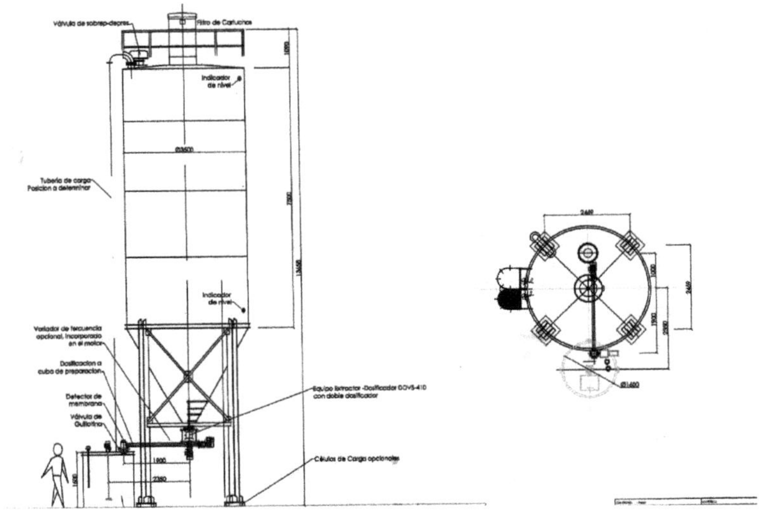

Figure 16.10. Lime storage silo.

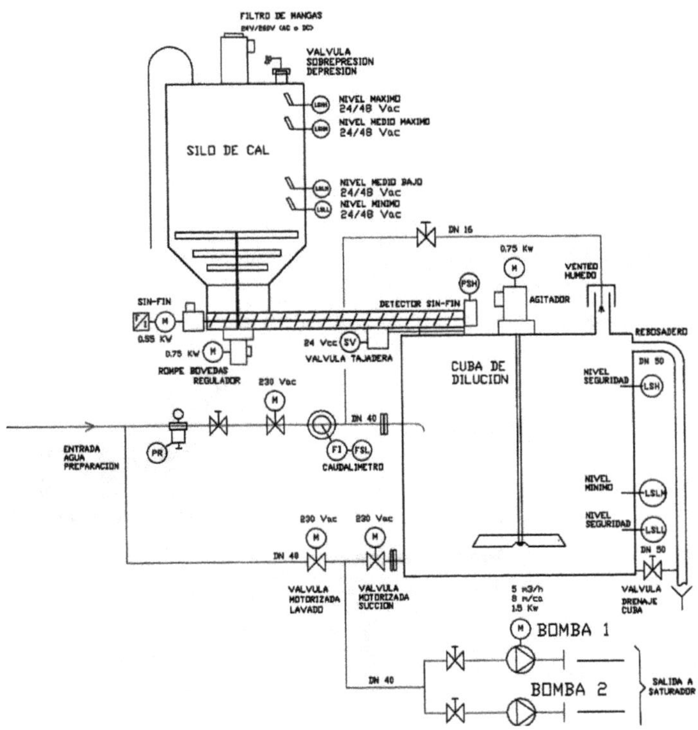

Figure 16.11. P&ID Lime slurry preparation.

- There are (2) ways of dosing lime slurry; with centrifugal or with dosing pumps.
- **Centrifugal pumps** (1+1). Pumps with a flow of 10-15 m^3/h manufactured in cast iron with semi-open impeller and a maximum speed of 1450 rpm. Sealing systems are available with either a standard packed gland arrangement with a hydraulic seal or a drop in cartridge mechanical seal.

In some installations, peristaltic pumps have been used instead of centrifugal pumps. Abrasive particles (sand) passing through peristaltic pumps will commonly increase wear on the internal surface of the tubing and lead to reduced life as a consequence. However regular tube replacement will avoid this potential problem.

Operation of this preparation system requires a constant flow of dilution water, so calcium hydroxide flow will be adjusted through a frequency converter. Calcium hydroxide dosage will be proportional to the treated water flow and dilution will be carried out with treated water. Consequently, there is no water consumption as water returns to the system saturated of lime.

- **Dosing pumps** (n+1). The pumps with have a metallic head (AISI 316) specially designed for lime milk. Pump flow is regulated with speed convertor or stroke electronic positioner.

- **Control.** Make-up batches of hydrated lime slurry to pre-set concentrations (3%) are required and the slurry feed rate is adjusted in response to the process demands (treated water flow) through a variable speed feeder.
  Once, the first mixing tank is empty, the second mixing tank (if available) is used and the solution can be prepared in the empty tank.
- **Piping and valves.** Lime slurry piping should be designed to ensure a minimum flow velocity of 4 m/second with a minimum line size of 1-1/2". Care should be taken to eliminate short radius elbows. Avoid in line dilution of the lime slurry. Pump suction lines should be as short as possible with no bends. Piping materials can be PVC, ABS or carbon steel.

Valves should be chosen with care to avoid problems. For pump isolation valving straight through diaphragm valves are recommended. Lime slurry take off valves should be of the pinch valve type, which are self-cleaning with a 1" minimum diameter. Pinch valves can be of the straight through, reduced port or tapered style, operating in a timer controlled on-off manner. Alternately they can be operated in a modulating manner provided an intermittent fully open pulse is provided by the control scheme.

16.3.4 Carbon dioxide storage.

Carbon dioxide is delivered and stored at the desalination plants in compressed liquid form. The liquid is converted into gaseous form in evaporators at 3 bars at 12°C dissolved in carrier water. A typical CO_2 installation consists of liquid CO_2 storage tanks equipped with refrigerator units; vaporizers; vapor heaters; carbon dioxide dosing systems with pressurized gas solubility panel and in line gas feed system.

Figure 16.12. Photography of different elements of remineralization.

EXAMPLE

To design a remineralization system, we need to find the RO projection with the best permeate quality (16°C with new membranes). Permeate analysis is TDS < 400 mg/L (TDS < 400 mg/L)

Cations	*Anions*
Ca^{++} 0.8 mg/L (as ion)	*Cl- 152.2 mg/L (as ion)*
Mg^{++} 2.5 mg/L (as ion)	*$SO_4^{=}$ 5.2 mg/L (as ion)*
K^{++} 4.6 m/l (as ion)	*HCO_3^- 2.0 mg/L (as ion)*
Sr^{++} 0.013 m/l (as ion)	
Na^+ 93.6 mg/L (as ion)	

PH: 5.8

$$5.8 = 6.3 + \log \frac{HCO_3^- \,(ppm\ as\ CaCO_3)}{CO_2\,(ppm\ as\ ion)}$$

$$5.8 = 6.3 + \log \frac{2 \times 0.82(ppm\ as\ CaCO_3)}{CO_2(ppm\ as\ ion)}$$

Then $CO_2 = 5.19$ *ppm*

Temperature: 16 °C

Total hardness (TH) =	Ca^{++} *0.8 mg/L x 2.5 as* $CaCO_3$	
	Mg^{++} *2.5 mg/L x 4.17 as* $CaCO_3$	
	12.425 mg/L	*as* $CaCO_3$

The values to be obtained are:

Sodium (Na^+)	< 150 *mg/L* Na^+
Cloride (Cl^-)	< 200 *mg/L* Cl^-
Total hardness (TH)	> 60 *mg/L* Ca^{++} <> *150 ppm* $CaCO_3$

Remineralization methods

A. CO_2*, lime and calcium carbonate*

Remineralization reaction

$$2CO_2 + Ca(OH)_2 \rightarrow (HCO_3)_2Ca$$

TH to add= 150 – 12.425 = 137.58 ppm $CaCO_3$*.*

<u>Ca(OH)₂ necessary</u>

$$137.58\ ppm\ CaCO_3 \times \frac{1\ ppm\ Ca^{++}}{2.495\ ppm\ CaCO_3} = 55.14\ ppm\ Ca^{++}$$

$$\frac{55.14}{40.1} = 1.375\ mols\ Ca^{++} <> 1.375\ mols\ Ca(OH)_2$$

$$Ca(OH)_2 = 1.375 \times 74.1 = 101.89\ ppm\ as\ Ca(OH)_2$$

$$Ca(OH)_2 = 1.375 \times 74.1 = 101.89\ ppm\ as\ Ca(OH)_2 \text{ at } 100\%$$

HCO_3^- formed.

1,375 moles Ca^{++} *correspond to 2.75 mols of* HCO_3^-

$HCO_3^- = 2.75 \times 61 = 167.75$ *ppm as* HCO_3^-

CO_2 necessary.

1,375 mols Ca^{++} *correspond to 2,75 mols of* CO_2

$CO_2 = 2.75 \times 44 = 121$ *ppm* CO_2

CO_2 *to add =* CO_2 *necessary-* CO_2 *permeate*

CO_2 *to add = 121 – 5.19 = 115.81 ppm as* CO_2

New treated water analysis is

CATIONS	Anions
Ca^{++} 55,94 mg/L (as ion)	HCO_3^- 169,39 mg/L (as ion)
Mg^{++} 2,5 mg/L (as ion)	$SO_4^=$ 5.2 mg/L (as ion)
K^{++} 4.6 m/l (as ion)	Cl^- 152.2 mg/L (as ion)
Sr^{++} 0.013 m/l (as ion)	
Na^+ 93.6 mg/L (as ion)	

pH of treated water must be between 6.5 and 8.5.

An extra dose of CO_2 will be necessary to keep an optimum pH

$$pH = 6.3 + log\frac{HCO_3^-(ppm\ as\ CaCO_3)}{CO_2(ppm\ as\ ion)}$$

$$6.5 = 6.3 + log\frac{139(ppm\ as\ CaCO_3)}{CO_2(ppm\ as\ ion)}$$

$$If\ pH = 6{,}5 \Rightarrow CO_2 = 87.7\ mg/l$$

$$If\ pH = 8{,}5 \Rightarrow CO_2 = 0.88\ mg/l$$

On the other hand, LSI must be above 0.

Ph	**LSI**
6.5	-1.31
8.5	0.69
7.9	0.09

CO_2 amount will be calculated on the pH of reference to have an LSI slightly positive

$$If\ pH = 7.9 \Rightarrow CO_2 = 3.5\ mg/l$$

DOSAGE

<u>CO_2</u>

To remineralize	115.81 ppm
To adjust Ph	3.5 ppm
TOTAL	119.31 ppm

A 10% margin correction will be added to the stoichiometric value to consider the reactions with feed water, so CO_2 consumption will be 131.24 ppm:

$Q = 27{,}000\ m^3/d$ <> 3543.5 Kg/d

Storage

Conditions of supply

Form......................................Liquid

Pressure.................................21 Kg/cm²

Temperature...........................-18 ºC

Density..................................1,02 Ton/m³

Transport..............................Containers of 14 Tn

Capacity................................15 days

Required capacity

$$C = \frac{3543.5\ kg/day \times 15\ days}{1} = 53.15\ tons$$

No de tanks............................1 of 50 Tn

<u>*Ca (OH)₂*</u>

$Ca(OH)_2$ (at 100%). consumption will be 101.89 ppm

The total $Ca(OH)_2$ (consumption with a 5% margin will be

Q = 27,000 m³/d <> 120,4 Kg/h $Ca(OH)_2$ 100%

Design value will be 120 Kg/h $Ca(OH)_2$ 100%.

Richness of calcium hydroxide: 95 %

$$\frac{120}{0,95} = 126.3kg/h <> 252.6l/h$$

<u>*LIME SLURRY PREPARATION*</u>

Lime milk flow at 5%:

$$\frac{252.6}{0,05} \times \frac{1}{1.000} = 5.05m^3/h$$

Two (1+1) pumps will be installed with a unit design flow of 5 m³/h

No of lime milk preparation systems: one (5 m³)

SATURATED LIME WATER PREPARATION

Concentration of lime water at 20ºC: 1.650 mg/L

Maximum flow of water

$$Q = \frac{27000\ m^3/day \times 101.89 \times 1.05/1\ kg/m^3}{1.650\ kg/m^3} = 1750.7\ m^3/day <> 72.9\ m^3/hour$$

LIME SATURATOR

No of decanters......................1

Type of decanters...................sludge recirculation

Design flow............................72.9 m³/h

Minimum retention time ......50 min.

Design upflow velocity...........3.5 m³/h/m²

Required settling surface....... $\frac{72.9}{3,5} = 20.8m^2$

Model selected........................Accentifloc no 7

Diameter................................5400 mm

Total height............................3575 mm

Cylindrical height..................2975 mm

Surface ....20.3 m²

Velocity ....3.45 m³/h/m²

Volume..................................74 m³

Retention time.......................61 min

Equipped with a turbine of 1 kW and a velocity of 20 rpm

Storage of $Ca(OH)_2$ *(Density = 0,5 Tn/m³)*

Storage time..........................15 days

$$\frac{120}{0,95} \times 24 \times 15 = 45.473,7\,kg$$

It will be installed (1) silo of 85 m³ of capacity

16.3.5 Limestone remineralization.

Limestone is a natural mineral made of calcite (calcium carbonate). Processing through calcite media allows us to dissolve this calcium source and in reaction with CO_2 to add both calcium hardness and bicarbonate alkalinity to water. Remineralization using calcite and carbon dioxide follows this equation:

$$CO_3Ca + CO_2 + H_2O \rightarrow (CO_3H)_2Ca \qquad \text{with calcite}$$

$$Ca(OH)_2 + 2CO_2 + H_2O \rightarrow (CO_3H)_2Ca \qquad \text{with lime}$$

Based on the equation, 1mg/L of calcite and 0.44 mg/L of carbon dioxide would need to be added in order to increase the desalinated water hardness and alkalinity by 1 mg/L as $CaCO_3$, so the use of calcite is usually more cost effective than the application of lime. The main advantage of the use of calcite is in the control of suspended solids.

In praxis, a parameter called empty bed contact time (EBCT) is used to describe the residence time inside the bed. It is calculated by dividing the volume of the limestone bed by the flowrate, and is given in minutes.

Due to the low solubility of calcite at the near neutral pH of treated water, pH should be reduced in the permeate to 4-4.5. Typically, they are designed to operate in the up flow mode as gravity or pressurized filters

- **Parameters in calcite contactor design**

- *Superficial velocity or loading rate*

The typical average design surface loading rate is in a range of 4 to 8 $m^3/m^2/h$ and the maximum 10 $m^3/m^2/h$. Higher velocities will lead to higher values of turbidity of permeate.

- *Empty bed contact time.*

It is in the range of 10 to 30 minutes. The needed EBCT increases with decrease in desalinated water temperature. The contactor should be designed for the minimum temperature. A contact time of 11 minutes is normally selected. For colder waters and below 14°C, 14-20 minutes are used.

- *Calcite contactor bed height.*

In practice the height of the calcite beds ranges from 1 to 3.5 m.

- *Calcite particle size and purity.*

Particle size affects the contact surface area and therefore the contactor performance

There are two possible solutions: It is important that the calcite be free from dust (diameter < 40 microns) and have a uniform particle size. This should be 1-3 mm, although plants are frequently found operated with particle size of 2-4 mm.

- **Type of calcite contactors.**

There are two types of calcite contactors available: upflow and down flow. Variations in the process configurations have their own advantages and disadvantages

Operation of down flow contactors require frequent calcite replenishment in order to maintain a constant bed height, as well as frequent backwashing to eliminate the fine particles from each calcite loading event. When a newly replenished contactor is brought online, backwashing is required to eliminate carryover of fine particles. The backwash waste stream generated will require an additional particle removal process prior to disposal into the sanitary sewer. Upflow calcite contactors minimize turbidity problems by not requiring the intermittent turbidity surges during the backwash process, and are favored in large installations. In up flow contactors, since less/no backwashing is required, dealing with a waste stream is a not an issue.

However a number of challenges have been detected with these filters:

- Turbidity problems in the effluent during reloading operations.
- Decreasing performance due to the continually diminishing height of calcite bed between loadings.

Researchers at the Canary Islands Water Center in Spain have developed an up flow contactor design (DrinTec™) to overcome the aforementioned difficulties by:

- 1. Maintaining a constant calcite bed height via a continuous feeding system.
- 2. Eliminating turbidity events during reloading by separating the filter bed surface from the built-in silo.

We will focus in this book on this system. The proposed upflow limestone contactor modules are built in concrete tanks. The operation of the up flow contactor is as follows: Water flows upwards. Once it has flowed through the bed, the now re-carbonated water enters a no-flow area and then leaves the cell via an internal perimeter overflow weir made of FRP. The overflow is then channeled to an outlet of the tank.

The tanks have a built in reserve silo in their upper part. A series of small feeding funnels are placed at the bottom of the silo which are supported by a steel structure. The funnels guide the calcite from the silo to the surface of the bed located under water. In this way, the product is dosed directly from the silo onto the bed by gravity, replenishing it continuously as it becomes dissolved. As a result, dosing occurs very gradually, approximately one granule of calcite per funnel per minute, under normal operating conditions. This steady dosing allows it to operate in the plant with a superficial velocity of up to 18 m/h, without turbidity surpassing 1 FNU.

The built in silo in the tanks allows it to operate autonomously for several weeks. The product in the reserve silos remains dry. The calcite is moistened only in the down pipes to the bed and, as the material is relatively inert, i.e. does not absorb water or dissolve easily, there are no problems of obstruction.

They can be divided into six parts, starting from the bottom:

1) the air injection system;
2) the water distribution zone;
3) the limestone bed,
4) the still-water zone,
5) the limestone feeding zone, and
6) the limestone silo. (see Figure 16.13 and fig 16.14 for details).

Part 1.- The air injection system consists of a network of PVC pipes embedded underneath the contactor concrete base. These pipes are collected into a manifold outside the concrete tank. The air blower is connected directly to this manifold.

Part 2.- The water distribution zone is located in the lower part of the tank. It consists of a 300 mm height infiltration platform. This platform consist of 996 x 996 mm^2 platforms that are placed against each other to conform an uniform infiltration surface. Contact surfaces between platforms, as well as platforms and walls, are sealed with a plastic resin. Each platform has three large openings on each sidewall to facilitate water distribution. They also have internal beams to withstand more than 6,000 kg/m^2. Each module has 81 nozzles, uniformly distributed on its surface.

Part 3.- The limestone bed zone (made of crushed limestone particles 1.5-2.5 mm in diameter) is located above the infiltration platform. The height of the limestone bed may vary according to specifications but it is normally between 1 and 2 meters.

Part 4.- The bed surface is approximately 800 mm below the perimeter of the spillway. This area is called the still-water zone. This distance reduces the possibility of particles elutriation even at high superficial velocities.

Part 5.- The limestone feeding zone is located above the still zone. It consists of a steel supporting structure and 996 x 996 mm^2 feeding plates. Each plate has nine (9) equally spaced funnels. These funnels allow distributed crushed limestone particles on to the limestone the bed surface, without creating turbidity. Limestone falls onto bed by gravity and is regulated by water.

Part 6.- The feeding structure in itself makes the floor of the limestone built in silo. The material stored in the in-built silo falls thru the funnels on to the limestone bed below and by gravity. The in-built silo is loaded from above thru strategically located openings. Material supplied in regular bags or big-bags can be transported above the silos with a moving crane and unloaded directly into them.

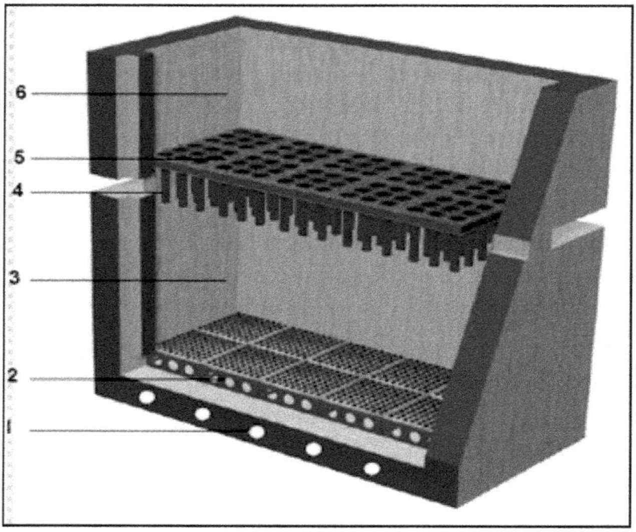

Figure 16.13. Section of limestone contactor I.

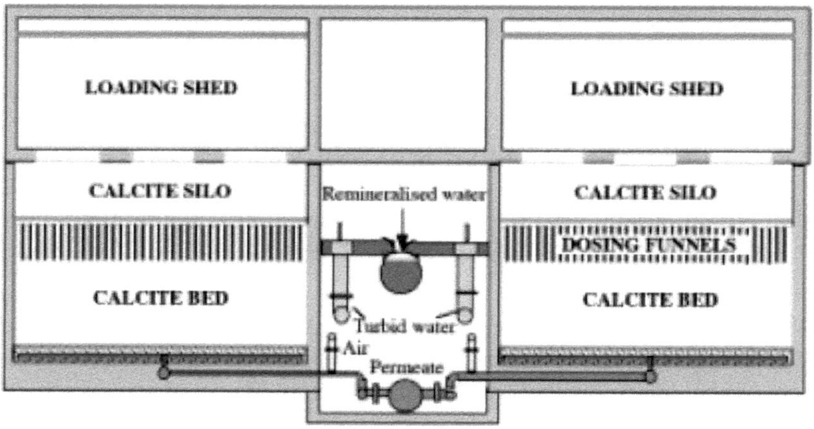

Figure 16.14. Lime stone contactor. II

16.4 Treated water disinfection.

The primary goal of water treatment is to ensure that the water is safe to drink and does not contain any disease-causing microorganisms. It will be necessary to add disinfectants to guarantee not only that the water is safe to drink but to prevent microbiological growth in storage deposits and water distribution networks.

The most commonly used disinfection alternatives are:

- Sodium hypochlorite.
- Calcium hypochlorite.
- Cloramines.

16.4.1 Sodium hypochlorite.

Usually 0.5 to 1 ppm as Cl_2 is dosed in the treated water deposit or in the pipe. Injection is carried out with a dosing pump and a chlorine analyzer is installed in the discharge of treated water pumps.

16.4.2 Calcium hypchlorite.

Calcium hypochlorite is water-soluble and comes in the form of powder, pellets, granules and tablets. Two preparation tanks with mechanical agitation are required. Maximum concentration of solution is of 20%. One tank will be in preparation/stand-by mode and the other tank will be used for dosage.

16.4.3 Chloramines.

Chloramines are produced when chlorine is added to water containing nitrogen compounds such as ammonia nitrogen. There are three types of chloramines that are produced, including monochloramines, dichloramines and trichloramines. From the water treatment perspective, the desired form of chloramine is monochloramine, because of the biocidal properties and minimal taste and odor production. Dichloramines and trichloramines are less desirable because of the chlorinous tastes and odors that they produce.

The major benefits of monochloramines include their tendency to produce fewer disinfection by-products such as TTHMs and HAA5s, minimal chlorination tastes and odors, persistence to reach distant areas of the distribution system.

Chloramines are less effective as biocide than free chlorine for inactivating pathogenic microorganisms. For that reason, chloramines are generally not used as a primary disinfectant. Chloramines are, however, an excellent secondary or final disinfectant because they form a very stable and persistent residual.

The reaction mechanism is:

$$NH_3 + ClOH \rightarrow ClNH_2 + H_2O\ (Monochloramines)$$

Dichloramines and trichloramines can also be produced with excess of chlorine and ammonium.

The equipment required to produce chloramines is essentially the same equipment required for all chlorination systems. Ammonium sulphate will be stored in a plastic material deposit and will be dosed with diaphragm pumps.

Residual chlorine control will be done with a chlorine analyzer, to analyze the free residual chlorine. Special analyzers will be required including buffers solutions (acetic acid, etc) to analyze total chlorine.

Although feedback control alone can be used to adjust for changes in process flow this is typically coupled with feed forward flow pacing. The primary control is feed forward flow pacing, which can react very quickly to changes in flow since there are not significant process delays associated with the feedback loop. The feed forward ratio is then slowly trimmed by the feedback loop. This is shown in Figure 16.15

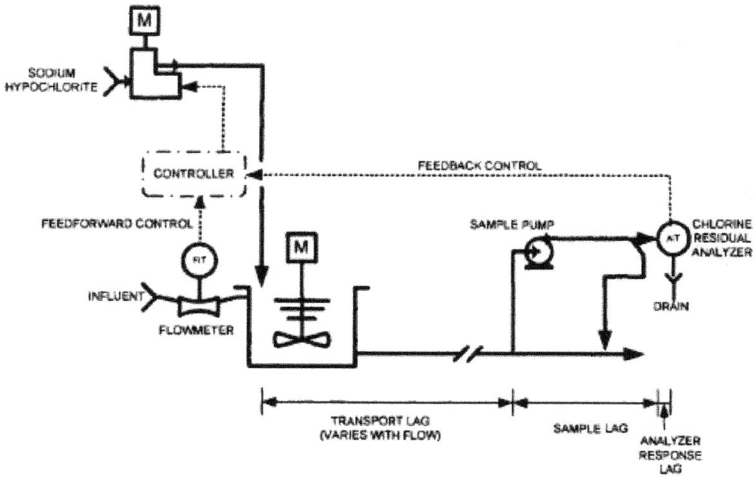

Figure 16.15. Control philosophy of chlorination.

16.5 Other products used to condition water.

Typically in Spain no additional chemicals are added in the post-treatment, but other chemicals can be used:

- **Calcium chloride $CaCL_2$** To add calcium hardness.
- **Magnesium sulphate $MgSO_4$** To remineralize and add permanent hardness. While it is acceptable for total hardness to be added only as calcium hardness for the purpose of protecting the water distribution system against corrosion, it could not be acceptable and provide somewhat limited human health protection and is of lesser agricultural value. This it is the reason that Magnesium is added in some desalination plants.
- **Magnesium Chloride** To mineralize and add permanent hardness.
- **Sodium bisulphite $NaHSO_3$ to** regulate the amount of residual chlorine in the water.
- **Phosporic acid H_3PO_4** to control the corrosion of piping.
- **Sodium Hydroxide Na(OH)** for final pH adjustment.

17

TREATED WATER DEPOSIT AND PUMPING.

Figure 17.1. Treated water pumping station.

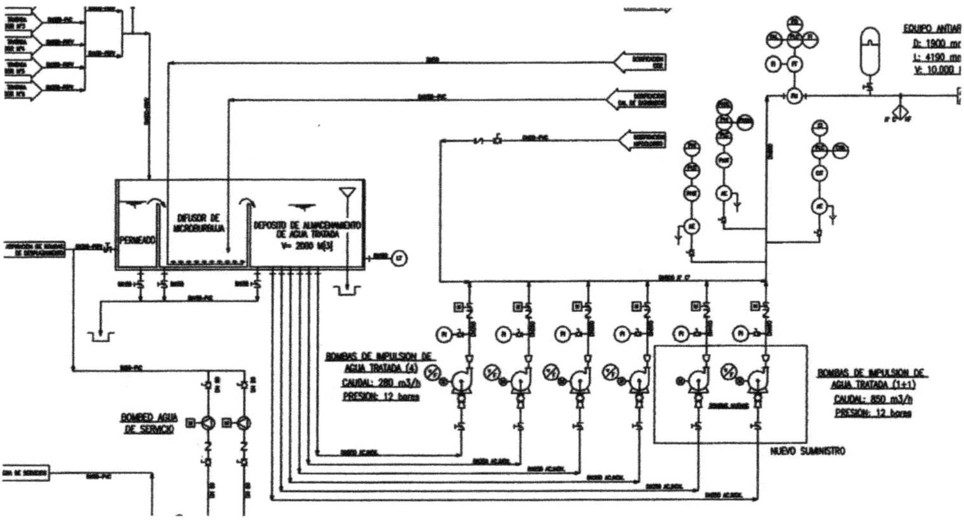

Figure 17.2. Treated water P&ID.

The capacity of a treated water deposit, the number of pumps and pressure is usually defined from Project specifications. The deposit should be, in theory, sized to pump the water during the lowest time periods of electricity cost. Consequently, the treated water deposits have a storage capacity of approximately 16 hours (in order to avoid the peak cost periods).

The main equipments we use are the following:

- Centrifugal pumps.
- Butterfly valves.
- Piping.
- Instrumentation.
- Water hammer deposit.

17.1 Centrifugal pumps and valves

Usually, working pressures are below NP 16 but in some cases can be up to NP 25

Split casing pumps are typically used. These pumps can handle large flows to medium pressures with a good efficiency (higher than 80%). The material of pumps should be AISI 316.

Butterfly valves. They are usually supplied in NP16 with the butterfly in stainless Steel. When the design pressure is 25 bars, metallic seats are used instead of EPDM seats.

Small valves (less than 300 mm NP10) are supplied with pneumatic actuators. For bigger valves or pressures of NP16 or higher, electrical actuators will be used.

17.2 Piping

Ductile iron+cement mortar as an internal lining is the predominant form used for treated water applications. The cement mortar comprises a mixture of cement and sand to a ratio of between 1:2 and 1:3.5. Portland cement is used for potable water; and for low pH it is common to use high alumina cements.

Carbon Steel+epoxy painting as an internal lining will be used for interconnecting piping. Helical seam carbon Steel pipe with neoprene for external and epoxy painting for internal painting is an alternative to ductile iron.

17.3 Instrumentation

The instrumentation required is:

- **PHIT.** A pH meter will be installed when remineralization chambers are installed at the outlet of a lime saturated water chamber. With static mixers, the pH meter to be installed is placed at the entrance to the reservoir.
- **LT.** Deposit level transmitter.
- **FIT.** Flow transmitter at the discharge side of pumps.
- **PHIT**. pH-meter at the discharge side of pumps.
- **Cl_2IT.** Residual chlorine analyzer at the discharge of pumps (in document attached it is possible to read the definitions of different chlorine measures).

There are instruments which measure total or free residual chlorine by injecting a solution of acetic acid and potassium iodide into the samples. Triple Validation chlorine monitoring systems are widely used. These systems process signals, all covering the same parameter and perform a check on the signals under one of a number of optional algorithms. The inputs can come from our own sensors or from existing third party products.

Figure 17.3. Chlorine measurement system.

17.4 Antiwater hammer deposit

There are two types of deposits. Bladder pressure tanks are a type of tank containing pressurized nitrogen and water separated by a membrane. They are pre-charged with nitrogen at the factory. They are used with seawater or small deposits of treated water. Tanks without a bladder (air in contact with water) are more common.

The system will consist of the following:

- **Tank** – a steel pressure vessel constructed in accordance with ASME, Section VIII, Div. 1 Pressure Vessel Code Requirements and to the working pressure necessary for the required service. It will be internally painted with an epoxy resin compatible with potable water.
- **Sight glass** – for visual indication of the tanks water level also for the full operating range of the system.
- **Level detector** – used to provide the feedback information necessary for the water level control in the tank. Several alternatives are available to transmit the water level information to the control panel including conductance probes, a differential pressure transmitter or a capacitance probe. Four (4) level switches will be installed and will give low and high level alarms.
- **Air Compressor** – an air cooled, two-stage, reciprocating type with an air receiver. The compressor works in conjunction with the unit control panel and provides the air for the add-air function of the control system.
- **Additional features** – valves are provided to complete the systems package. Butterfly valves will be installed in the feed, discharge and drainage. A check valve will be installed in the discharge.

18

NEUTRALIZATION, EFFLUENTS TREATMENT AND BRINE DISCHARGE

The effluents from SWRO result from:

- physical pretreatments: backwash water from filtration units. They may have ferric chlorine dosing.
- Reverse osmosis lines: Rinsing, Brine.
- Membrane cleaning units: dirty chemical solutions. Cartridge cleaning.

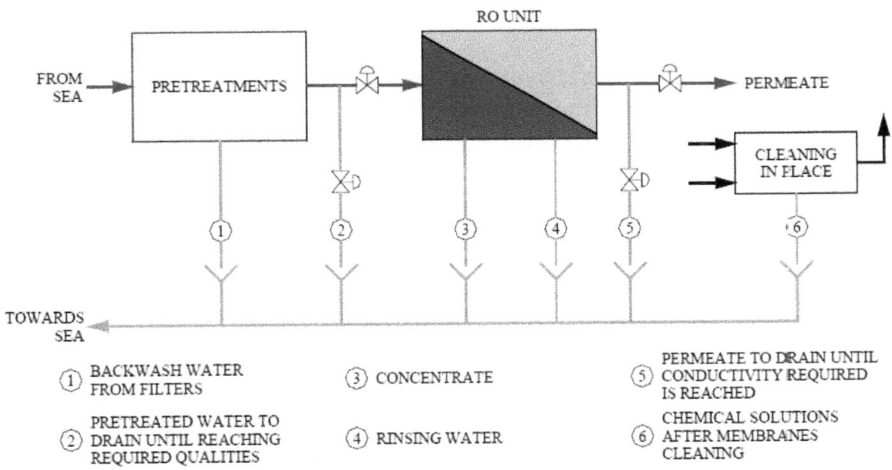

Figure 18.1. Effluents of SWRO.

These effluents are characterized:

- For the first ones, by the high loads of suspended solids containing biological, mineral and organic matters;
- For the second ones, regarding the concentrate, by a high content in dissolved salts (generally from 1.7 to 2.5 times that of the seawater content) and by an acid pH (pH about versus a seawater pH close to 8).

- For the third ones, by a pH that may be very alkaline (pH 11) or very acid (pH 3) and strong loads in biological, mineral and organic matters. Generally, a chemical cleaning of the membranes is performed with two types of solutions, first with an acid solution and then with an alkaline solution. Neutralization of membranes washing is done with sulphuric acid or with lime saturated water for acid cleanings. However, acid cleanings are not normal. Alkaline cleaning is the normal procedure for these kind of applications.

Therefore, it is advisable to discharge both of them in a stirred buffer tank in order to obtain auto-neutralization and then convey the mixture at a slow flow rate into the concentrate drain, to strongly dilute.

With regard to the quantities of produced effluents:

- For the physical pretreatments, they vary from 2% to 10% of the seawater pumped, according to: the selected filtration process: filtration on granular materials (sand, anthracite + sand), microfiltration using washable elements, ultrafiltration; the microbiological, organic and physical-chemical characteristics of the seawater to be treated.
- The concentrate produced by the RO units vary from 40% to 60%. The RO units discharge, in addition to the concentrate:
- At each start-up, pretreat water until the required quality is obtained (especially with respect to turbidity), as well as permeate until its conductivity matches the required value;
- At each shutdown, rinse water in order to reduce the salinity of the water contained in the concentrate zone of the membranes, so that the direct osmosis phenomenon is limited or avoided, since it might dry the membranes. This rinsing can be performed with seawater or permeate, each one containing, or not, a biocide such as sodium bisulfite.
- Rejects from chemical cleanings mainly depend on the frequency of these cleanings.

Treatment of effluents can be divided in two (2) groups:

1. **Neutralization without suspended solids treatment.**
2. **Neutralization with suspended solids treatment.**

18.1 Neutralization without suspended solids treatment

This treatment is used in plants with sand filters without coagulation (ferric chloride). Effluents have a low concentration of suspended solids. Such as water coming from deep vertical or horizontal wells.

Washing frequency of filters is infrequent (more than 48 hours) and membrane washing frequency also (every three months).

A neutralization tank which collects all effluents and has an storage capacity equivalent to the volume of (1) sand filter backwashing unit will be used. The tank is usually buried to collect all the effluents using gravity.

(1+1) centrifugal pumps will be installed to recirculate and convey the mixture at a slow flow rate into the concentrate in (1) or (2) hours. These pumps can be centrifugal and manufactured of plastic material with automatic priming or submerged and made of painted carbon steel with sacrifice anode.

To guarantee full neutralization of effluents, pumps will recirculate the water into the tank while pH is continuously monitored. From a set of automatic valves, water is discharged to the drain network when pH is acceptable.

As commented above, a pH meter must be installed. In addition, a flow transmitter may be required to control the discharge rate. Here are the details of P&ID neutralization

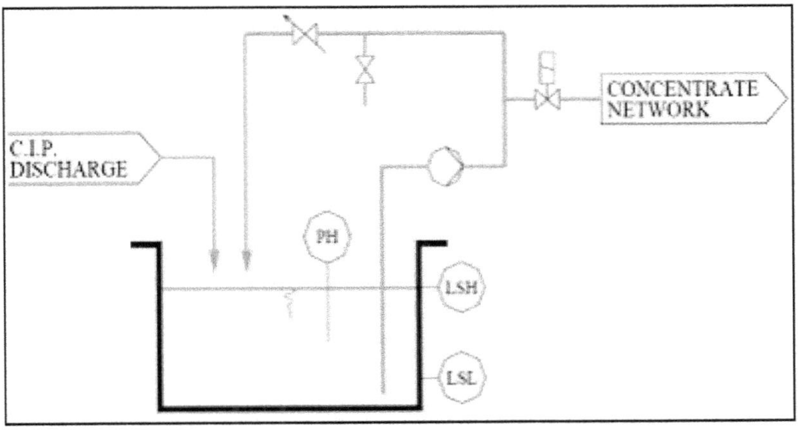

Figure 18.2. Neutralization deposit.

18.2 Neutralization with suspended solids treatment

This treatment is used in plants where sand filters+coagulation (ferric chloride) is used as a pretreatment. Effluents present a high suspended solids content and a reddish color associated with ferric hydroxide. When precoat filters are installed, the diatomaceous discharge will need to be treated.

Two (2) independent treatment systems are installed in these plants. One treatment for sand filters back washing, and another treatment for membranes and cartridge filters backwashing.

18.2.1 Filters back washing treatment.

It is necessary the have a tank with a minimum storage capacity equivalent to the volume of two or three backwashes of filters (better if for three).

It is necessary to take into consideration that the filters can be washed every 24 hours (especially first stage filters) and every 48 hours for second stage filters. In larger plants, the interval between washes can be of one hour (1) or less.

For filters of 50 m^2, the tank must have a volume of approx. 1050 m^3 and must have an agitator installed to keep the solids suspended. Some aspects must be considered in the design:

1. **If the backwash water is to be recovered because seawater is used instead of brine**

The treatment must include a coagulation-flocculation followed by a clarification. For coagulation-flocculation, an anionic polymer is used. Hydrated lime will be used if pH must be raised. Flocculation is performed in a tank equipped with an adjustable low speed turbine, while the clarification is performed in a settling tank, of a lamellar type. The treated water contains less than 10 mg/L of suspended solids, which is quite similar to seawater. Upflow design velocity of decanters will be between 7-10 m/h. With higher velocities, there is a potential spillage of sludge, even though high rate decanters can operate at velocities up to 15-18 m/h.

The lamella settling tank will have a sludge recirculation system and a thickener system. The design capacity of decanter will depend on the number of filters and frequency of backwashing. For a 100,000 m^3/h production plant, the maximum flow will be 350 m^3/h with a double filtration (24+16).

In waters with a high content of organic matter, it may be not recommended to recirculate the water from the effluents treatment plant to the intake of swro, in spite of water savings. It must be considered to mix the water with rest of effluents and discharge it in order to minimize the potential biofouling of membranes.

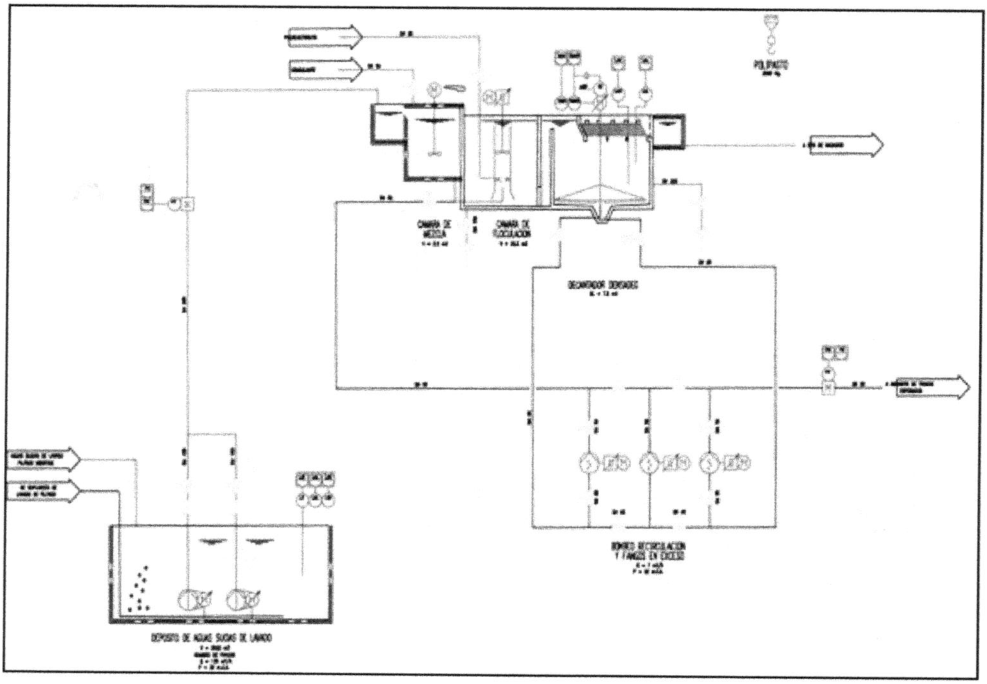

Figure 18.3. Suspended solids treatment system.

2. **Backwashing water is not recovered (Brine is used as backwash water).**

- <u>*Treatment of suspended solids using a thickener - decanter.*</u>

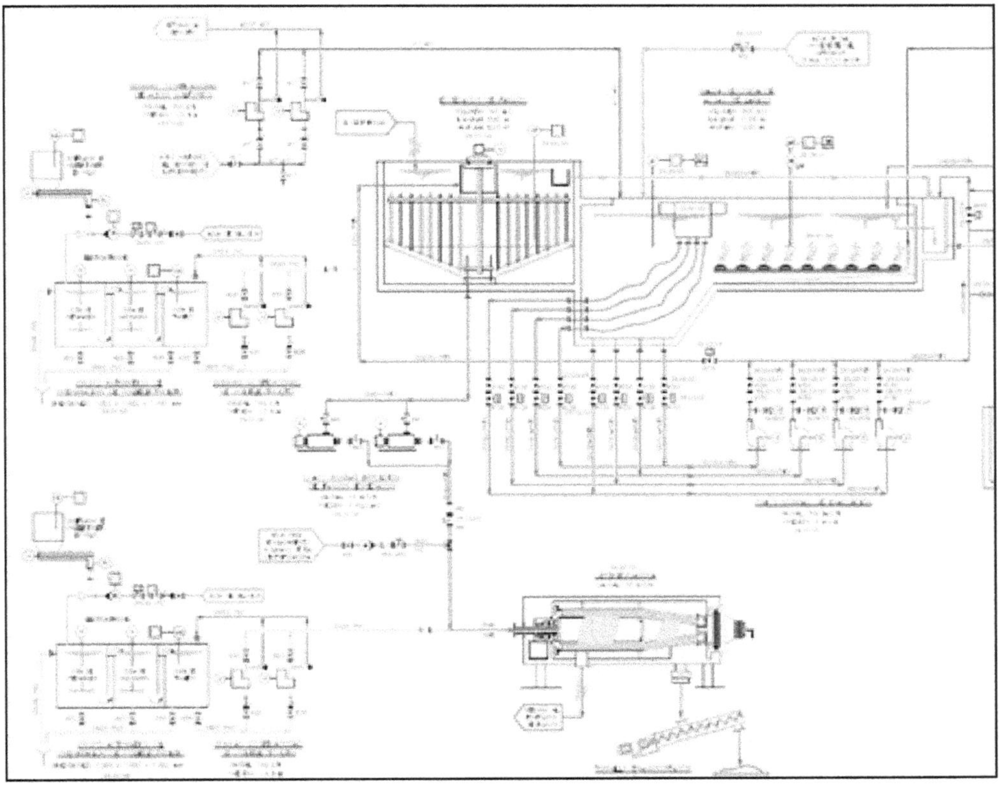

Figure 18.4. Treatment of suspended solids using a thickener-decanter.

In this case, backwash water must be treated in a decanter thickener (without sludge recirculation) with quick mixing chamber for reactives (polyelectrolyte and/or lime slurry) and designed for a velocity of 2 to 3 m/h. It must be taken into account that treated effluent is mixed with brine in a ratio of 1 to 12 so water with a concentration of suspended solids of 50 ppm when diluted with brine is reduced to 5 ppm.

Parts in contact with water in decanter-thickener setup should be manufactured out of polyester except for the shaft which should be rubber lined or ebonite covered.

If lamellas are installed in the decanter, the upflow velocity could increase up to 10 m/h, but it will force us to increase the elevation of the decanter in order to have enough storage capacity.

- *Treatment of backwash water with a DAF.*

Backwash water filters must be washed in a DAF. The equipment should be manufactured out of plastic with a maximum design capacity of 160 m^3/h, in cells of dimension of 7x2.7 meters, and a velocity of 3.3 m/h.

For example, for a flow of 350 m^3/h, two (2) units would be required including (2) water pressurization pumps and one (1) common flocculator.

This type of treatment is more expensive that the alternative of a decanter-thickener. The drawback of this system is that you have to use a sludge tank for storage before delivery to drying beds. The deposit must include a mixing chamber to mix with lime and submerged agitator.

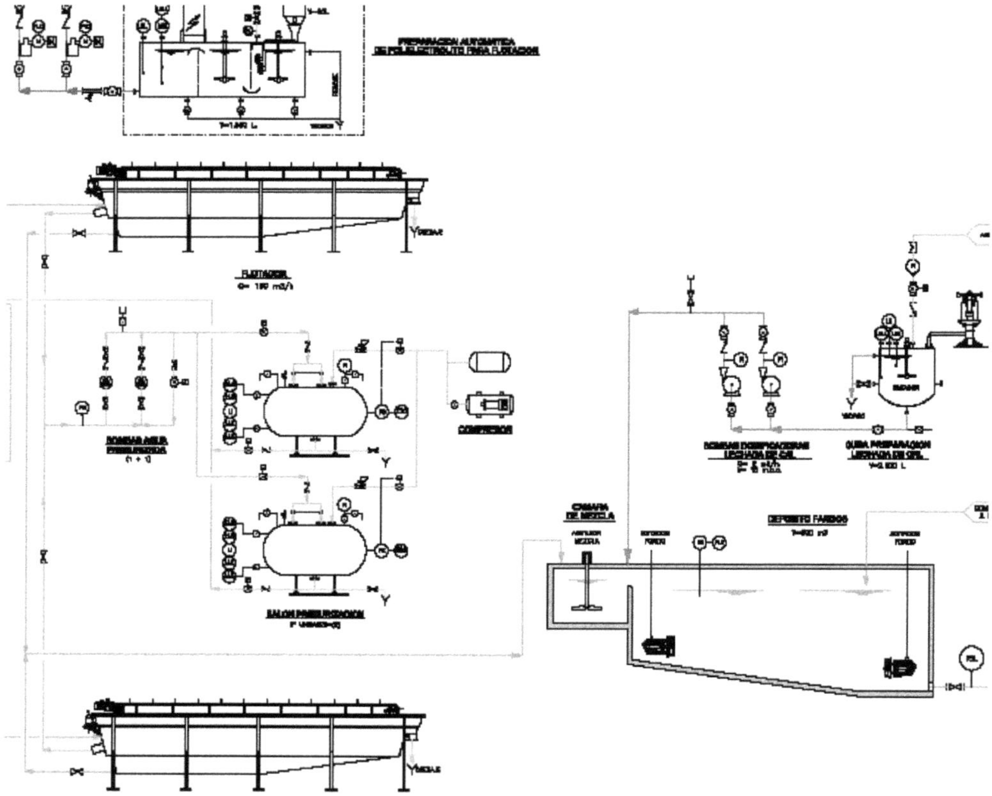

Figure 18.5. Treatment of suspended solids with DAF.

In both cases, the sludge treatment can be carried out with a centrifuge or belt filter.

An additional factor to consider if a centrifuge is selected as the method to dewater is that the sand and anthracite which escape from filters can damage the dewatering unit as they are highly aggressive.

Duplex for centrifuges and AISI -316 for belt filters should be the materials to be used. Screw pumps in both circumstances will be with a rotor in duplex.

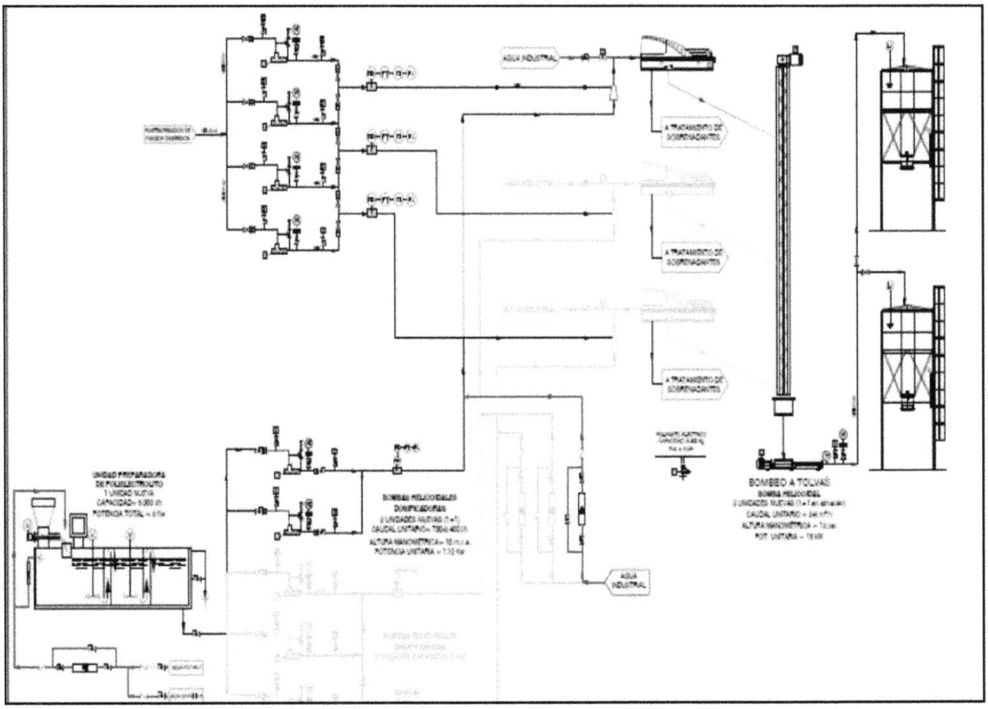

Figure 18.6. P&ID of dewatering system.

18.2.2 Membranes and cartridge filters washing.

In this case, a neutralization tank will be used with a storage capacity equivalent to the volume of on rack backwashing. In both cases, pumps will be used for recirculation and neutralization.

18.3 Brine discharge

The various density differences between the brine and the receiving water represented by the buoyancy flux causes different flow characteristics of the discharge. The dense RO effluent flow has the tendency to fall as a negatively buoyant plume. Figure 18.7 illustrates the typical behavior of positively or negatively buoyant jet discharging into the receiving water through a submerged single port outfall.

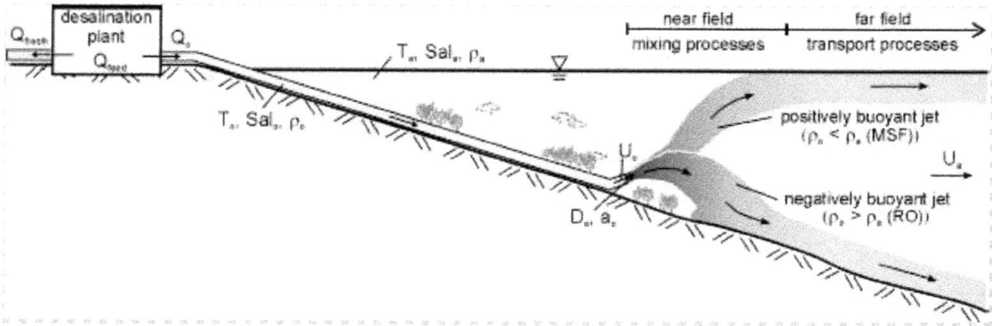

Figure 18.7. Brine discharge detail.

Ocean outfalls are classified according to their location (onshore surface discharges / offshore submerged discharges), by their mixing features (single port / multiport) and their effluent characteristics (positively buoyant, or negatively buoyant). However, such discharges should be analyzed carefully and generally be avoided due to their limited mixing characteristics, high visibility, and their need for large scale coastal constructions.

Shoreline discharges may cause shoreline impacts by causing high concentrations accumulating in the near-shore region due to the limited mixing characteristics of these discharges. Further direct impacts are caused by the often necessary large scale discharge and protection structures (wave protection, stilling basins, etc.), and their effect on coastal currents and sediment transport characteristics.

Therefore, it is recommended to apply modern efficient mixing devices, which overcome the limitations of the traditional surface onshore discharges. Such single or multiport submerged diffuser systems are characterized by their flexible location and their high mixing rates (Figure 18.8). These discharge technologies follow two main principles, aiming for enhanced effluent dispersion in the receiving environment and providing an adequate discharge sitting to avoid pollutant accumulation, to protect sensitive regions and to utilize natural purification processes.

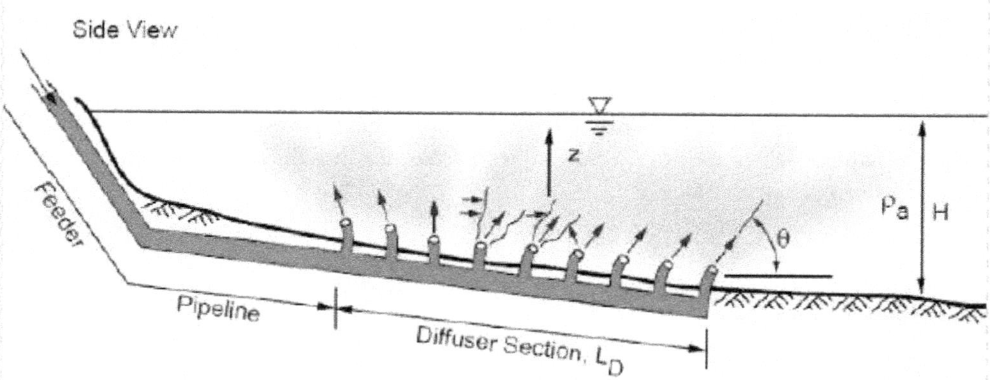

Figure 18.8. Brine discharge details II

There are different materials and designs available, depending on the construction method (Figure 18.9) applied, and the prevailing ambient conditions.

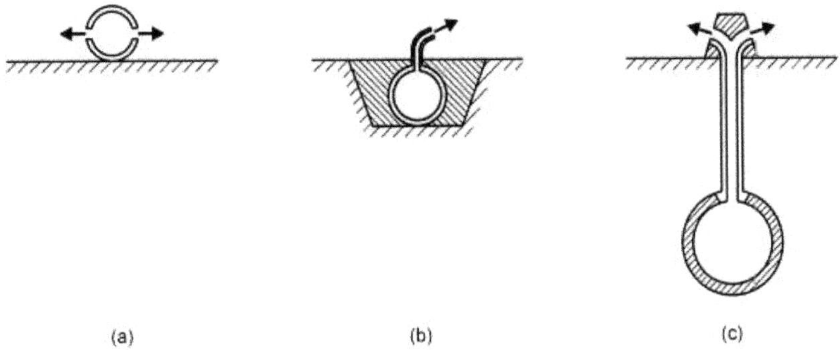

Figure 18.9. Typical construction details for multiport diffusers in water bodies: (a) Diffuser pipe on bottom with port holes, (b) diffuser pipe buried in trench with short risers, (c) deep tunnel.

Diffuser configuration:

For the optimization of the diffuser, the following criteria must be met:

- The diffuser section should be placed perpendicular to the prevailing ocean current. If there is no one prevalent current direction, then a T-port diffuser can be used.
- Design flows must be discharged satisfactorily through the ports to ensure continuity of flow. Generally, the total cross-sectional areas of the ports should be less than 0.7 times the cross-sectional area of the main pipe at any point in the diffuser. A port diameter of less than 75 mm is susceptible to blockage.
- Maintain sufficient flow in each port to prevent the intrusion of seawater by gradually increasing port sizes towards the end of the pipe and ensure the densimetric Froude Number for each port is greater than unity.
- Ensure an even distribution of flows, through all the diffuser ports, because the flow is directly related to the achievable initial dilution, and the worst performing port will be considered as representative of the performance of the diffuser.
- Maintain scouring flows within the diffuser section.
- Optimum dilution will be obtained with diffusers discharging between 45° and 60° to the horizontal and with alternate ports directed in opposite directions.
- The distance between any two ports must be such that the plumes do not merge during the rise of the buoyant plumes.
- The jet velocity will be not less than 3,5 m/sec.

The ports can be equipped with a check valve made of reinforced elastomer (Figure 18.10). These valves operate easily and reliably. They enable an operator to isolate the outfall if the discharge stops. In addition, the pressure required to open them can be selected in order to homogenize the distribution of the flow on all the diffusers. Thus, the upstream valves, which are the most advantaged, will require a higher differential pressure than those located downstream.

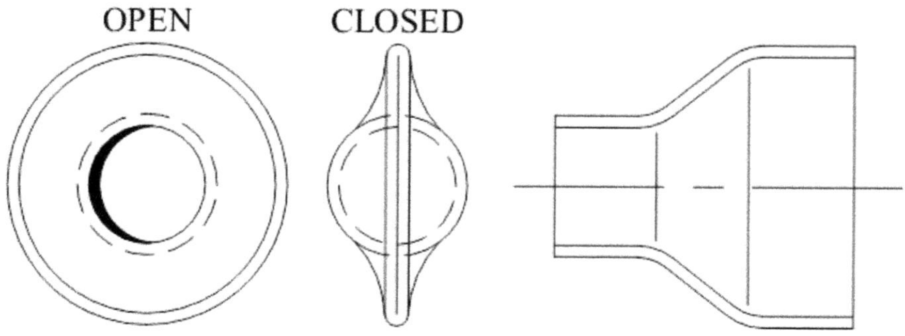

Figure 18.10. Check valve installed in a diffuser.

A multi-port diffuser consists of a header pipe containing two or more ports (with or without risers) which inject a series of turbulent jets at high velocity into the ambient receiving water body. These ports or nozzles may be connected to vertical risers attached to an underground pipe or tunnel or they may simply be openings in a pipe lying on the bottom. Multi-port diffusers are typically used where:

- Maximizing dispersion is imperative;
- Effluent flow rates are greater than 157 .73 m^3/h.
- Bathymetry is not extreme and
- Underwater slope stability is good.

Intermittent flow could cause sediment ingress into the diffuser pipe during times of no-flow. Appropriate port valves could be used to prevent this problem.

Outfall configuration.

The optimization of the main pipe diameter depends on the present and future flow conditions and available or practical head (gravity or pumps). The design should ensure:

- A main pipe velocity of greater or equal to 0.7 m/sec. is maintained to prevent deposition of solids and provide scouring abilities;
- The maximum flows are discharged with available gravity head or cost-effective pumping, taking into account the increase in roughness during the lifetime of the outfall and all losses at fittings in the main pipe and diffuser.

19 ELECTRICAL EQUIPMENT

Electrical equipment and services:

1. Electrical distribution.
2. Transformation centers.
3. Medium voltage motors.
4. Motors.
5. MCC.
6. Speed converters.

19.1 Electrical distribution

Electrical network main supply voltages are:

- 123 kV.
- 66 kV Used when energy demand is important > 10 MW.
- 20 kV for small and medium SWRO <10MW.

Supply voltage will depend not only on energy demand but on location and country.

19.1.1 Connection to the utility network

The most commonly used configurations for connection are as follows, in the order of increasing power demand and service reliability:

- o MV duplicate supply service, (power demand < 25 MVA).
- o MV duplicate supply service with double busbar. (> 25 MVA).

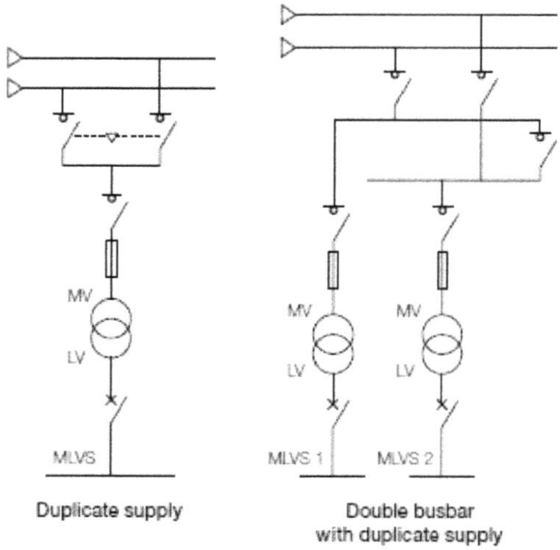

Figure 19.1. Connection to the utility network configurations.

19.1.2 Electrical architecture

The recommended electrical architecture for large desalination pumps is:

- An MV duplicate supply service from the Utility substation, and a double MV distribution configuration.
- The different process units are supplied at LV by separate transformers, with a double ended LV distribution and interconnected MV switchboards.
- The largest motors can be supplied directly at MV.
- Power factor correction is provided by MV capacitor banks.

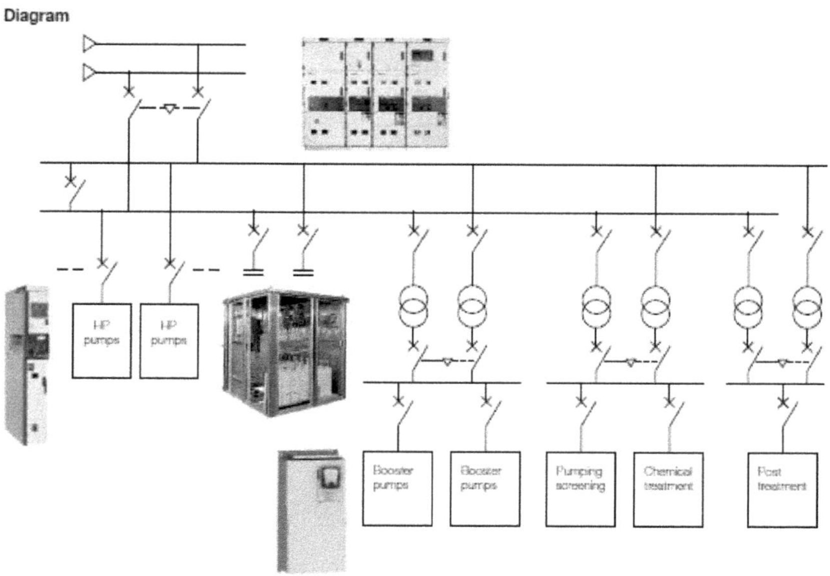

Figure 19.2. Electrical architecture of SWRO.

19.2 Electrical power requirements

The electrical power requirement of the SWRO plants comprising different pre-treatment regimes and different seawater feed salinity is listed in Table 19.1

Table19.1. Electrical power demand of typical SWRO plants

Region@selected seawater design parameters	SWRO Plant capacity (m^3/h)	Pretreatment	Specific Energy consumption (kWh/m^3)
Mediterranean sea and Atlantic ocean @TDS 39,000 mg/L & 15-30 °C	200,000	FF1 /UF	3.5 / 4
	100,000	FF1/UF	3.5 /4
	20,000	Beach Wells	3.8
Red sea and Indian ocean @ TDS 43,000 mg/L & 20-35 °C	200,000	FF1	3.7
	100,000	FF1	3.7
	20,000	Beach Wells	4.2
Arabian Gulf@TDS 46,000 mg/L & 20-35 °C	200,000	DAF+FF2	4.2
	100,000	DAF+FF2	4.2
	20,000	Beach Wells	4.3

Including SWRO, post-treatment and WWTP without potable water pumping The electrical power demand ranges between 3.2 kWh/m^3 and 4.3 kWh/m^3, the highest value corresponding to the Gulf Region, which clearly reflects the influence of the increasing pre-treatment scope and higher seawater TDS on the desalination plant energy demand for the production of 1 m^3 desalinated water. The energy demand of typical desalination plants in the Red Sea remains quite moderate in general. The related figures are quite comparable with plants equipped with MF/UF pre-treatment in the Mediterranean. The minimum registered values have been about 3 kWh/m^3 with a consumption in the HP pump/ER system of 1.6 m^3/h.

A typical value for an open intake with an advanced pretreatment system and a two pass RO system would be approx. 3.6-3.7 kWh/m^3 in the Mediterranean Sea.

The power installed in a desalination plant is the maximum instantaneous energy demand with a variation of ±5%,. It must add the power consumption of the last HP pump start. In medium installations, this value is so high that the type of starter is fixed by the electrical company. In big desalination plants > 100 MW at 66 kV, it should not be a problem with a direct start.

19.3 Transformation centers.

The standard voltages in seawater desalination plants

- High pressure pumps: 6 kV, in USA 4,6 kV is used and in the Middle East 11 kV.
- Rest of equipment: 690 V for motors > 400 kW and 440 V for the rest.

- Control system: 230 V monophasic with UPS.
- System of security and protection: 125 VCC.

In big plants (higher than 10 MW) with power distribution to the plant at 66 kv or bigger, voltage transformation will be to 6 kV for high pressure pumps. In smaller installations (lower than 10 MW), the transformation to 6 kV will be only for high pressure pumps and 400 V/690 V for rest of equipments

The number of transformers installed is usually one (1) per line. Line size will be up to 100000 m^3/day depending on power demand. If a stand-by transformer is installed it is not common to connect it.

Big transformers are usually placed outdoors. Smaller ones can be installed indoors. In both cases, oil cooled transformers, with or without an expansion vessel are the first and cheapest option. Dry-type transformers are an alternative and more expensive but with advantages.

The voltage transformers shall be protected by suitably rated circuit breakers in the primary and in the secondary winding.

19.4 Medium voltage switchgears

– The MV switchgears and motor control panels are specified to be metal clad and completely enclosed with a class of enclosure protection of IP52 according to IEC298. The enclosure protection of IP52 is also maintained when the circuit breaker is in the test position. The cubicles, which are installed outside, are required to be of protection class of IP55 with sunshades if required. The voltage is usually 6Kv.

The SF6 hermetically sealed metal chamber encapsulates all the life components and offers a high degree of protection against humidity contamination from corrosive gases, dust, moisture and rodents

– The floor openings under the cubicle panels desk are covered and/or sealed to be both fire- and vermin-proof. Individual switchgear panels are segregated into four or six compartments with metallic partitions:

- The circuit breaker compartment.
- The bus bar compartment.
- The control and monitoring compartment.
- The power cable compartment.

– The degree of protection of these zones is at least IP4X according to IEC298.

The control voltage is usually 125 VCC.

19.5 Motors

Medium voltage motors are used for high pressure pumps. Low voltage will be used for the rest of the motors.

19.5.1 Medium Voltage motors

Voltage ranges from 11 kV a 4,6 kV.

Service voltage of motors is usually 11 kv for motors above 400-500 kW (if VFD is installed an LV motor should be considered as an alternative) and 6,6 kv for motors from 250 kW to 400 kW. However, some specifications make requests for medium voltage motors above 200-250 kW.

In the desalination sector, MV voltage motors are only used for high pressure pumping. The rated power of these pumps range from 1000 to 2500 kW or even higher depending on rack size.

Medium voltage motors are designed for high efficiency (> 97%) or normal efficiency (96<>95%).

Forced lubrication of bearings will be required for large motors (power > 1500 -200 kw). Forced lubrication systems will include instrumentation, pumps, valves.

The insulation of all motors is specified to be class "F" according to IEC85 with class "B" temperature rise during the starting as well as at the steady state operation at full load. MV motors controlled by a vacuum type switchgear are equipped with lightning arrestors for suppression of over voltages generated at switching operations. It is worth mentioning that all motors are specified to be a totally enclosed fan cool type protection class IPW54 at a minimum due to severe ambient conditions. Velocity will be 2980 rpm at 50 Hz and 3580 rpm at 60 Hz

PT-100 in bearings and wiring and resistance-type detectors (RTD) for bearing temperature protection shall be supplied on each bearing brought out to a separate terminal box.

Figure 19.3. Medium voltage motors.

19.5.2 Low voltage motors.

690 V and 400 V are the standard voltages. 690 V for power higher than 400 kW, and 400 V for the rest.

All motors are specified to be totally enclosed fan cool type protection class IP 54 as minimum due to severe ambient conditions. The insulation of all motors is specified to be class "F" according to IEC85 with class "B" temperature rise during the starting as well as at the steady state operation at full load.

The following protection and accessories will be installed. For motors bigger than 50 kW, heating resistance and thermistors in winding. Above 250 kW, PT-100 in bearings. For motors above 400 kW, also PT-100 in winding.

When a VFD is installed, it must be specified the range of velocities and the power required at each velocity. Bearings special isolation and bigger frame could possible be necessary.

19.6 Motor Control center

A motor controller is a device or group of devices that serves to govern the electric motor and it is uses for LV and MV motors.

19.6.1 Medium voltage

MV cabinets are similar to the transformation center cabinets (one for each motor) with an automatic circuit breaker. Motor protection and control systems must be able to send real-time data to control centers and receive the resulting commands.

Figure 19.4. Medium voltage MCC.

Direct starter is acceptable for big desalination plants >10 MW (start-up intensity will be 6 In), but in medium size SWRO it will be necessary to install soft starters (3,5 In) Speed converters also act as soft starters reducing the stress on the network, motors and pumps. During the starting process, the speed converters progressively increase the motor speed and smoothly accelerate the load to its rated speed. Soft starters eliminate high starting currents and voltage dips which can cause process trips. With soft starters, maintenance costs will be reduced and the lifetime of the equipment extended.

One soft starter will be used in this scenario for all the motors. The motor will be connected to the specific cell once it has started.

It is recommended to install an electrical energy counter to measure the energy consumption of the high pressure pumps.

19.6.2 Low voltage

A motor control center (MCC) is an assembly of one or more enclosed sections having a common power bus and principally containing motor control units (cubicles).

Each motor controller in an MCC can be specified with a range of options such as separate control transformers, pilot lamps, control switches, extra control terminal blocks, various types of thermal or solid-state overload protection relays, or various classes of power fuses or types of circuit breakers.

Cubicles can be fixed or extractable. Fixed cubicles are usually installed for big motors or with frequency converters. Extractable cubicles are used for small motors.

Motor protection and control systems must be able to send real-time data to control centers and receive the resulting commands. This data should be transferred via a commonly-used industrial communications protocol

A motor starter has four basic functions:

- Isolation.
- Short-circuit protection.
- On/off control.
- Overload protection.

Motor starters can be enhanced with additional features depending on the requirements of a particular motor.

- Power features: speed adjustment, progressive starting, phase inversion, etc.
- Control features: auxiliary contacts, time delay, communication, etc.

A motor starter can be designed in one of three different ways according to the desired function.

Motor starter design and functions

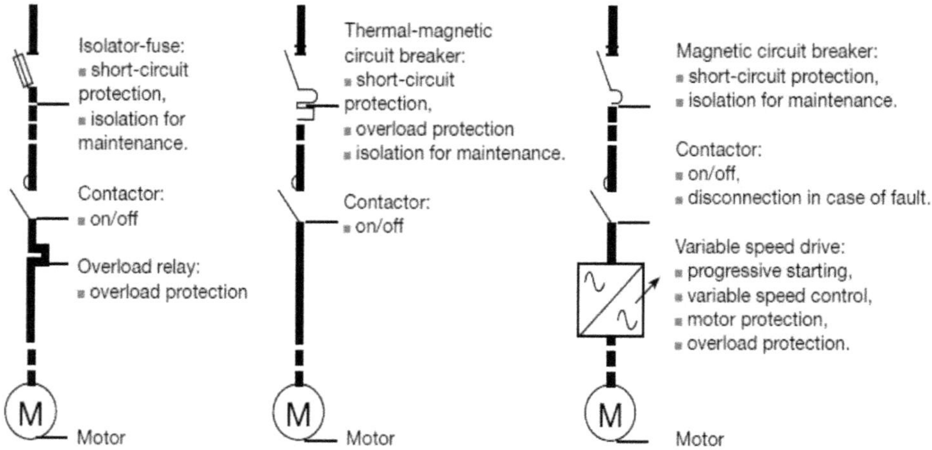

Figure 19.5. Types of motor starters.

There are different types of motor starters: direct on line, start-delta starting, soft starting, variable speed drive.

Three-phase motor with low to medium power rating. For large motors with rated power bigger than 75 kW, soft start-up will be used.

19.7 Variable frequency drives

Fixed-speed motor load applications that are supplied direct from AC line power save energy when they are operated at variable-speed, by means of VFD. Fixed-speed operated loads subject the motor to a high starting torque and to current surges that are up to eight times the full-load current. AC drives instead gradually ramp the motor up to operating speed to lessen mechanical and electrical stress, reducing maintenance and repair costs, and extending the life of the motor and the driven equipment.

Variable frequency drives are used in seawater pumps, recovery system booster pumps and in booster pumps of two passes and two stages systems.

- Type of variable frequency drives.

Variable frequency drives will be classified in function of power, voltage and type of rectifier installed.

VFD installed in dosing pumps and pumps with a power bigger than 75 kW (usually 400/460 V) will be considered as small VFD.

Rest of VFD > 75 kW will be used in centrifugal pumps. Up to 300 kW, voltage will be 400 V and 690 V will be used above 300 kW.

They are usually mounted in MCC, except for big motors (> 400 kW) which will be in separate cabinets.

20
CONTROL SYSTEMS

Control systems include software and hardware. The hardware can be local or can be installed in the control room

Software and Hardware are both elements of a control system. Depending on the selected control system, the manufacturer of both elements could be the same.

20.1 Types of control systems

20.1.1 Overview of DCS or PLC systems

Two (2) types of control systems can be used in desalination plants. Systems based on distributed control (**DCS**) or in **PLCs**.

DCS are integrated as a control architecture containing a supervisory level of control overseeing multiple, integrated sub-systems that are responsible for controlling the details of a localized process. Product and process control are usually achieved by deploying feed back or feed forward control loops whereby key product and/or process conditions are automatically maintained around a desired set point. To accomplish the desired product and/or process tolerance around a specified set point, specific PLCs are employed in the field and proportional, integral, and/or derivative settings on the PLC are tuned to provide the desired tolerance as well as the rate of self-correction during process upsets. DCS are used extensively in large scale desalination plants

PLCs are computer-based solid-state devices that control processes. While PLCs are control system components used throughout SCADA and DCS systems, they are can be used as the primary components in smaller control system configurations. However, PLCs based ICS is getting more powerful and reliable approaching to the standards of DCS systems for large scale plants.

20.1.2 Control systems operation

The basic operation of an ICS is shown in Figure 20.1 Key components include the following:

- **Control Loop.** A control loop consists of sensors for measurement, controller hardware such as PLCs, actuators such as control valves, breakers, switches and motors, and the communication of variables. Controlled variables are transmitted to the controller from the sensors. The controller interprets the signals and generates corresponding manipulated variables, based on set points, which it transmits to the actuators. Process changes from disturbances result in new sensor signals, identifying the state of the process, to again be transmitted to the controller.
- **Human-Machine Interface (HMI).** Operators and engineers use HMIs to monitor and configure set points, control algorithms, and adjust and establish parameters in the controller. The HMI also displays process status information and historical information.
- **Remote Diagnostics and Maintenance Utilities.** Diagnostics and maintenance utilities are used to prevent, identify and recover from abnormal operation or failures.

A typical ICS contains a proliferation of control loops, HMIs, and remote diagnostics and maintenance tools built using an array of network protocols on layered network architectures. Sometimes these control loops are nested and/or cascading–whereby the set point for one loop is based on the process variable determined by another loop. Supervisory-level loops and lower-level loops operate continuously over the duration of a process with cycle times ranging on the order of milliseconds to minutes.

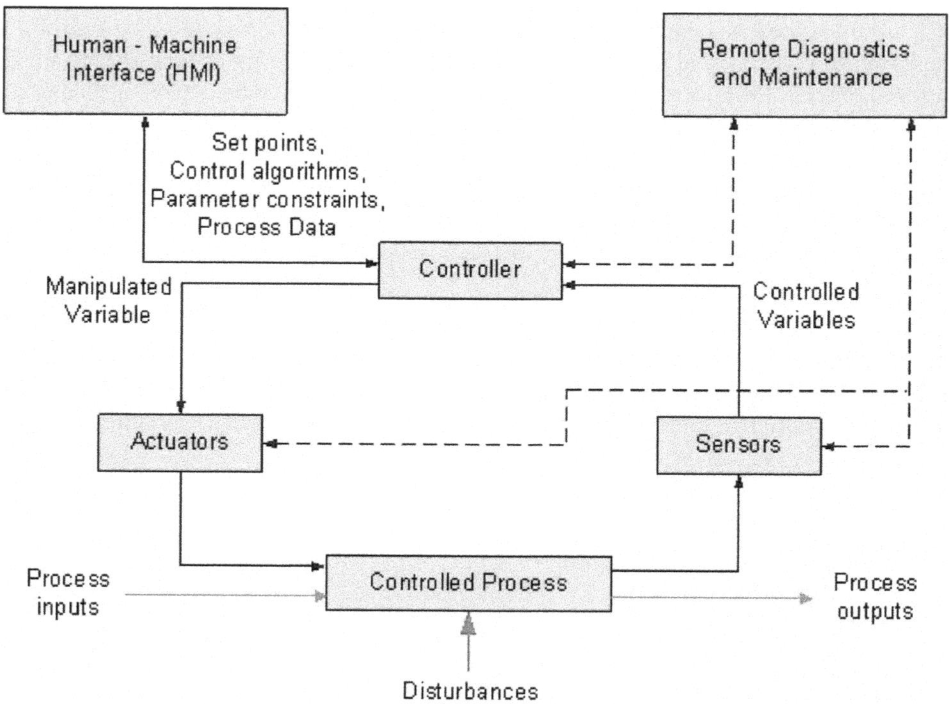

Figure 20.1. Components of control system.

20.1.3 Distributed control systems

A DCS uses a centralized supervisory control loop to mediate a group of localized controllers that share the overall tasks of carrying out an entire production process. By modularizing the production system, a DCS reduces the impact of a single fault on the overall system. In many modern systems, the DCS is interfaced with the corporate network to give business operations a view of production.

An example implementation showing the components and general configuration of a DCS is depicted in Figure 20.2. This DCS encompasses an entire facility from the bottom-level production processes up to the corporate or enterprise layer. In this example, a supervisory controller (control server) communicates to its subordinates via a control network. The supervisor sends set points to and requests data from the distributed field controllers. The distributed controllers control their process actuators based on control server commands and sensor feedback from process sensors.

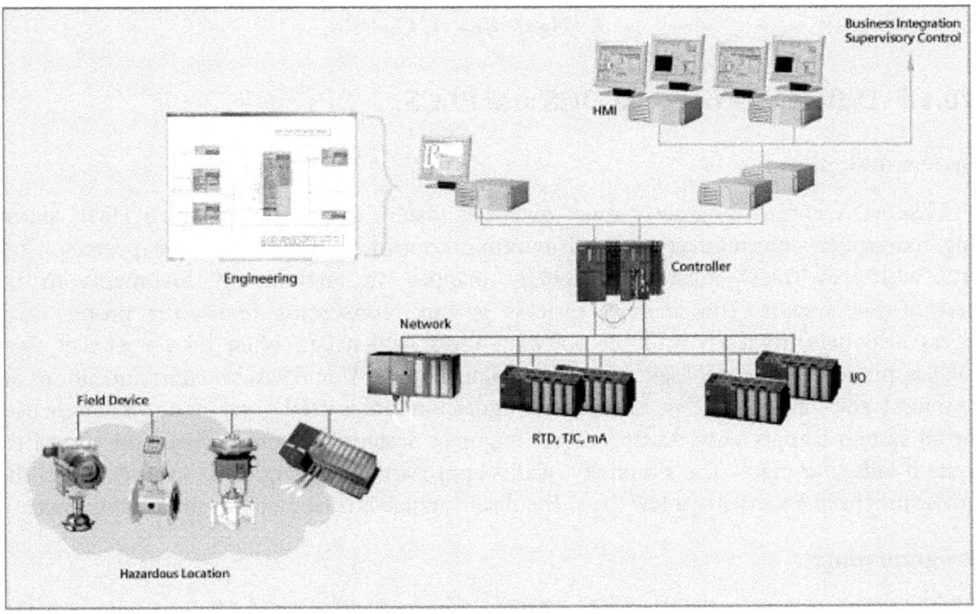

Figure 20.2. DCS.

20.1.4 Programmable logic controllers

PLCs are used in both SCADA and DCS systems as the control components of an overall hierarchical system to provide local management of processes through feedback control as described in the sections above. In the case of SCADA systems, they provide the same functionality of RTUs. When used in DCS, PLCs are implemented as local controllers within a supervisory control scheme. PLCs are also implemented as the primary components in smaller control system configurations. PLCs have a user-programmable memory for storing instructions for the purpose of implementing specific functions such as I/O control, logic, timing, counting, three mode proportional-integral-derivative (PID) control, communication, arithmetic, and data and file processing.

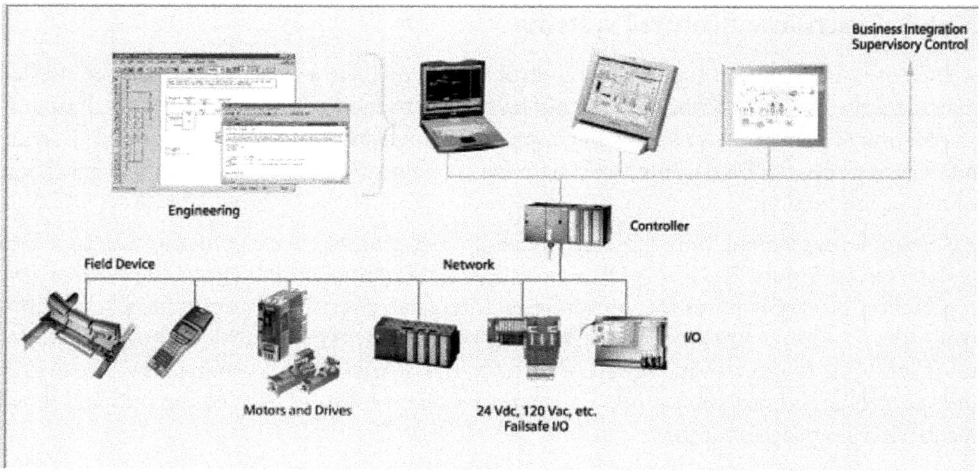

Figure 20.3. PLC.

20.1.5 Differences between DCS and PLCS

System design.

PLC/SCADA control engineers must map out system integration between HMI, alarming, controller communications and multiple controllers for every new project. Control addresses (tags) must be manually mapped in engineering documents to the rest of the system. This manual process is time consuming and error prone. Engineers also have to learn multiple software tools, which can often take weeks of time. DCS approach: As control logic is designed, alarming, HMI and system communications are automatically configured. One software configuration tool is used to set up one database used by all system components. As the control engineer designs the control logic, the rest of the system falls into place. The simplicity of this approach allows engineers to understand this environment in a matter of a few days. The disadvantage is the quality of sinoptics is lower

Programming.

PLC/SCADA control logic, alarming, system communications and HMI are programmed independently. Control engineers are responsible for the integration/linking of multiple databases to create the system. Items to be manually duplicated in every element of the system include: scalability data, alarm levels, and Tag locations (addresses). Only basic control is available. Redundancy is rarely used with PLCs. One reason is the difficulty in setting it up and managing meaningful redundancy for the application.

The DCS way: When control logic is developed, HMI faceplates, alarms and system communications are automatically configured. Redundancy is set up in software quickly and easily, nearly with a click of a button.

Commissioning and start-up.

Testing a PLC/HMI system is normally conducted on the job site after all of the wiring is completed and the production manager is asking "why is the system not running yet?" Off line simulation is possible, but this takes an extensive effort of programming to write code which will simulates the application you are controlling. Owing to the high cost and complex programming, this is rarely done. DCS benefits: Process control systems come with the ability to automatically simulate the process based on the logic, HMI and alarms that are going to be used by the operator at the plant.

This saves significant time on-site since the programming has already been tested before the wiring is begun. Potential savings are 10 - 20% depending on the complexity of the start up and commissioning.

20.2 Control system equipment

The following is a list of the major control components of an ICS:

- **Control Server.** The control server hosts the DCS or PLC supervisory control software that communicates with lower-level control devices. The control server accesses subordinate control modules over an ICS network.
- **SCADA Server or Master Terminal Unit (MTU).** The SCADA Server is the device that acts as the master in a SCADA system. Remote terminal units and PLC devices (as described below) located at remote field sites usually act as slaves.
- **Remote Terminal Unit (RTU).** The RTU is a special purpose data acquisition and control unit designed to support SCADA remote stations. RTUs are field devices often equipped with wireless radio interfaces to support remote situations where wire-based communications are unavailable. Sometimes PLCs are implemented as field devices to serve as RTUs; in this case, the PLC is often referred to as an RTU.
- **Programmable Logic Controller (PLC).** In SCADA environments, PLCs are often used as field devices because they are more economical, versatile, flexible, and configurable than special-purpose RTUs.
- **Intelligent Electronic Devices (IED).** An IED is a “smart” sensor/actuator containing the intelligence required to acquire data, communicate to other devices, and perform local processing and control. An IED could combine an analog input sensor, analog output, low-level control capabilities, a communication system, and program memory in one device. The use of IEDs in SCADA and DCS systems allows for automatic control at the local level.
- **Human-Machine Interface (HMI).** The HMI is software and hardware that allows human operators to monitor the state of a process under control, modify control settings to change the control objective, and manually override automatic control operations in the event of an emergency. The HMI also allows a control engineer or operator to configure set points or control algorithms and parameters in the controller. The HMI also displays process status information, historical information, reports, and other information to operators, administrators, managers, business partners, and other authorized users. The location, platform, and interface may vary a great deal. For example, an HMI could be a dedicated platform in the control center, a laptop on a wireless LAN, or a browser on any system connected to the Internet.
- **Data Historian.** The data historian is a centralized database for logging all process information within an ICS.

- **Input/Output (IO) Server.** The IO server is a control component responsible for collecting, buffering and providing access to process information from control sub-components such as PLCs, RTUs and IEDs.
- **Control Server.** The control server hosts the DCS or PLC supervisory control software that communicates with lower-level control devices. The control server accesses subordinate control modules over an ICS network.
- **UPS**
- **Printer**

20.3 Communication instruments and cards

Fieldbus Network. The fieldbus network links sensors and other devices to a PLC or other controller. Use of fieldbus technologies eliminates the need for point-to-point wiring between the controller and each device (2 cables per instrument for analogic signals 4-20 mA). The devices communicate with the fieldbus controller using a variety of protocols. The messages sent between the sensors and the controller uniquely identify each of the sensors. Querying a smart field sensor can provide the primary measurement information on sensor failure, instrument, manufacturer, calibration and any required data for control and command.

The most popular and accepted fieldbus choices in the desalination sector are PROFIBUS and FUNDATION FIELDBUS.

FUNDATION FIELDBUS allows control strategies to be programmed in the smart filed device as well as the linking device. For example, a mart flow control valve positioner may have the flowmeter signal as an input and may be programmed with a PID flow control loop. This improves reliability It is a system of communication in series, fully digital, working in both directions with maximum distance of approx. 1900 m and a data speed of 31,35 Kbits/sec.

20.4 Control systems suppliers

It must be distinguished between hardware manufacturer and equipment integrator who is responsible for the software and correct operation including start-up.

YOKOWAWA, EMERSON (Rosemount), ABB, are equipment suppliers. These companies can supply hardware and software including start-up.

With PLC based systems, Siemens (PCS 7), Rockwell (Control Logix) or Schneider (Modicon) are the most important suppliers.

In SCADA, the alternatives are Intouch (Wonderware), RSview (Rockwell) and Winn cc (Siemens).

21

VARIOUS EQUIPMENT

21.1 Control valves.

Different types of control valves are used in desalination. They will be described in this chapter.

- *Valve sizing calculation (general).*

Control valve sizing is based on flow coefficient Cv calculation. Flow coefficient Cv calculation is made for required flow rate and for related pressure drop in control valve. With the flow coefficient Cv calculated, the size of control valve can be selected, or two control valves from different manufacturers can be compared in terms of flow capacity for certain pressure drops and the same control valve size.

Flow coefficient Cv is determined experimentally by control valve manufacturers.

$$Q = \frac{C_v \times \Delta P^{1/2}}{\delta}$$

Q = Flow gpm

ΔP = headloss in psig

δ = fluid density

If Kv coefficient is used instead of C_v, units will be m^3/h for flow and bars for headloss.

Usually, it is recommended the control valve have a headloss of 25% of the total headloss in line where valve is installed. It is also recommended that the final selection of the control valve to be carried out by the valve manufacturer due to the importance of noise and cavitation analysis at high head losses.

Variables to analyze are:

- Maximum and minimum flow.
- Type of fluid, Maximum temperature and density.
- Maximum and minimum headloss.
- Size and diameter of pipe and material.
- Maximum level of noise.

- *Materials.*

AISI 316 is used for treated water pumps and duplex or superduplex is used for 1st pass brine and second pass brine depending on salinity.

- *Actuators.*

Pneumatically operated control valve actuators are the most popular type in use, but electric actuators can be used. The spring-and-diaphragm pneumatic actuator is most commonly specified due to its dependability and simplicity of design. Pneumatically operated piston actuators provide high stem force output for demanding service conditions.

Adaptations of both spring-and-diaphragm and pneumatic piston actuators are available for direct installation on rotary-shaft control valves Electric and electro-hydraulic actuators are more complex and more expensive than pneumatic actuators. They offer advantages where no air supply source is available, where low ambient temperatures could freeze condensed water in pneumatic supply lines, or where unusually large stem forces are needed.

Pneumatically operated valves depend on a positioner to take an input signal from a process controller and convert it to valve travel. These electro-pneumatic positioners can use electrical current (usually 4-20 mA) or digital signal

Typically, spring return (single acting) actuators, fail open, are used.

21.2 Butterfly valves

Butterfly valves are installed in the low pressure areas of a desalination plant, typically in the early stages of production upstream of the membranes and then at the end on the permeate lines. By keeping the butterfly valves open, each valve creates a pressure loss and with the centered disc the differential pressure is much lower than in offset disc valves. Where there are a high number of valves, say 1000 units, installed in a desalination plant the accumulation of pressure saved on each valve will result in energy savings. In order to minimize the effects of corrosion, all KSB's butterfly valves contain a thick rubber liner encapsulated within the valve body and discs are offered in a choice of Halar and hardened rubber surface finishes. It is only the disc and liner that come into contact with the fluid.

AISI 316 will be used in treated water line, reverse osmosis backwashing, drainage.

In the following positions, the butterfly valve actuators will have positioners.

- Raw water inlet of pressure filters. Head loss of 1 bar.
- Filtered water outlet of pressure filters. Headloss of approx 0.5 bars.
- Drain valve in suction side of high pressure pumps. Headloss of 2-3 bars. If pressure is higher, other type of valves must be installed.
- Permeate outlet of RO rack.

Butterfly valves are commonly used as control valves in applications where the pressure drops required of the valves are relatively low. They should limit to operate between 30%-60% maximum opening.

21.3 Rotary motion control valves

They are modified ball valves or butterfly valves for big sizes which can control functions from 10% to 100% of flow.

Control valves are used for the following applications:

- **Permeate water control**. Treated water production at rack outlet is controlled with a valve (production varies with water temperature and age).

- **2nd pass brine pressure control.** Rack operation pressure is controlled in function of flow.
- **1st pass brine pressure control:** 1st pass operation pressure is controlled during start-up in two-pass and common collector brackish water installations.

One of the most typical control valves used in the desalinations sector is the CAMFLEX model valve from Massonelian. It is equipped with an electropneumatic positioner for 4-20 mA signal. Connections are flanged ANSI-RF-300 lbs and the body if installed at the permeate outlet is manufactured in AISI-316. Concentrate control valves must be open during start-up of RO racks. Starting against a closed or almost closed concentrate valves could cause the recovery to be exceeded which may lead to scaling.

Fig. 1 Type 3310 Segmented Ball Valve with Type BR 31a Pneumatic Rotary Actuator

Fig. 2 Cutaway view of Type 3310 Segmented Ball Valve with Type BR 31a Pneumatic Rotary Actuator and positioner

Figure 21.1. Photography of control valves.

21.4 Agitators

They are used to mix liquids and solids. The design and characteristics will depend on the mixing volume.

For tanks up to 1000 liters, agitator speeds will be 1500 rpm. In general, if mixing volume is larger than 1000 liters, gear reducers will be required.

Normally, the agitators are vertical, mounted at the center of tanks, and vertical baffle plates are attached on the inner wall of the tank to obtain vertical convection and turbulence in the direction of axial flow. Reinforcement could be required at the top of tanks.

Horizontal agitators could be used for some applications

In seawater reverse osmosis desalination plants, agitators will be used in:

- Sodium bisulphite mixing.
- Reverse osmosis cleaning tank.
- Lime dilution tank.
- Quick mixing and floculation.
- Neutralization tank (horizontal).

Standard material for the construction of agitators is AISI 316, except for coagulant and cleaning system. In these cases, rubber lining or additional protection will be required in the parts in contact with water.

The most important manufacturers of agitators are MILTON ROY, TFB, GRUNDFOS and CHEMINEER.

21.5 Compressors

Compressors produce the compressed air for automatic valves and rest of pneumatic equipment. There are three types of compressors:

- **Piston compressors**: For low flows. Robust equipment.
- **Centrifugal**: For medium flows.
- **Rotary Screw:** For big flows.

The compressor, takes the air from outdoors, and compress it up to 6.5-7 kg/cm^2, elevating the temperature with the same moisture content. When the air is cooled, water condensation is produced, so a refrigerator is installed at the compressor outlet to reduce the temperature.

The complete system usually includes (1) dryer, (1) filter, (1) deposit to regulate potential service fluctuations and two (2x100%) compressors.

The technical criteria used to determine the air consumption is indicated below:

- On-off valves: 280 l/min and 5% of use.
- Regulation valves: 25 l/min and 100% of use.
- Piping and accessories are manufactured in copper.

21.6 Water pressure group

It usually consists of two (2) centrifugal pumps and pneumatic deposit (diaphragm) to keep the pressure constant. A standard capacity of deposits in SWRO plants is from 5 to 15 m^3.

21.7 Static mixer

It is recommended to use static mixers to guarantee proper mixing of chemical products The two positions where the static mixer will be installed are.

- High pressure pumps suction side (antiscalant, sodium bisulphite, sulphuric acid)
- Remineralization (CO_2, lime).

They are manufactured in GRP with independent connections for the injection of each reactive.

Figure 21.2. Photography of static mixer.

21.8 Centrifugal pumps

The typical pumps used in a seawater reverse osmosis desalination plant are:

- Split case pumps (horizontal or vertical).
- Multistage.
- DIN.

Different type of impellers can be used depending on the application:

- **Closed**: To enclose the impellers, two circular plates are attached to both sides of the blades. It is used with clean water.
- **Semi open**: The semi-open impeller is constructed with a circular plate attached to both sides. The open impeller is less likely to clog with solids, but if it does, it is easy to clean. In desalination, it is used in the water with lime pumps.
- **Vortex:** Used for sludge
- **Channel**: Used for sludge

Velocity

- For high pressure: 3000 rpm.
- For low head, sludge: 1500 rpm.
- For sludge: 1000 rpm.

Below, find typical seawater reverse osmosis (SWRO) with location of critical centrifugal pumps.

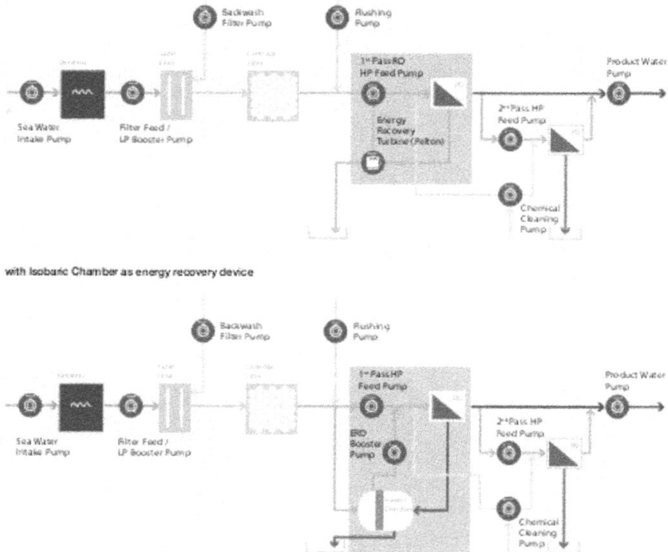

Figure 21.3. Position of centrifugal pumps in a SWO.

The high pressure pumps (multistage axially split and multistage horizontal ring pumps) and seawater intake pumps (vertical pumps) have been studied in detail in previous chapters, so we will only refer to the rest of pumps.

Application matrix

Technology	Intake	Filter feed/ LP booster	ERD booster	Filter backwash	Chemical cleaning	Product transfer
End suction pumps		X	X	X	X	X
Axially split pumps	X	X				X
Multistage pumps						X
Vertical pumps	X	X				

The most important manufacturers in the seawater desalination sector are FLOWSERVE, SULZER and TORISHMA.

- **Axially split pumps.**

They have single stage double entry impellers arranged either horizontally or vertically. The casing is axially split with the suction and discharge branches located in the lower half, which allows the complete pump rotor with bearings to be removed for maintenance without disturbing the main pipework and valves. Shaft sealing is performed by optional mechanical seals.

The capacity of these pumps can be up to 10000-15000 m^3/h and the total head up to 200-250 meters of column of water.

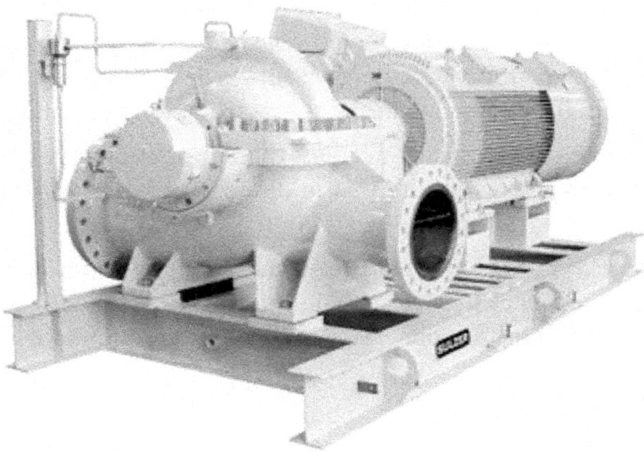

Figure 21.4. Photography of axially split pumps.

- **End suction pumps.**

They are single suction and single stage centrifugal pumps. They are designed to comply with ISO and ANSI standards and as indicated in application matrix, can be used for different applications.

The capacity of these pumps can be up to 1500 m^3/h and a total head up to 10-15 bars.

Figure 21.5. Photography of end suction pumps.

21.9 Supports and piping

Piping stress calculations must be carried out in order to check that the stress is within the maximum allowable for the concerned material and at the connection to equipment nozzles.

Once the routing is confirmed to be acceptable, the piping isometric drawings are issued for construction.

Piping stress analysis is systematic for high temperature/high pressure service and large diameter lines. When such a line comes into service, if it can not expand, because it is constrained, it be will be subject to internal stress, which could exceed the strength of the material and make the line come off its position and even fall from it supports. Allowance for displacement of the line must be provided in the design, by means of expansion loop and guiding supports.

Special attention must be paid at the connections to rotating equipment, where the flexibility of the inlet and outlet lines must ensure that minimum loads are transferred to equipment nozzles.

Indeed excessive forces on the connected equipment could result in its displacement. Typically, the pumps, will be displaced, resulting in its misalignment with the driver (motor). Supports are provided to minimize the connecting lines loads on equipment nozzles.

Positions of supports must also allow, during installation of the pipe work, the effortless fitting of the pipes connecting to nozzles of rotating machinery. Very stringent tolerances are imposed by vendors in terms of parallelism between pipe flanges and rotating machinery nozzles. Forced connection of piping flanges to machine nozzles using hydraulic tools can generate forces on nozzles and induce shaft misalignment.

Pipe supports are otherwise standardized to the maximum extent, in order to allow their mass pre-fabrication. Two types of supports can be distinguished:

- Fixed: Installed in the elbows, to protect pumps.
- Sliding supports: Slide supports should be use particularly on long pipelines to allow thermal expansion of pipes. Distance between fixed supports must me be enough to allow pipe flexion.

Figure 21.6. Photography of low pressure piping supports.

21.10 Verification unit

Desalination plants may include a verification and testing unit which allows it to verify the operation and performance of membranes, when received during the operation of the plant.

This equipment gives the opportunity to test different chemical products and design parameters.

This verification unit includes a high pressure pump which feeds a vessel with one membrane. The plant also includes a cleaning system with tank, cartridge filters and independent pump.

Figure 21.7. Verification unit.

22

COST EVALUATION OF DESALINATION PLANTS

22.1 Capex

In tables presented below CAPEX are calculated for different macro-regions, plant capacities and pre-treatment technologies

- 200,000 m^3/d comprising pre-treatment of FF1 in table 22.1
- 200,000 m^3/d comprising pre-treatment of MF/UF in table 22.2

The specific CAPEX varies between 1750 US$/$m^3$d and 2250 US$/m^3d depending on plant capacity and pre-treatment applied. The specific annual CAPEX ranges between 0.37 US$/$m^3$ corresponding to the largest capacity of 200,000 m^3/d plant with one-stage filtration (FF1) and 0.52 US$/$m^3$ comprising a plant capacity of 100,000 m^3/d and more intensive pre-treatment of DAF and two-stage filtration (FF2). The option of small plants with 20,000 m^3/d and beach wells intakes come up to the high specific CAPEX costs with 0.52 US$/$m^3$. The economy of scale is significant.

Capital cost of 200,000 m^3/day production SWRO plant with open intake

Pretreatment consisting of gravity filters and post treatment of lime +CO_2

Plant life time of 25 years with an interest rate of 6%

Table 22.1. 200,000 m^3/d comprising pre-treatment of FF1

Systems	System cost $	Cost partition %	Specific cost $/$m^3$day
Intake, pumping station and outfall	50000000	14.3	250
Pretreatment system	40000000	11.4	200
Membranes (in case MF/UF)	-	-	-
Pretreatment without membranes	40000000	11.4	200
Reverse osmosis	130000000	37.2	650

Systems	System cost $	Cost partition %	Specific cost $/m³day
Reverse osmosis without membranes	116000000	33.2	580
Membranes w.o vessels	14000000	4	70
Potable plant	16000000	4.6	80
Drinking water storage and pumping	16000000	4.6	80
Wastewater treatment	10000000	2.9	50
Mechanical equipment without membranes	248000000	70.9	1240
Auxiliary systems	11000000	3.1	55
Civil works	25000000	7.2	125
Electrical works	23000000	6.6	115
I&C works	12000000	3.4	60
Total	330000000	100	1665
Contingencies	5%	4.8	83.3
SWRO plant total	349650000	10	1748.3
	$/year		
Annual capital cost	**27351972**		**0.37**

Capital cost of 200,000 m^3/day production SWRO plant with open intake

Pretreatment consisting of MF/UF and post treatment of lime +CO_2

Plant life time of 25 years with an interest rate of 6%

Table 22.2. 200,000 m^3/d comprising pre-treatment of MF/UF

Systems	System cost $	Cost partition %	Specific cost $/m³day
Intake, pumping station and out-fall	55000000	15.1	275
Pretreatment system	50000000	13.7	250
Membranes (in case MF/UF)	6000000	1.6	30
Pretreatment without membranes	44000000	12	220

Systems	System cost $	Cost partition %	Specific cost $\$/m^3day$
Reverse osmosis	130000000	35.6	650
Membranes w.o vessels	14000000	3.8	70
Reverse osmosis without membranes	116000000	31.7	580
Potable plant	16000000	4.4	80
Drinking water storage and pumping	16000000	4.4	80
Wastewater treatment	13000000	3.6	65
Mechanical equipment without membranes	260000000	71.2	1300
Auxiliary systems	11000000	3	55
Civil works	20000000	5.5	100
Electrical works	24000000	6.6	120
I&C works	13000000	3.6	65
Total	348000000	100	1740
Contingencies	5% 17400000	4.8	87
SWRO plant total	365400000	100	1827
Annual capital cost	**28584043**		**0.39**

22.2 Opex

22.2.1 Staff

For large desalination plants, the number of operation staff is practically unrelated to the plants production capacity. The number of staff required to operate a plant will depend on the plant size and operation complexity. Office cleaning is usually entrusted to an external specialized firm. The average costs would be between 0.015 to 0.025 $\$/m^3$.

Table 22.3. Cost of staff.

Plant capacity (m^3/day)	**Plant automation, complexity and labor skill level**	
	High	**Low**
100000	14-18	20-24
200000	18-28	30-40
300000	35-50	60-80

22.2.2 Fixed maintenance costs.

Certain maintenance operations must always be performed whether the desalination plant is in operation or not:

- Civil works maintenance.
- Gardening.
- Pipes and surrounding space checking.
- Machines painting.
- Transformation center maintenance.
- Routine (preventive maintenance).
- Emergency (reactive) maintenance.
- Emergency (Reactive) Maintenance.
- Spare Parts and Equipment.
- Rebuilding High Pressure Pump Internals

Every 5 Years;

- Rebuilding ERDs every 10 Years

The annual cost of these operations for a large plant is the 2-4% of the direct capital costs This represents from 0,035 $\$/m^3$ to 0,060 $\$/m^3$

22.2.3 Miscellaneous

Fixed costs for miscellaneous items include the following:

- Work clothes.
- Health and safety equipment (gloves, helmets, auricular, glasses, etc.).
- Office furniture, equipment and materials.
- Telephone and ADLS, bills, photocopy machine.
- Vehicles.
- Desalination plant insurance (all risks insurance for materials and machine damage).
- Staff training.
- Licenses, taxes and permits.
- Fungible laboratory materials.

The repercussion per m^3-desalinated water would be: 0.01-0.1 $\$/m^3$.

22.2.4 Power connection

Water desalination plants operate on a full time basis, that is, 100% capacity every day of the year, except for the 7 to 15 days needed for general revision (transformers, pipe and pumps clearing and maintenance), once a year.

In these circumstances, the best option is to subscribe an electric power rate giving access to a basic supply.

Since the subscribed power depends on the desalination plants specific power consumption, its cost will depend on:

- The plant specific power consumption.
- Its branching tension.

The repercussion of this item in every m^3-desalinated water, would be approx.: 0.01 $/m3

22.2.5 Energy.

The greater portion of the OPEX costs is attributed to the electrical energy consumption in SWRO plants. The specific electricity consumption for the cost implementation of an SWRO plant at the Arabian Gulf coast is higher being related to the more demanding seawater conditions in this area. This implies the requirement to undertake additional water pre-treatment such as two-stage floc-filtration and DAF, which results in additional power consumption.

The electrical power requirement of the SWRO plants comprising different pre-treatment regimes and different seawater feed salinity are shown in detail in chapter 19 and will range from 3.5 to 4.5 kWh/m^3. The tariff will be variable.

22.2.6 Reagent cost

The chemicals normally used in desalination plants may be classified in various large groups: Pretreatment chemicals and post-treatment chemicals. Dosage varies according to temperature and depends on whether the water will be mixed with water of different precedence before distribution or not. The chemicals used to clean the membranes and the cleaning frequency depend on the intake water quality. Dosage and frequency also depend on seawater quality, reagents demands increasing as the water quality gets worse and vice versa

For a conventional plant dealing with relatively clear, pollution free water, and reagent costs may be approximately 0.025 \$/m^3. This value could exceed up to 0.05-0.06 \$/m^3 for complex pretreatment with waters which require intensive pretreatment.

22.2.7 Membranes and cartridge filters replacement

For the SWRO plant, a membrane replacement rate has to be assumed depending on the efficiency of pre-treatment. Lower membrane exchange values are correlating with higher costs for increased pre-treatment effort.

Considering open intake SWRO plants, designed and operated to the state of-the-art, an AMLT (average membrane life time) of about 3.5 years for the first pass and about 4.5 years for the second pass should be considered. Annual membrane replacement rates are 15%/yr and 12%/yr. for the first and the second pass RO respectively and the cost may be approximately 0.035 \$/m^3. For beach wells, the cost could be as low as 0.020 \$/m^3 and maximum values could be approx. 0.05 \$/m^3.

22.2.8 Environmental and performance monitoring

Environmental Monitoring Costs – US\$0.005 to 0.020/m^3

Plant Performance Monitoring Costs – US\$0.005 to 0.015/m^3

Environmental Monitoring Costs Include:

- Plant Discharge Sampling.
- Offshore Data Collection.

Plant Performance Monitoring Costs Include:

- Source and Product Water Sampling.
- Process Data Collection and Analysis.

Attachment I

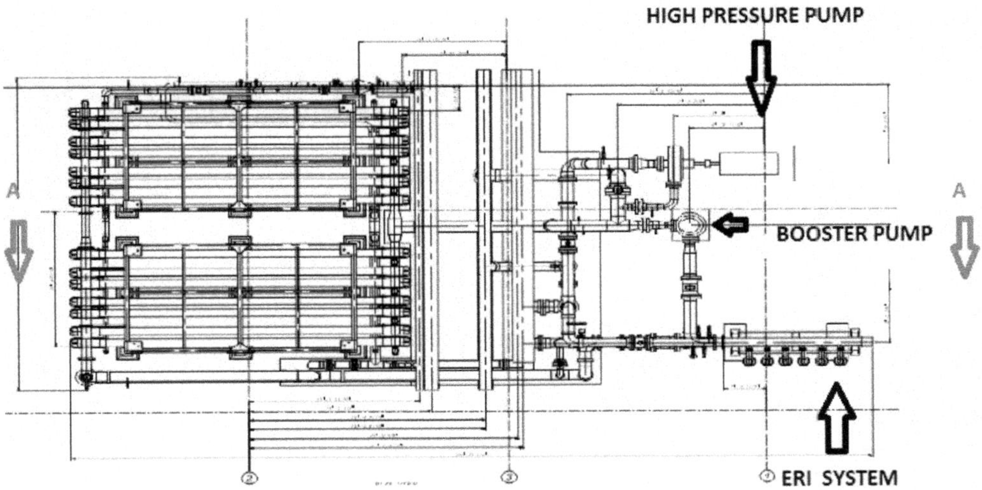

RO FIRST PASS PLAN VIEW

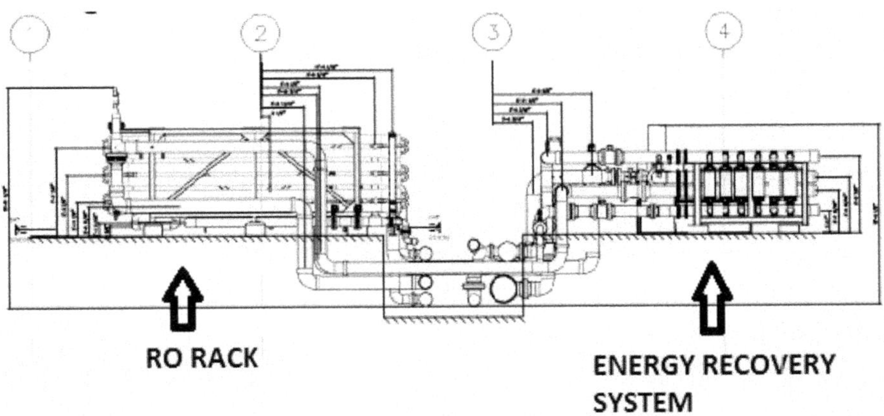

RO FIRST PASS SECTION A-A

BIBLIOGRAPHY

"Advances in Water Desalination"
Noam Lior (Editor)
ISBN: 978-0-470-05459-8
712 pages
January 2013

"Basic Principles of Membrane Technology (Paperback)"
J. Mulder
ISBN: 07923 4248-8

"Beach well seawater intake as feed for an RO desalting system"
Abdel-Jawad, M.; Ebrahim, S
Source: Desalination v 99 n 1 Nov 1994. p 57
(A Desalination Primer)

"Criteria and procedure for selecting a site for a desalination plant"
Tsiourtis, N.X. (2008),
Desalination, vol. 221, January, pp. 114-125.

"Desalination Cost Assessment and Management"
Nikolay Voutchkov.
ISBN: 978-616-90836-65
Water Treatment Academy Publications 2011

"Desalination and Advanced Water Treatment. Economics and Financing"
Corrado Sommariva
ISBN: 0-86689-069-6
Balaban Desalination Publications
2010

Dow, The Dow Chemical Company, "FILMTEC Membranes Product Information Catalog", Form 609-00204-1108, 2009.

"Environmental Issues of Desalination".
Younos, T. (2005)
Journal of Contemporary Water Research & Education. Universities Council on Water Resources UCOWR 132.

"Fundamental chemistry and engineering aspects of post-treatment processes for desalinated water"
Birnhack, L., Voutchkov, N., & Lahav, O. (2011-A review. Desalination 273(1), 6-22. doi: 10.1016/j.desal.2010.11.011

"Liquid Separations with Membranes. An Introduction to Barrier Interference
Author: Karl W. Böddeker:a
ISBN: 978-3-540-47451-7
Springer
2008

"Membrane pretreatment of reverse osmosis: Long-term experience on difficult waters
Author: Durham, Bruce; Walton, Andy
Source: Desalination v 122 n 2 1999. p 157-170.

Metcalf and Eddy (2003) Wastewater Engineering : Treatment and Reuse 4th edition
Mc Graw Hill

"MF/UF pretreatment in seawater desalination: applications and trends"
R.P. Huehmer
Proceedings of World Congress in Desalination and Reuse, International

"Optimizing Design And Cost Of Seawater Reverse Osmosis Systems."
Van Dijk, J. C.; de Moel, P. J.; van den Berkmortel, H. A.
Source: Desalination v 52 n 1 Dec 1984 p 57-73
Publication Year: 1984

'Pre-treatment strategies for seawater desalination by reverse osmosis system', Prihasto, N., Qi-Feng, L. and Seung-Hyun, K. (2009)
Desalination, no. 249, pp. 308-316.

"Reverse Osmosis and Nanofiltration"
Edited by Kathleen A. Faller
ISBN: 0 89867 978 8
American Water Works Association 1999 AWWA Manual of Water Supply Practices

United States Bureau of Reclamation (USBR) (2009): Treatment of Concentrate.
Desalination and Water Purification Research and Development
Program Report No.155, May 2009.

'Seawater intakes for desalination plants',
Gille, D. (2003) *Desalination*, vol. 156, August, pp. 249-256.

"Seawater Pretreatment"
Nikolay Voutchkov
ISBN: 978-974-4017956
Water Treatment Academy Publications 2010

"Seawater Desalination Impacts of Brine and Chemical Discharge on the Marine Environment"
by Sabine Lattemann and Thomas Hoepner
ISBN: 0-86689-062-9
Desalination Publications and DEStech Publications 2003

'State-of-the-art of reverse osmosis desalination', The Fritzmann, C., Lowenberg, J., Wintgens, T. and Melin, T. (2007) *Desalination,* vol. 1, no. 216, December, pp. 1-76.

"Guidebook to Membrane Desalination Technology : Reverse Osmosis, Nanofiltration and Hybrid Systems Process, Design, Applications and Economics"
Mark Wilf; Leon Awerbuch; Craig Bartels; Mike Mickley; Graeme Pearce; Nikolay Voutchkov (Author)

Zenon environmental Inc (2003) Operation and maintenance Manual Ultrafiltration of wastewater treatment plant, City of Woostock Georgia

Dow (2011) Product Manual. Ultrafiltration . Version 3

www.ingramcontent.com/pod-product-compliance
Ingram Content Group Australia Pty Ltd
76 Discovery Rd, Dandenong South VIC 3175, AU
AUHW011414220626
428909AU00010B/136

9 780820 602004